MW01077770

# GENETICS and CRIMINALITY

The LAW AND PUBLIC POLICY: PSYCHOLOGY AND THE SOCIAL SCIENCES series includes books in three domains:

*Legal Studies*—writings by legal scholars about issues of relevance to psychology and the other social sciences, or that employ social science information to advance the legal analysis;

*Social Science Studies*—writings by scientists from psychology and the other social sciences about issues of relevance to law and public policy; and

*Forensic Studies*—writings by psychologists and other mental health scientists and professionals about issues relevant to forensic mental health science and practice.

The series is guided by its editor, Bruce D. Sales, PhD, JD, University of Arizona; and coeditors, Stephen J. Ceci, PhD, Cornell University; Norman J. Finkel, PhD, Georgetown University; and Bruce J. Winick, JD, University of Miami.

***

*The Right to Refuse Mental Health Treatment*
Bruce J. Winick
*Violent Offenders: Appraising and Managing Risk*
Vernon L. Quinsey, Grant T. Harris, Marnie E. Rice, and Catherine A. Cormier
*Recollection, Testimony, and Lying in Early Childhood*
Clara Stern, William Stern, and James T. Lamiell (translator)
*Genetics and Criminality: The Potential Misuse of Scientific Information in Court*
Jeffrey R. Botkin, William M. McMahon, and Leslie Pickering Francis

# GENETICS and CRIMINALITY

## The Potential Misuse of Scientific Information in Court

EDITED BY

**Jeffrey R. Botkin**

**William M. McMahon**

**Leslie Pickering Francis**

AMERICAN PSYCHOLOGICAL ASSOCIATION

WASHINGTON, DC

Copyright © 1999 by the American Psychological Association.
All rights reserved. Except as permitted under the United States Copyright Act of 1976, no part of this publication may be reproduced or distributed in any form or by any means, or stored in a database or retrieval system, without the prior written permission of the publisher.

Published by
American Psychological Association
750 First Street, NE
Washington, DC 20002

Copies may be ordered from
APA Order Department
P.O. Box 92984
Washington, DC 20090-2984

In the UK and Europe, copies may be ordered from
American Psychological Association
3 Henrietta Street
Covent Garden, London
WC2E 8LU England

Typeset in Times Roman by GGS Information Services, York, PA

Printer: United Book Press, Inc., Baltimore, MD
Cover Designer: Berg Design, Albany, NY
Editor/Project Manager: Debbie K. Hardin, Reston, VA

**Library of Congress Cataloging-in-Publication Data**
Genetics and criminality : the potential misuse of scientific information in court / edited by Jeffrey R. Botkin, William M. McMahon, Leslie Pickering Francis.
p. cm. — (The law and public policy)
Includes bibliographical references and index.
ISBN 1-55798-580-4
1. Forsenic genetics—United States. 2. Forensic pyschiatry—United States. I. Botkin, Jeffrey R. II. McMahon, William M. III. Francis, Leslie, 1946– . IV. Series.
KF8964.G4 1999 99-20607
614′.1—dc21 CIP

**British Library Cataloguing-in-Publication Data**
A CIP record is available from the British Library

*Printed in the United States of America*
*First Edition*

# CONTENTS

# CONTRIBUTORS

**Jeffrey R. Botkin,** Department of Pediatrics, University of Utah

**Dan W. Brock,** Philsophy and Biomedical Ethics Department, Brown University

**Allen E. Buchanan,** Philosophy Department, University of Arizona

**Mary Coombs,** School of Law, University of Miami

**Hilary Coon,** Department of Psychiatry, University of Utah

**Mary Crossley,** Hastings College of Law, University of California

**Rebecca Dresser,** Law and Medicine Department, University of Washington/St. Louis

**Leslie Pickering Francis,** Law and Philosophy Department, University of Utah

**Gerald N. Grob,** Department of History, Institute for Health, Health Care Policy, and Aging Research, Rutgers University

**Samuel B. Guze,** Washington University School of Medicine, St. Louis

**Creighton C. Horton II,** Office of the Attorney General, State of Utah

**Jeffrey A. Kovnick,** Psychiatry Department, University of Utah

**Edward J. Larson,** History and Law Department, University of Georgia School of Law

**Mark Leppert,** Human Genetics Department, University of Utah

**William M. McMahon,** Child Psychiatry Department, University of Utah

**Steven O. Moldin,** Genetics Research Branch, Division of Neuroscience and Basic Behavioral Science, National Institutes of Health

**Lisa S. Parker,** Department of Human Genetics, University of Pittsburgh School of Medicine

**Michael L. Perlin,** New York Law School

**Robert F. Schopp,** University of Nebraska College of Law

**Mark A. Small,** Institute for Families in Society, University of South Carolina

**Daniel A. Summer,** Summer & Summer Law Office, Gainesville, Georgia

**Robert M. Wettstein,** Department of Psychiatry, University of Pittsburgh School of Medicine

# ACKNOWLEDGMENTS

A number of our colleagues have made essential contributions to this project. Jean Nash spent countless hours developing and coordinating the project at all of its stages. Keri Gould provided valuable insight on legal issues and was key in coordinating the participation of several authors. Margaret Battin was an energetic and insightful member of the planning committee from the beginning and a valuable contributor to the discussion. Kimberly Avery and justice Michael Zimmerman provided consistent support through the Snowbird Institute. Ray Gesteland provided valuable ideas and encouragement from the Department of Human Genetics.

Other valuable colleagues to whom we wish to express our thanks include Mellanee Kilpack, Deb Dutson, Dorothy Dart, Lara Taub, and Bill Byerley. Additional funding for this project was provided by the Snowbird Institute, the Department of Psychiatry, and the Division of Medical Ethics at the University of Utah School of Medicine.

We also are deeply grateful for the patience and understanding of our families and for their sacrifices in our pursuit of this project.

# PREFACE

Stephen Mobley walked into a Domino's Pizza store in 1991, robbed the manager, and then shot the young man in the head as he pleaded for mercy. Despite Mobley's ability to commit such a cold-blooded crime, he has been described as a bright, witty man who is well-read and can speak intelligently on a number of topics. Mobley has a history of violence and antisocial behavior from early in his life, as do other members of his family, going back several generations. Mobley confessed to the crime, and his conviction for the murder was prompt. However, prior to his sentencing, scientific information emerged from a research study that gave Mobley and his lawyer some hope—hope that he might be spared the death penalty. A mutation was identified in a gene in a large family in the Netherlands that was associated with violent, antisocial behavior.[1] What if Mobley's family had this same mutation? What if his behavior was influenced by some aberrant brain chemistry induced by an abnormal gene? Could he be held responsible for his actions—*should* he be held as responsible as the rest of us? Perhaps this would not justify an acquittal, Mobley's lawyer reasoned, but it might be enough mitigating evidence to spare his life.

The judge refused the motion to test Mobley for the genetic mutation. Mobley now sits on death row. In the aftermath of this highly publicized case, Mobley's lawyer had a flood of calls from attorneys who wanted to know more about this new avenue for criminal defense.

Scientific developments and social forces are converging that will foster the use of genetic information in courts of law. The courts and prisons are swamped, and the justice system is eager for an objective evaluation that may shed light on an individual's future behavior. Scientific research is actively pursuing the biological underpinnings of animal and human behavior. There is strong evidence that genes play a role in many types of mental illness. Perhaps, as the Netherlands study suggested, there will be genetic factors identified in some instances of violent or antisocial behavior. If such genetic factors exist, they may be relevant for only a few people and even fewer acts of criminal violence. But even if genetics sheds only a little light on violent behavior, the stakes are high in criminal law, and thus genetic analysis and interpretation may become an integral feature of the criminal and juvenile justice systems.

If there are specific genes that predispose people to violence, the potential for appropriate use of this genetic information in the courts is unclear. The fundamental question is whether we as a society care *why* a criminal behaved the way he or she did. For the purposes of prosecution, does it matter if there is a genetic predisposition? Does it matter if there was an environmental predisposition such as, say, an impoverished and abusive home? Should we consider such factors in determining guilt or sentencing? Should it matter whether the accused is an adult or a child in our consideration of "causation"?

Although questions of the appropriate use of genetic information in court are important, the potential abuse of genetic information in courts of law must also be considered carefully. The justice system has authority over the freedom and the very lives of those under its scrutiny. Misunderstandings and misuse of genetic information may have disastrous effects. The eugenics era is a clear reminder of how fragmentary scientific information can be misappropriated for social policy campaigns. It is not difficult to imagine that genetic analysis in the hands of prosecutors could be used to influence decisions about guilt and sentencing: ''Ladies and gentlemen of the jury—the plaintiff has been accused of molesting young children placed under his care and supervision. Evidence will show that as a kindergarten teacher he had ample opportunity to commit these vile crimes. Further, hard scientific evidence will show that he has the genetic makeup of a sexual deviant. . . .'' Such a hypothetical claim is at the complex intersection of law, philosophy, medicine, and science. Lawyers tend to understand little of science or medicine, and scientists and physicians tend to understand little of the law and the U.S. legal system. With these forces in play, the potential for the misuse or abuse of contemporary genetics in the legal system is substantial.

## About This Book

This book is the product of a project funded by the Ethical, Legal and Social Implications Branch of the National Human Genome Research Institute at the National Institutes of Health. The chapters have been written across disciplinary boundaries to address the potential applications of genetic testing in the context of criminal and juvenile law. One purpose of this volume is to assist judges and attorneys in understanding some of the basic science and clinical medicine relevant to mental health disorders and how they might relate to criminal behavior. This information may be increasingly relevant to scientific and medical aspects of criminal cases. In addition, the volume explores the legal philosophy concerning the use of mental health conditions as exculpatory or mitigating factors in criminal proceedings. Judges and attorneys will need to have a clear understanding of these medical and legal issues to adequately and fairly address the use of genetic information on behavioral conditions in criminal law.

This volume is also addressed to scientists, clinicians, and social scientists with an interest in the ethical, legal, and social implications of genetic research. Scientists who develop new technologies have an obligation to consider the potential application in society—both good and bad. The respect with which scientists and physicians are held in society gives them a measure of power in influencing how new technologies are used, even if the applications are beyond their realm of expertise. This book provides an orientation to those without legal education or experience as to how the courts may consider new genetic tests and medical information in the context of criminal and juvenile law.

The book is organized into four parts. The first three parts include an introduction, primary chapters on key issues, and secondary chapters that we have termed ''elaborations.'' The fourth part is made up of primary chapters in the form of conclusions and recommendations. Primary chapter authors sent drafts of their manuscripts to authors of the secondary chapters. The secondary authors were asked both to comment on the chapter drafts and to develop aspects of the subject not covered by the primary chapter authors.

Part I addresses three foundational subjects: the history of efforts to understand mental health disorders; the history of mental health disorders in criminal law, specifically the insanity defense; and the philosophy of free will. This part illustrates that the basic issues involved in the ''genetics defense'' are not new. Genetics offers a new twist to our struggle to understand abnormal behavior and to assign the proper treatments, blames, and punishments.

Part II focuses on important clinical information about mental health disorders and addresses progress and difficulties in genetic research on these conditions. Although there have been many advances in our understanding of the biological origins of behavior, it is becoming increasingly clear that complex behaviors have extensive roots. Simple genetic tests that will provide clear explanations of past behavior or clear predictions of future behaviors are unlikely to emerge.

Part III addresses how the law may deal with new genetic information about mental health problems. More specifically, if there *are* genetic tests for mental disorders, how might the law treat the "genetics defense" for criminal behavior? This part suggests that new genetic discoveries are unlikely to shake the basic assumptions of responsibility on which the criminal justice system is based. Given the less rigid procedures under which the juvenile justice systems operates, there may be wider possibilities for the use (and abuse) of genetic information in this system.

Part IV summarizes the conclusions drawn from the analysis in the first three parts and offers recommendations for future work. The chapters by William McMahon and Leslie Pickering Francis emerged from collaborative discussions between the authors and editors. Mary Coombs was invited by the editors to address the important racial issues raised by the use of genetics in the legal system. The recommendations emphasize the tentative nature of the scientific information, question the potential relevance of genetic links to antisocial behavior in criminal and juvenile law, and advocate interdisciplinary education to foster an informed approach to these issues.

The majority of authors in this volume are profoundly skeptical that a "genetics defense" will find a secure justification in criminal law. Clearly any changes in public policy should arise only after an extended interdisciplinary analysis of the relevant science, law, philosophy, and medicine—a dialogue that this volume is designed to foster.

## Reference

1. Brunner, H. G., Nelen, M., Breakfield, X. O., Ropers, H. H., and van Oost, B. A. "Abnormal Behavior Associated With a Point Mutation in the Structural Gene for Monoamine Oxidase A." *Science* 262 (1993):578–580.

# Part I

# History of Genetic Research and the Philosophy of Free Will and Determinism

# INTRODUCTION TO PART I

Part I establishes a historical and philosophical foundation for the subsequent discussions of applications of genetic testing in criminal and juvenile law. In the first chapter, Gerald Grob, a historian of medicine, describes the cycles of professional enthusiasm and disappointment over the past two centuries engendered by efforts to understand mental health disorders. New insights have repeatedly offered prospects for effective treatments or preventive measures for these devastating conditions. Once again, genetic research is rekindling hope that modern science is "on the threshold" of success. Grob's analysis illustrates that how society and the medical profession conceptualize the nature of psychiatric conditions has a profound effect on how society organizes and manages the care of these chronically ill individuals. Ed Larson, a historian and legal scholar, picks up this theme with a focus on the eugenics era in which the distinction between mental illness and criminal behavior was blurred. During the early decades of this century, the belief in hereditary causes of abnormal and criminal behavior was influential, fostering the widespread use of forced sterilization to combat "feeble-mindedness" and criminal insanity. Explanations of social problems swung to environmental and cultural factors after World War II, but Larson notes that contemporary excitement over human genetics heralds a return to biological explanations of and remedies for mental health disorders.

Michael Perlin, a legal scholar, provides a fascinating history of the "insanity defense," including the popular misconceptions about this defense and the complex political influences that have shaped its application in the courts. Perlin describes the profound hostility of the public toward this defense, noting that it has become one of the most controversial aspects of the criminal law. Rather than providing a slick escape for clever defendants, however, the insanity defense is infrequently used, often unsuccessful, and, when "successful," typically leads to a longer incarceration than convictions following standard defenses. Perlin argues that little about the criminal justice system's approach to mental illness has been based on careful attempts to understand the science and medicine of these conditions.

In chapter 3, Dan Brock and Allen Buchanan, both philosophers and bioethicists, step back from debates in the legal system to consider the broader philosophical implications of genetic factors in human behavior. What do genes that are associated with specific kinds of behavior mean for our concept of free will? Freedom or agency are central defining concepts for persons in Western thought. Brock and Buchanan emphasize, however, that whether our actions are determined by a host of complex internal and external factors is ultimately a metaphysical question and therefore not open to proof through empirical data. Although challenges to the concept of free will are not new, Brock and Buchanan develop the implications of new genetic information on our common *beliefs* and *justifications* concerning our freedom to act according to our wills. If, indeed, genes are found that predispose individuals to certain behaviors, the implications are several: There is potential for society to vigorously protect itself from those perceived to have diminished responsibility, while at the same time offering prospects of expanding control of those genetic factors that limit our freedom.

Lisa Parker, a philosopher based in a department of human genetics, develops a different set of implications of genetic determinism in her elaboration. If genes are associated with certain behaviors, genes are likely to exert their influence through behavioral predispositions rather than through any direct triggers for action. If so, and if individuals become aware of their predispositions through genetic testing, then Parker argues that perceptions of individual responsibility will *increase* rather than decrease. If one knows one's genetic weaknesses, then presumably there will a responsibility to avoid the social or environmental circumstances that can lead to trouble. Parker also argues that society may gain new responsibilities as we better understand what social and environmental factors cause problems for those with certain biological predispositions.

Robert Schopp, a scholar of law and psychology, strengthens this theme of enhanced individual responsibility through genetic information in his elaboration. Schopp explores our notions of free will and responsibility through a number of case vignettes. These cases illustrate how we often make complex intuitive judgements about responsibility for actions, depending on a variety of potentially mitigating factors. Schopp suggests that we need to develop more sophisticated conceptions of free will and responsibility to account for numerous internal and external influences on behavior and our rational judgements about those behaviors. If responsibility flows from the decisions of competent, practical reasoners, Schopp argues, then increased knowledge of genetic predispositions to certain behaviors ultimately will increase responsibility to manage personal behavior in conformity with society's moral and legal rules.

# Chapter 1
# ON THE THRESHOLD:
## Illusion and Reality in American Psychiatric Thought

Gerald N. Grob

As human beings we generally inhabit two different worlds simultaneously. The first is characterized by contingency, indeterminacy, and an inability to comprehend or control the numerous variables that shape our environment. Our judgments, analyses, and actions often represent a pragmatic response to a seemingly intractable and partially incomprehensible universe. The second is an imaginary and idealized world—one characterized by certainty and clarity, where pure and precise knowledge leads to a kind of understanding that enables human beings to cope with or to solve perennial problems. The static nature of this idealized world fosters the illusion that the creation of a veritable utopia is within reach.

This characterization can be applied to the specialty of psychiatry as well as to medicine in general. As clinicians, psychiatrists are compelled to deal with individuals whose pathology is rarely amenable to therapeutic clarification or simple prescriptions. To be sure, the specialty has developed elaborate nosological systems that group together seemingly like diseases and conditions. These classification systems, however, are neither inherently self-evident nor given. On the contrary, they emerge from the crucible of human experience; change and variability, not immutability, are characteristic. In a classic work in 1838, Isaac Ray conceded that no classification could be "rigorously correct." Disease categories often overlapped and differences over diagnoses were common. "But such is the order of nature," he added, that "we must make the most of the good it presents and remedy its evils in the best manner we can."[1] At a quite different level, psychiatrists provide elaborate theoretical explanations of the nature and etiology of mental disorders. These explanations usually include a blend of psychological, environmental, physiological, and genetic factors, although the relative importance of each changes over time. Successive generations generally perceive their own explanations to be a closer approximation to truth. If theories are not immediately verifiable by empirical data, they are justified by the claim that in the near future it will be possible to cross a ubiquitous threshold and arrive at a full understanding of those mental disorders that have plagued humanity throughout recorded history, and probably beyond.

In offering such a simplified caricature, I do not mean to denigrate psychiatry in particular or medicine in general. Every discipline—whether in the physical, natural, social, or behavioral sciences—has similar characteristics; its members are invariably committed to a search for final and irreducible truths. The need to provide overarching explanations, moreover, is not only a function of the internal dynamics of disciplines. On the contrary, all societies demand explanations to account for the unexpected or abnormal. Like their other medical brethren, psychiatrists traditionally provided both an explanation for abnormal behavior and a rationale for treatment and prevention. Had they rejected such a role, they might have impaired their professional legitimacy and prepared the way for other groups willing to meet perceived social needs. In this chapter I should like to illustrate some of these generalizations by examining the evolution of American psychiatric thought and (to a lesser extent) practice over the last two centuries, and to conclude by offering some personal observations. This discussion will serve to place contemporary thinking about biologic underpinnings of psychiatric illness and antisocial behavior in its broader social context.

## Holistic Concepts of Mental Health in the Nineteenth Century

Before 1820 the literature dealing with mental illnesses in the United States was relatively sparse. Insanity was peripheral to the lives of laypeople and physicians alike; the primary concern of both was with endemic and epidemic diseases that posed a threat to life itself. A few physicians, including the redoubtable Benjamin Rush, had written about insanity, but they tended to be exceptions. The creation of mental hospitals beginning in the second quarter of the nineteenth century, however, introduced a new element. Physicians associated with these institutions limited their activities to a single group of patients rather than a representative cross-section. The physical health problems of mental hospital patients were not novel. Yet the reasons for their confinement were specific and unique, and asylum doctors quickly began to distinguish themselves from their colleagues in general practice. Beginning in the 1830s and accelerating in succeeding decades, physicians employed in mental hospitals (either as superintendents or assistant physicians) created one of the earliest medical specialties and produced an extensive literature dealing with mental illnesses. In search of professional legitimacy, the young specialists insisted that they could merge theory and data to clarify phenomena that in the past had eluded human understanding. Not only did they seek to confer legitimacy on their institutions and specialty, but they hoped that their rational explanations of insanity would foster public confidence. The resulting atmosphere of trust would then remove the negative stigma often associated with mental disorders and facilitate the early commitment of insane persons by their friends and families.[2]

Most psychiatrists, like physicians in general, conceived of disease in individual rather than general terms. Health was a consequence of a symbiotic relationship among nature, society, and the individual. Disease represented an imbalance that followed the violation of certain divine natural laws that governed human behavior.[3] Insanity occurred when false impressions were conveyed to the mind because the brain or other sensory organs were impaired. Mental illnesses were perceived to be somatic and to involve lesions of the brain, the organ of the mind. To have argued otherwise would have approached blasphemy. If the mind itself (often equated with the soul) could become diseased, it could conceivably perish. The immortality of the soul, on which Christian faith is based, would thereby be denied or negated.[4] Most American psychiatrists were reared in a Protestant culture, and instinctively rejected a model of disease that threatened traditional moral and religious values.

Such reasoning provided asylum physicians with a model of mental illnesses that was especially compatible with the belief that such illnesses were precipitated by psychological, environmental, and somatic factors interacting with the constitution or predisposition of the individual. Excepting accidental causes (e.g., a blow to the head), insanity was generally brought on by a frequently unknowing violation of certain natural laws that governed human behavior. Immorality, improper living conditions, or unnatural stresses that upset the natural balance of the individual could also precipitate mental disorders. Mid-nineteenth-century psychiatrists managed to conflate moral values and science into their model of insanity.

The holistic concept of mental illnesses that saw physical disorder of the brain as the cause of the disorder required an act of faith by most U.S. psychiatrists. Except for a few cases in which autopsies revealed the presence of a brain tumor or other gross abnormality, the link between the brain and madness remained a mystery. Given their inability to demonstrate a relationship between anatomical changes and behavior, psychiatrists identified mental disorders by observing external signs and symptoms. In this respect they were no different from other physicians who defined pathological states in terms of visible and external signs (such as fever). Although disagreeing on the diagnosis of diverse signs and symptoms, few physicians questioned this approach, if only because they could conceive of no alternative. Psychiatrists accepted disease as a given; the inability of patients to function, combined with severe behavioral symptoms, was sufficient evidence of the presence of pathology. An act of faith notwithstanding, most psychiatrists were enthusiastic at the prospect of future progress; they believed themselves poised to cross the threshold and enter a new era of enlightenment. "We may therefore reasonably expect within a short period valuable additions to our knowledge of the nervous system," Amariah Brigham, a prominent hospital superintendent and one of the leaders of the young specialty, wrote in 1840.[5]

Most asylum doctors were vaguely aware of the intellectual and logical difficulties that followed from their somatic view of insanity. Their references to lesions generally lacked empirical support, and acceptance of their presence rested on faith rather than observation. Classification of mental disorders (nosology) was equally problematic, for there was no way to relate brain physiology to specific behavioral patterns. Indeed, nomenclature and classification aroused little enthusiasm among mid-nineteenth-century U.S. psychiatrists, if only because diagnosis was not related to specific therapies or to specific causes. No system of classification, conceded Brigham in 1843, appeared to be of "much practical utility." All categories based on symptoms "must be defective, and perhaps none can be devised in which all cases are arranged."[6] Nor were Brigham's views unique. Samuel B. Woodward, head of the Worcester Hospital, had observed earlier that insanity was a "unit, indefinable . . . easily recognized . . . [but] not always easily classified." He believed that therapy was independent of any nosological system and had to reflect the unique circumstances presented by each individual case.[7] And when Pliny Earle, one of the most important asylum physicians of the nineteenth century, was approached toward the end of his long career on the possibility of developing a universally accepted classification of mental disorders, he took issue with the idea. "In the present state of our knowledge," he observed,

> No classification of insanity can be erected upon a pathological basis, for the simple reason that, with but slight exceptions, the pathology of the disease is unknown. Hence psychiatrists were still forced to fall back upon symptomatology—"*the apparent mental condition,* as judged from the outward manifestations." Nineteenth-century nosologies—unlike their twentieth-century counterparts—were simple, and employed such categories as mania, monomania, melancholia, dementia, and idiocy.[8]

Recognition of the difficulties that blocked the formulation of a comprehensive nosology, however, did not impede discussions of the origins and causes of mental disorders. Despite an inability to demonstrate meaningful relationships between physiology and the presence of particular behavioral signs or symptoms, the social and cultural role of medicine required that physicians—psychiatrists and others—provide some explanation of disease processes.

The causes of mental illnesses, most psychiatrists agreed, could be subsumed under two general headings—physical and "moral" (i.e., psychological). The physical causes—a blow to the head, a disordered organ other than the brain, various somatic illnesses—affected the structure of the brain, thereby impairing cerebral functioning. Most psychiatrists, however, were far more concerned with the "moral" causes of insanity, if only because of their inability to influence presumably unknown physiological processes. The "moral" causes of insanity (which seemed to account for the majority of cases) were in some respects related to inappropriate behavioral patterns; individuals who ignored the natural laws that governed human behavior placed themselves at risk. Lifestyle thus became a key analytic category in causal explanations. When referring to "laws," psychiatrists were not endorsing a deterministic universe; rather, law was a moral construct that stipulated, but did not necessarily require, ideal behavioral forms. The moral causes of insanity included—to cite a few examples—intemperance, masturbation, overwork, domestic difficulties, excessive ambitions, faulty education, personal disappointments, marital problems, jealousy, and pride. Nor was individual lifestyle the only determinant of illness and health. Social relationships and institutions also played a role. There was general agreement that insanity and civilization were linked; the pressures of an urban and commercial society influenced prevailing etiological patterns.[9]

Within this protean model heredity played a minor role at best. The term most frequently employed in mid-nineteenth-century literature was *hereditary predisposition.* The structure of the brain, to be sure, was transmitted from parents. But even insane parents did not bequeath to their offspring a defective brain that would lead to insanity. Without "the agency of moral causes," wrote Brigham, predisposition would not by itself lead to disease.[10] In this sense there was no appreciable difference between insanity and other diseases. All were subject, wrote Isaac Ray (perhaps the most important psychiatrist of his generation) to the "ordinary relations of cause and effect." Those who inherited a predisposition to mental disease had it within their power to avoid precipitating "incidents and conditions." There was reason to believe, he added, that many persons "have warded off an attack of [mental] disease, by looking the evil firmly in the face, and resolutely shunning, in their diet, regimen, habits, occupations, amusements, mental and bodily exercise of every description, whatever might be supposed likely to produce unhealthy excitement."[11]

Nineteenth-century hereditarian ideology was extraordinarily fluid. The Lamarckian view that acquired characteristics could be inherited prevailed; genetic endowment involved tendencies and predispositions rather than fixed or immutable qualities. Environment, therefore, remained the decisive element. In one of the most important mid-nineteenth-century demographic investigations of mental illnesses, Edward Jarvis employed an environmental explanation to account for the high rates of insanity among Irish immigrants to the United States, as compared with far lower rates among native-born citizens. "There is good ground for supposing," he noted, "that the habits and condition and character of the Irish poor in this country operate more unfavorably upon their mental health, and hence produce a larger number of the insane in ratio to their numbers than is found among the native poor." Being in a strange culture, ignorant of American customs, and lacking education and flexibility, they proved unable to adapt easily to their new environment. Their intemperate drinking habits, Jarvis added, only compounded their personal frustrations, and the result was high rates of insanity.[12] As late as 1883 Henry P. Stearns conceded the existence of an "insane" or "nervous diathesis," but also identified two channels of relief. The first involved the union of a healthy parent with one with a predisposition toward disease; the result was the neutralization of the latter's morbid tendencies. The second was an emphasis on education, which could "modify proclivities toward those morbid neuroses which result from hereditary influences."[13] The inability to substantiate claims about heredity in no way inhibited definitive pronouncements of one sort or another. Thus William A. Hammond would argue that the tendency toward certain diseases "is derived from the seminal fluid of the male, and in an equal or perhaps greater degree from the ovaries of the female, does not admit of a reasonable doubt."[14]

Race was an equally problematic category. Although the mid-nineteenth-century medical and scientific communities engaged in an extended debate over the relative mental capabilities and the unity or separate origins of the different races, psychiatrists remained distant from such discussions. As hospital administrators, they were involved with managerial issues and devoted little time to speculation about racial abstractions. As a consequence, their attitudes and practices were shaped by prevailing community mores and sentiments, and they concurred with the belief that patients of different races could not be kept together in integrated facilities. To be sure, psychiatrists from the American South—more so than their Northern brethren—believed that susceptibility toward insanity was in part shaped by race. Although such attitudes became somewhat more common toward the close of the nineteenth century, they never gained the allegiance of the specialty as a whole.[15] Even before World War II racial explanations of mental disorders found relatively little support among psychiatrists.[16] Although race clearly shaped patterns of care and treatment, it never became a distinctive element in psychiatric thought; racial theory drew its strength from other sources.

To define the nature and etiology of mental illnesses was only a beginning, not an end. The goal of psychiatry—like that of medicine generally—was the alleviation and cure of disease. Despite their recognition that disease processes remained shrouded in mystery, most psychiatrists believed that mental illnesses could be cured. Because insanity was in large part precipitated by improper behavioral patterns and psychological reactions associated with a deficient environment, it followed that treatment had to begin with the creation of a new and presumably appropriate environment. Home treatment was ineffective, for the physician had no means of controlling the domestic environment in which the disease began or eliminating undesirable influences. Institutionalization, therefore, was a sine qua non. Once in a hospital, the patient could be exposed to a judicious mix of medical and moral (i.e., psychological) treatment. Although moral treatment was seen as conceptually distinct, mid-nineteenth-century psychiatrists did not distinguish in practice between it and medical therapy, although the former was generally regarded as more effective than the latter.[17] The indissoluble linkage of mind and body required that careful attention be paid to both; to neglect one for the other would only perpetuate the continued presence of the disease process. In effect, early psychiatrists were reiterating the ancient belief that health involved both mind and body.[18] Heirs to a body of traditional medical thinking that dated to antiquity, asylum physicians perceived of disease as a holistic entity that could not be disaggregated into specific categories. Health represented a physiological equilibrium; disease occurred because of imbalance. The goal of treatment was to restore the balance.

In the 1830s and 1840s mental hospital superintendents (i.e., psychiatrists) claimed striking therapeutic successes. Mental illnesses, they averred, if treated in the early stages, were more easily cured than most diseases. Some claimed a recovery rate of between 80 and 100 percent. In the second edition of his classic work on mental hospitals (first published in 1854), Thomas S. Kirkbride reiterated his faith in the curability of mental diseases. Nor did recidivism temper the prevailing optimism. A recurrence of insanity—such as a recurrence of an infectious disease—represented a new case. Mental hospitals, Woodward wrote in 1842, "are blameless for the numerous recommittals, while those who have recovered from insanity will throw themselves into these channels of excitement, and seek rather than avoid these known causes of disease. This should not be so. If those who are predisposed to insanity would avoid these and many other known causes, they might safely pass on, and, in most cases, continue well."[19]

In annual reports, articles, speeches, and books, the founding generation of American psychiatrists offered an extraordinarily optimistic vision. Insanity was both preventable and curable. Americans simply had to follow a particular lifestyle to avoid disease. Those who had the misfortune to become ill could be cured, assuming that states met their ethical responsibilities by providing sufficient resources to create institutional systems capable of providing care and treatment to all those in need. Mental hospitals, Kirkbride insisted, "can never be dispensed with,—no matter how persistently ignorance, prejudice, or sophistry may declare to the contrary—without retrograding to a greater or less extent to the conditions of a past period, with all the inhumanity and barbarity connected with it."[20]

## Emerging Biologic Concepts in Mental Health

The fortuitous circumstances that created the conditions conducive to apparent therapeutic successes and the provision of care, however, proved relatively short-lived. By the latter half of the nineteenth century the structure and functions of mental hospitals—and therefore of the specialty of psychiatry—had begun to undergo a gradual transformation. Intended as small curative institutions that fostered close relationships between the medical and lay staffs and patients, hospitals grew in size and complexity. Considerations of order and efficiency began to conflict with therapeutic imperatives. The vision of a harmonious institution proved difficult to implement. The realities presented by an increasingly diverse patient population that included individuals who often behaved in bizarre and disruptive ways led to friction with the medical and lay staff. In theory all patients were to receive the same quality of care. In practice the variables of class, race, ethnicity, and gender resulted in internal distinctions. Conceived as self-contained and independent institutions, asylums in fact were shaped by legal, administrative, political, and fiscal environments in which they existed, and as a consequence faced problems similar to those at other public and social institutions. In short, major differences developed between the ideology of asylum care and the realities of institutional life.[21] The publication of Pliny Earle's book on curability in 1887 was indicative of the shift. One of the founding patriarchs of the specialty, Earle in his later career repudiated the curability claims of the 1830s and 1840s even while defending the legitimacy of and need for mental hospitals.[22]

Increasingly large numbers of aged and chronic patients after 1890 hastened the transformation in the character of public mental hospitals. Because of their desire to shift fiscal responsibilities from the community to the state level, local officials redefined senility in psychiatric terms. This redefinition facilitated the closing of almshouses, which in the nineteenth century served in part as old age homes, and led to the transfer of aged individuals without financial means to mental hospitals.[23] Internally, the change in the nature of the patient population created a more depressing environment. To cure and discharge patients was associated with an aura of optimism and achievement; it reflected a predominantly male ethos that valued effective therapy above all other functions. To care for those who rarely manifested improvement and ultimately died was hardly consistent with modern images of medical and scientific progress; the caring function was identified with females and with hopelessness and ineffectuality and thus was often devalued. For psychiatrists the rise of custodialism created negative images of themselves, their work, and their institutions. Ultimately this situation led them to reexamine their position and to become receptive to new roles that shattered their hitherto inseparable links with mental hospitals, which then began to lose some of their social legitimacy.

Nowhere was the eclectic and divided nature of the specialty better revealed than in the famous trial of Charles J. Guiteau, the assassin of President James A. Garfield in 1881. In a trial that drew national attention, the mental status of the defendant became the key issue. John P. Gray, superintendent of the Utica Asylum and editor of the *American Journal of Insanity,* testified for the prosecution, and Edward C. Spitzka, a young German-trained neuroanatomist and bitter foe of hospital psychiatry, appeared on behalf of the defense.[24]

Gray had rejected the extremes of idealism and materialism; the former separated mind and body, and the latter reduced all phenomena to purely physical explanations that led to an absolute determinism. An adherent of the Scottish common sense or moral philosophy school, Gray accepted the concept of an objective physical reality and the belief that human beings possessed innate moral faculties independent of the material world. Insanity required the presence of cerebral disease, and he emphatically rejected any hereditarian explanation on the grounds that it abrogated individual responsibility; free will and sin were realities in human affairs. Paradoxically, Gray was among the first to appoint a pathologist at his hospital to conduct postmortem examinations in the hope of illuminating the pathology of insanity.[25]

Gray's testimony was a logical extension of his psychiatric thinking. He insisted that Guiteau (whom he had examined during a lengthy pretrial interview) was rational and sane. There was a fundamental distinction between insanity—the "offspring of disease"—and "mere demoralization of character"—"the offspring of education in vicious lines of thought and conduct and from the indulgence of passions." Heredity, moreover, involved merely the transmission to children of "susceptibility." Gray argued that Guiteau's act was assuredly not the product of insanity, but rather of depravity and immorality as evidenced by his previous life.[26]

Spitzka, by contrast, was influenced by the European concept of "degeneration," which rested on a vague hereditarian foundation. Guiteau's behavior, he averred, could only be understood as a product of heredity, possessing as he did a brain that represented a structural atavism. The taint of insanity could be seen by an examination of the Guiteau family genealogy—a view subsequently echoed by George M. Beard, the American popularizer of the concept of neurasthenia (a disease that grew out of the unique conditions of American life and whose symptoms included mental, and physical exhaustion and an inability to perform mental or physical work). In a text published a few years after the trial, Spitzka argued that individuals could be born with such a "defective or perverted anatomical foundation" that they "may properly be said to be born insane."[27]

The contradictions between theory and practice slowly but surely fostered changes in psychiatric thinking. Older patterns never disappeared completely. Instead they underwent subtle modifications as the balance between the relative importance of environment, heredity, and physiology shifted, and the faith in the efficacy of older therapies diminished. This is not to suggest that psychiatrists abandoned their efforts to illuminate the nature and etiology of mental disorders, or that they adopted pessimistic or fatalistic explanations. What occurred instead was a reformulation of theories of mental disorders that appeared more consistent with concepts that were transforming general medical thought and practice in the late nineteenth century.

After 1870 American medicine underwent a fundamental transformation. Increasingly it became identified with bacteriology and other biological and physical sciences. At the same time the general hospital began to assume its modern form, and authority slowly shifted from lay trustees to physicians. This trend mirrored the transformation of the general hospital from an institution providing care for socially marginal groups to one reflecting a new emphasis on science and technology and catering to more affluent groups capable of paying the high costs involved.[28]

The emergence of modern scientific medicine had a profound impact on the specialty of psychiatry. In the mid-nineteenth century asylum physicians enjoyed high status, and even rebuffed efforts by the fledgling American Medical Association to induce what is today the American Psychiatric Association to affiliate.[29] By the turn of the century, the respective roles of physicians and psychiatrists had been reversed. Because psychiatrists were employed overwhelmingly in public mental hospitals and seemed far removed from the institutions that defined the character of the new scientific medicine (laboratories, medical schools, and teaching hospitals), they were perceived by their medical colleagues to be hopelessly anachronistic and out of touch. Indeed, in 1894 S. Weir Mitchell, the renowned neurologist, castigated institutional psychiatrists for their isolation. "You were the first of the specialists," he noted, "and you have never come back into line. It is easy to see how this came about. You began to live apart and you still do so. Your hospitals are not our hospitals; your ways are not our ways. You live out of range of critical shot; you are not preceded and followed in your wards by clever rivals, or watched by able residents fresh with the learning of schools." Mitchell went on to deplore the absence of a spirit of scientific research, the distrust of asylum therapeutics, and the obsession with administrative and managerial concerns.[30]

Reacting to widespread criticism, psychiatrists began to redefine the intellectual foundations of their specialty in the hope of narrowing the gap between themselves and their medical colleagues. There was general agreement that the managerial and administrative functions could no longer define the nature of the specialty. Edward Cowles, superintendent of the oldest private mental hospital, paid homage to the founders of American psychiatry, but insisted that the time had come to move in a new direction. In his presidential address in 1895, he defined the new psychiatry in terms far removed from the founding generation. The alienist, he noted, must be "a student of neurology, and uses its anatomy and physiology." Equally important, psychiatry would increasingly draw on contributions from organic chemistry and bacteriology. "Thus it is that psychiatry is shown, more than ever before, to be dependent upon general medicine." Indeed, Cowles anticipated a theme that would become more significant—namely, the "toxic causation of [mental] disease."[31]

The new psychiatry that took shape after 1900 tended to be eclectic. Explanations of the nature and etiology of mental disorders continued to incorporate somatic, psychological, environmental, and genetic factors even though sharp disagreements persisted over the relative merits of each. At one extreme were purely somatic explanations that had the virtue of blurring any distinctions between psychiatry and medicine. In his presidential address in 1901, Peter M. Wise noted that clinical experience and laboratory findings were leading psychiatrists to the conclusion that cell integrity was dependent "more upon chemical processes than upon structure." Mental pathology, therefore, was due to chemical changes even though present techniques could not validate this claim.[32] At about the same time Henry J. Berkley, a clinical professor of psychiatry at Johns Hopkins, divided mental disorders into two parts; functional maladies and organic–degenerative disorders. Although he identified heredity as "the most important factor . . . [in] the pathology of the main types of insanity," he mentioned the subject on only 3 of the book's nearly 600 hundred pages.[33] Stewart Paton, also of Hopkins, wrote in 1905 that there was too much "glib talk about the problems of heredity," and that the uninitiated were led to believe that much was known. The inheritance of mental traits was "far more complex than those encountered in dealing with the transmission of mere physical qualities." Much of the data purporting to prove the case for heredity, he added, was virtually worthless.[34]

The older somatic tradition, however persistent, nevertheless faced new challenges after the turn of the century. "Dynamic psychiatry" (the name by which it was known) offered an alternative to purely somatic explanations by creating new models of mental disorders. Generally speaking, nineteenth-century psychiatrists made a fundamental distinction between health and disease. The presence of mental illnesses was indicated by dramatic behavioral and somatic signs that deviated from the prior "normal" behavior of the individual. The new model of psychic distress, by contrast, suggested that behavior occurred along a continuum that commenced with the normal and extended to the abnormal. Such an approach elevated the significance of the life history and prior experiences of the individual, thereby blurring the clear demarcation between health and disease. Indeed, psychiatric intervention began to emerge as a distinct option well short of the acute stage of the mental illness. From here it was but a short step to suggest that early outpatient treatment either in offices or clinics might prevent the onset of the severe mental disorders that up to that time had required institutionalization. Indeed, the psychiatric jurisdiction began to expand to include not simply pathology, but the psychology of the normal as well.[35]

Dynamic psychiatry was indissolubly linked with the career of Adolf Meyer, who shaped several generations of American psychiatrists following his move to the newly created Henry Phipps Clinic at the Johns Hopkins School of Medicine in 1910. Trained originally in neurology, Meyer shifted his interests to psychiatry following his migration to the United States from Switzerland in 1892. Rejecting Cartesian dualism, he posited a ''biological conception of man'' that integrated somatic and psychological elements. All mental reactions had physiological counterparts; conversely, purely psychical (i.e., functional) disorders, just as organic lesions, were disorders of the life of the brain. The goal of psychiatry was to integrate the life experiences of the individual with physiological and biological data. Because every individual constituted ''an experiment of nature,'' an understanding of life history was crucial. Mental disorders were largely behavioral in that they involved defective habits whose origin lay in the experiences of early life.[36]

Within Meyer's genetic–dynamic framework, heredity occupied a relatively minor role. Aware of the early eugenics movement that took shape around the turn of the century, he was nevertheless skeptical of its claims. Existing data made it virtually impossible to apply Mendelian rules, if only because ''definitions of unit characters'' were nonexistent. We do not know enough, Meyer wrote in 1896, ''of the facts that underlie the acknowledged influence of heredity and the laws of growth and development.'' Indeed, his belief in the crucial role of the previous life history of the individual ensured that he would never be overly committed to a genetic explanation of mental disorders.[37]

Dynamic psychiatry never became a closed ideological movement that insisted on allegiance to a single set of beliefs; diversity rather than unity was characteristic. Nevertheless, it had a transforming effect on the specialty and its institutions. It expanded the role of psychiatry to include psychologically troubled individuals as well as allegedly dysfunctional social structures and relationships. It began the process of breaking the hitherto inseparable links between psychiatry and mental hospitals; outpatient institutions and private practice in the future would begin to increase in significance. ''I regard the future of mental medicine as filled with golden promise,'' prophesied Charles G. Wagner in his presidential address in 1917. ''Serious, thoughtful students of psychiatry are busily at work on problems of vital importance, and I venture to predict that within the period of a decade or two their labors will result in a much better understanding of the etiology, pathology, diagnosis and treatment of mental diseases than we now possess.''[38] The specialty, in other words, would cross the ubiquitous threshold within a generation and thus clarify age-old problems.

Wagner's optimistic comments notwithstanding, American psychiatry remained a divided and ambivalent specialty, if only because the problems of explaining human behavior, normal or abnormal, seemed beyond human capacity. The very concept of mental disease, for example, could not be separated from the deeper and more profound problem of explaining the nature of human beings in general and their behavior in particular. At one extreme were certain deterministic systems that reduced behavior to physiological mechanisms and ruled out independent thought or actions that did not have specific causal antecedents. Most psychiatrists, however, adhered to eclectic models that posited a link between mental and biological factors. But the nature of such links remained shrouded in mystery, and the very concept of mental phenomena posed seemingly unresolvable theoretical difficulties.

In other branches of medicine, by contrast, the demonstration of a relationship between the presence of certain symptoms and a specific bacterial organism had led to the development of a new classification system based on etiology rather than symptomatology. The inability to pursue a parallel course left psychiatry with a classification system based on external symptoms that tended to vary in the extreme. Conclusive evidence that paresis was actually the tertiary stage of a disease that began with a prior syphilitic infection offered one possible model for psychiatric diseases; the same was true of pellagra, a disease with a dietary etiology. Nevertheless, neither psychiatrists nor pathologists were able to identify other specific disease entities in comparable terms. Indeed, Simon Flexner, director of the Rockefeller Institute for Medical Research, was dubious about even undertaking neuropsychiatric research; he virtually insisted ''that there were no problems in a fit state for work.''[39] Flexner's position was not accepted by members of the specialty; to concede its validity might undermine their legitimacy.

Nevertheless, the striking successes in identifying specific diseases with specific organisms led some psychiatrists to adopt parallel explanations of the etiology of mental disorders. By the second decade of the century the focal infection theory already adopted in general medicine had begun to gain some acceptance among institutional psychiatrists. Its most enthusiastic proponent was Henry A. Cotton, who had begun his professional career as an intern under Meyer at the Worcester State Hospital in Massachusetts. Cotton was subsequently converted to the view that chronic, masked, or focal infections played a ''very important role in the etiology of the psychoses.'' Persuaded that there was a definitive link between infections and mental disorders, Cotton embarked on a system of therapy at a New Jersey state hospital during the 1920s that included the extraction of all teeth, tonsillectomy, colostomy, and colectomy.[40]

The ferment within psychiatry in the early part of the twentieth century was also marked by institutional change. By the turn of the century, two innovations had appeared: the research institute and the psychopathic hospital. The creation of such institutions was related to changes in medicine in general. The specific germ theory of disease suggested an explanation that was empirically verifiable and that seemed to point the way toward the development of specific therapies. Scientific and technological innovation also created conditions that made possible the emergence of the general hospital in its modern form. Why could not psychiatry emulate the successes of biological medicine and create new institutional forms? By the early part of the new century the New York Psychiatric Institute, the Henry Phipps Psychiatric Clinic, and Boston Psychopathic Hospital had come into existence. Unlike traditional mental hospitals, these institutions had multiple functions; the treatment of individuals in the early or acute stages of psychological distress, research, and training. One of their goals was to provide short-term and immediate therapy, thus, it was hoped, obviating the need for commitment to state hospitals.[41]

Paradoxically, the optimistic expectations of psychiatrists and the institutional innovations of the early twentieth century did not significantly alter the lives of the thousands of patients in traditional mental hospitals. The number of strictly psychopathic institutions was never large, and most dealt with relatively small numbers of young patients being treated for conduct disorders rather than psychotic conditions. For the majority of patients in traditional mental hospitals—including the aged and paretic—there were no known therapeutic interventions to eliminate the source of their difficulties. Yet most psychiatrists never conceded their inability to deal with the kinds of patients found in public institutions. On the contrary, the illusion of power and knowledge created its own perceptions of rapid change. Such was the case in the early twentieth century when a vision of a scientific psychiatry integrated with modern medicine held sway.

Another visible symbol of change was the creation of a mental hygiene movement after 1900. Reflecting a commitment to science, mental hygienists saw disease as a product of environmental, hereditary, and individual deficiencies; its eradication required a fusion of scientific and administrative action. As members of a specialty that they believed was destined to play an increasingly central role in the creation of a new social order, psychiatrists began to broaden the scope of their discipline. The emphasis on scientific research rather than care or custody, on disease rather than patients, and on alternatives to the traditional mental hospital was merely a beginning. More compelling was the utopian idea of a society structured in such a way as to maximize health and minimize disease. In 1917, Thomas W. Salmon, a key figure in the reorientation of psychiatry, spoke about the future of the specialty. The new psychiatry, he insisted, had to reach beyond institutional walls and play a crucial part "in the great movements for social betterment." Psychiatrists could no longer limit their activities and responsibilities to the institutionalized mentally ill. On the contrary, they had to lead the way in research and policy formulation and to implement methods in such areas as mental hygiene, care of the feeble-minded, eugenics, control of alcoholism, management of abnormal children, treatment of criminals, and to help in the prevention of crime, prostitution, and dependency.[42]

The emphasis on mental hygiene took programmatic form with the founding of the National Committee for Mental Hygiene in 1909. This organization was the creation of Clifford W. Beers, who had spent time in several mental hospitals in his early adulthood. In 1908 he published his classic *A Mind That Found Itself*.[43] Beers's initial goal was to create a national society dedicated to improving conditions for insane individuals. But as a result of his contacts with several leading figures in the field—notably Meyer—Beers's original emphasis on institutional improvement was replaced by a more ambitious, but amorphous, goal of promoting mental health. The attractiveness of mental hygiene for psychiatry was obvious; it opened new vistas and occupational roles for practitioners while shifting attention away from a custodial role and inability to cure admittedly vague disease entities. Mental hygiene also had the virtue of hastening the integration of psychiatry and medicine, because it provided the former with the rhetoric of a biologically oriented specialty and thus seemed to put it in step with other medical prevention movements. When Salmon became medical director in 1912, he began to reshape the focus of the National Committee for Mental Hygiene and further diminish its preoccupation with mental hospitals. The surveys conducted under its auspices, once limited to institutionalized mentally ill people, began to deal with truancy, sexual immorality, juvenile delinquency, and other similar categories. During the 1920s the National Committee also became involved with the child guidance movement.[44]

The growing preoccupation with environmental explanations of mental disorders and preventive strategies did not mean that hereditarian thinking had disappeared. Indeed, by the turn of the century a eugenics movement had taken shape. Its roots lay in the European fascination with theories of degeneration associated with B. A. Morel, Cesare Lombroso, and Max Nordau. To be sure, a figure like George M. Beard developed his own uniquely American adaptation in which neurasthenia became the primary stage of disease that in later life became transformed into more serious diseases, which in turn could be transmitted to offspring. Similarly, Richard L. Dugdale offered his study of the Jukes as an example of familial degeneration, although he conceded that environmental change could reverse genetic decline.[45] After 1900, however, the neo-Lamarckianism that dominated earlier theories was seemingly discredited by the work of such figures as August Weismann and the rediscovery of Mendelian genetics. Fearful that an alleged increase in degeneracy in general and mental disorders and feeble-mindedness in particular threatened the biological well-being of the American people, a nascent eugenics movement began to support a variety of interventionist measures, including marriage regulation, immigration restriction, and involuntary sterilization.[46]

In general, psychiatrists—precisely because they were not wedded to a deterministic genetics that made serious mental disorders the product of an inviolate germ plasm—were only marginally involved. The data relating insanity and newly arrived immigrants were weak, and the development of more sophisticated statistical techniques after 1900 undermined the claim that immigrants had higher rates of mental disorders than native-born Americans. In theory, institutional psychiatrists were supportive of eugenical concepts (as were social and behavioral scientists and physicians in general), but in practice there was far less enthusiasm for such invasive measures as sterilization of defective persons, including the mentally ill. In his presidential address in 1903, G. Alder Blumer recommended that individuals with family histories of insanity and alcoholism be prohibited from marrying.[47] A few—including William D. Partlow (superintendent of institutions in Alabama caring for the retarded and insane) and E. E. Southard (superintendent of Boston Psychopathic Hospital) favored sterilization. The majority, however, remained silent, and a small group, including William A. White (a leading institutional psychiatrist who also helped to disseminate psychoanalytic concepts), indicated their opposition. Between 1907 and 1940 about thirty states enacted sterilization laws, and at least 18,552 persons underwent surgery. The number of those sterilized was not equally distributed; more than half of all such operations occurred in California, and Kansas and Virginia accounted for an additional 25 percent. Opposition from a variety of sources in most states rendered enforcement of laws difficult if not impossible.[48] After World War I support among American psychiatrists for such invasive practices as sterilization declined precipitously.[49]

The absence of strong support for an activist eugenic program did not imply that American psychiatrists rejected hereditarian formulations. Most continued to accept the belief, at least in theory, that heredity or predisposition played a role even if its precise character was shrouded in mystery. The clinical nature of the specialty, however, tended to weaken any affinity for genetic explanations. If mental disorders were simply the product of an inviolate germ plasm, the role of the psychiatrist would be limited to diagnosis, if only because genetic therapy was inconceivable at that time. White, for example, noted that children were born with a biological inheritance. Heredity provided human beings with certain essential organs. Yet it could not explain the formation of character traits or personality. Indeed, that such traits could be modified suggested the primary importance of environment. Given his affinity for psychoanalytic and psychodynamic formulations, it was not surprising that White emphasized the crucial significance of infancy and early childhood experiences in shaping character. "Heredity," he wrote in the published version of the Thomas Salmon lectures delivered at the New York Academy of Medicine in 1935, "represents the limitations which past experience, as it has been laid down in the phylum, imposes upon the growing, developing, evolving organism. Environment represents the possibilities, or their absence, dependent upon the stimuli which may or may not be projected upon the growing organism and result or not in the stimulation of potentialities which are resident therein."[50] In a widely used textbook, Edward A. Strecker and Franklin G. Ebaugh declared that heredity as an etiological factor was "overvalued as an aid in discussing either diagnosis or prognosis." "Insanity," they added, "is not a unit character whose transmission can be traced."[51]

With the exception of those involved in the eugenics movement, most psychiatrists dealt with hereditarian formulations in such general terms as to be clinically and experimentally useless. One exception was Abraham Myerson, a figure who straddled the line between neurology and psychiatry. In 1925 Myerson attempted to sum up in book form existing knowledge about the inheritance of mental disorders and at the same time to avoid the platitudes that often accompanied discussions of the subject. Prior to the discovery of the tubercle bacillus, he observed at the outset, tuberculosis was held to be an inherited disease, a fact that should serve as a warning to those who assigned causality in the absence of evidence. Indeed, Myerson attributed genetic explanations of mental disorders to *"an inferiority complex."* Because such disorders did not respond to contemporary therapies, *"We throw off the feeling of inferiority that their existence forces on us, and take refuge in heredity as a cause, since no one can rationally expect us to eliminate 'heredity.'"*[52]

Myerson emphasized the absence of supporting data and the circular reasoning of those who espoused a hereditarian etiology. "*Because* one believes, *in advance,* in hereditary causes of mental diseases, *therefore* any striking disease or condition in the ancestors or collaterals of a patient with mental disease . . . [becomes] the hereditary cause of that disease." He was particularly critical of the methodology employed in gathering family histories by Charles B. Davenport and colleagues at the Eugenics Record Office. The role of heredity in the etiology of mental disorders was a problem to be solved by "clinical-experimental methods." Perhaps in the future laws could be formulated, "but that day is far off." "As a working hypothesis," Myerson added, "it seems to me more logical to search the environment for the causes of family mental disease than to fall back on 'pure heredity.'"[53] The protean—if not confused—nature of hereditarian thinking in psychiatry was confirmed a few years later in a study under the auspices of the National Research Council to determine the actual state of psychiatric knowledge. Although hopeful in tone and optimistic about the possibilities of future progress, this collaborative report constituted a veritable catalogue of what was unknown.[54]

A decade after the publication of his book Myerson chaired a committee of the American Neurological Association charged with the task of evaluating eugenical sterilization, a technique supported by Harry H. Laughlin of the Eugenics Record Office and just written into law in Nazi Germany. Myerson and his colleagues denied that the incidence of schizophrenia or manic-depressive disorders had risen, or that racial deterioration was a reality; environmental influences were more significant that genetic ones. They found no basis for sterilizing mentally ill individuals, and supported only voluntary sterilization for certain diseases for which a genetic basis had been established (e.g., Huntington's chorea).[55]

Pedigree and family research, which had originally been identified with the eugenics movement, began to acquire more respectability in psychiatry as a result of the work of Franz J. Kallmann. A German Jew who emigrated to the United States in 1936, Kallmann had studied with Ernst Rudin, who had begun to develop psychiatric genetics at the University of Munich and subsequently played an important role in the Nazi eugenics movement. Kallmann accepted the genetic basis of many forms of mental disorder, and in the United States began to study family histories to confirm the belief that schizophrenia had a genetic etiology. He became affiliated with the prestigious New York State Psychiatric Institute and Columbia University's College of Physicians and Surgeons. In a series of influential books and articles that examined family pedigrees and particularly the experiences of monozygotic twins, Kallmann attempted to demonstrate genetic transmission of psychopathology. Attributing schizophrenia and related disorders to a single recessive gene, he developed data sets that supposedly demonstrated the accuracy of his claim. To be sure, there were, at least by contemporary standards, severe methodological and substantive shortcomings in his work. He often began with hospitalized patients, and thus tended to find supportive data from retrospective research. In presenting statistics, he did not explain his techniques; he was vague about the nature of his subjects; and controls were entirely absent. Equally important, much of his as well as other data suggested that mortality was higher and fertility lower among individuals suffering from schizophrenia and manic-depressive disorders, as compared with the general population. Yet there was no evidence that the incidence of these disorders was decreasing (which should have been the case). Nevertheless, his work was regarded with respect by his colleagues, and ultimately contributed to the reemergence of biological psychiatry after mid-century.[56]

Kallman's conclusions were implicitly challenged by Horatio M. Pollock, a pioneer psychiatric epidemiologist who directed the statistical division of the New York State Department of Mental Hygiene. Employing genealogical techniques similar to Kallman, Pollock concluded that it was not possible to apply simple Mendelian laws to explain the inheritance of mental disorders. Although there was a higher *probability* for offspring to develop a mental disease if there was a history of the disorder in the family, its transmission was "not a fatalistic process." A diathesis or predisposition was insufficient; environment probably played an equally compelling role. Pollock offered no definitive conclusions but called for more research to shed light on the admittedly complex problem of the respective roles of heredity and environment.[57]

Oddly enough, clinical practice during the 1930s was far removed from the theoretical debates of this decade. Even as "dynamic psychiatry" gained legitimacy, institutional psychiatrists demonstrated a marked affinity for new somatic therapies that were justified in empirical rather than theoretical terms. Fever therapy, insulin and electroshock therapy, and prefrontal lobotomy were quickly absorbed into the psychiatric armamentarium. They offered to psychiatrists what appeared to be efficacious interventions. These treatments could also be understood by their medical colleagues, thus hastening the integration of psychiatry into medicine, and could facilitate the behavioral adjustment of the mentally ill. For patients and their families the obvious gain was the possibility of leaving the mental hospital and residing in the community. Hospital officials were equally enthusiastic; successful interventions would both relieve overcrowding and restore their therapeutic role, which had been eclipsed because of the presence of large numbers of chronically mentally ill persons requiring custodial care. The rapid acceptance of these therapies was also expedited by the vast publicity accorded them in the popular media. Newspapers and magazines as well as radio disseminated information about these therapies and created the impression that the specialty had crossed a threshold and now possessed demonstrably effective therapies.[58]

It is, of course, difficult to evaluate the impact of these aggressive somatic interventions. The obstacles impeding the evaluation of their efficacy remained formidable. A major investigation[59] of the outcome of all first admissions to a single Pennsylvania hospital from 1916 to 1950 yielded some surprising results. First admissions (which during the entire period totaled 15,472) were divided into four chronological periods: 1916–1925, 1926–1935, 1936–1945, and 1946–1950. The probability of release of functional psychotics increased from 42 to 62 percent between 1926–1935 and 1946–1950. Nevertheless, the authors of this study conceded that it was impossible to specify the causal elements responsible for the changes; a variety of factors, including novel therapies, had to be taken into consideration. It was entirely possible, for example, that better-risk patients were being admitted; that new administrative practices within the hospital were involved; and that changes in familial and community practices enhanced the possibility of release. They therefore called for more adequate clinical trials.[60] Whatever the case, the therapeutic claims of the 1930s were for the most part not grounded on a foundation of authoritative data. These claims for the most part reflected the confident outlook of psychiatrists, who believed in the efficacy of their therapies and who were unwilling to accept a caretaker role for chronic patients. In the end, faith in science and the scientific method shaped the activism of practitioners in the interwar years.

## The Rise of Psychodynamic Psychiatry

World War II represented a watershed for American psychiatry. Between 1940 and 1945 the specialty doubled in size. More significantly, wartime experiences transformed both the structure and nature of psychiatry. Psychiatrists became familiar with the high rejection rate for neuropsychiatric disorders during the operations of Selective Service as well as the disabling psychological effects of both training and combat situations. Slowly but surely they learned what their predecessors had observed between 1917 and 1919—namely, that environmental stress played a major role in the etiology of mental maladjustment. The psychiatric lessons gleaned from wartime experiences had significant policy implications. The greatest successes in treating soldiers with psychological symptoms occurred at the battalion aid station. Conversely, the therapeutic success rate declined in rear echelon units. A logical conclusion followed; treatment in civilian life, as in the military, had to be provided in a family or community setting rather than in a remote or isolated institution. The implication for psychiatry was clear; community and private practice would replace employment in mental hospitals. Concern with social and environmental determinants, at the same time, implied a radically different role for psychiatry. Not only would the community become the focal point of psychiatric practice, but practitioners would become active in promoting appropriate social and environmental changes that presumably optimized mental as well as physical health. Moreover, psychiatrists believed that it was possible to identify individuals experiencing psychological distress and provide early treatment, thus preventing the onset of more serious mental disorders.[61]

The new psychodynamic (or psychoanalytic) psychiatry that took shape after 1945 reflected the triumph of environmentalism in American social thought in general and the social sciences in particular. Increasingly the attention of psychiatrists shifted away from a concern with institutionalized severely mentally ill patients toward a preoccupation with those interpersonal, social, and environmental factors that promoted mental illnesses and maladjustment. The dominance of psychodynamic concepts was institutionalized and perpetuated by departments of psychiatry chaired by individuals within the psychodynamic tradition.[62] Mental illness, Menninger insisted, was "personality dysfunction and living impairment." Patients should not be seen as

> individuals afflicted with certain diseases but as human beings obliged to make awkward and expensive maneuvers to maintain themselves, individuals who have become somewhat isolated from their fellows, harassed by faulty techniques of living, uncomfortable themselves, and often to others. Their reactions are intended to make the best of a bad situation and at the same time forestall a worse one—in other words, to insure survival even at the cost of suffering and social disaster.[63]

The new orientation was evident as well in the work of the National Institute of Mental Health (NIMH), which came into being as a result of the National Mental Health Act of 1946. Mental hygiene, observed Robert H. Felix (its first director) and R. V. Bowers, had to be concerned "with more than the psychoses and with more than hospitalized mental illness." Personality, after all, was shaped by socioenvironmental influences. Psychiatry, in collaboration with the social sciences, had to emphasize the problems of the "ambulatory ill and the preambulatory ill (those whose probability of breakdown is high.)'' The community, not the hospital, was their logical habitat. Indeed, as early as 1945 Felix argued that psychiatry had an obligation to ''go out and find the people who need help—and that means, in their local communities.''[64]

The shift in the nature of psychiatric practice was reflected in a variety of developments. One was the movement away from careers in mental hospitals. Another was the rise of social psychiatry, which embodied a touching if naive faith that the specialty had an important role in resolving pressing social and economic problems. The creation of the Group for the Advancement of Psychiatry in 1946 and its subsequent career was indicative of the activism that was characteristic of the specialty in the postwar decades. Equally suggestive was the emergence of a new kind of psychiatric epidemiology. Before World War II statistical analysis had been largely limited to the study of the pattern of mental disease among hospital populations; such studies were presumably of interest to legislators and officials responsible for making appropriations and setting policy. After 1945, by way of contrast, there was a literal explosion of community and demographic studies of mentally ill individuals. These new studies reflected not only changes in data-gathering capabilities but also a preoccupation with the role of socioenvironmental variables and the relations between social class, diagnosis, treatment, and mental disease. To move from a concern with mental disease in institutional populations to the incidence in the general population and the role of socioenvironmental variables represented an extraordinary intellectual leap. Those involved with epidemiological studies of mental illnesses, however, generally ignored some of the methodological problems that followed. That such problems were rarely raised was suggestive of the degree to which those who were concerned with mental illnesses were emphasizing the centrality of an environment that had no clearly defined boundaries.[65]

Although psychodynamic psychiatrists conceded that heredity played a role in mental health, they tended to minimize its significance. Indeed, *genetics* had become a term to be abhorred after 1945 because it was identified with Nazi ideology and genocide. ''Inheritance,'' Strecker wrote in 1952, ''in the past was grossly overrated.'' Nor would eugenics ''solve the problems of psychiatry.''[66] Most psychiatric textbooks briefly alluded to heredity. Inborn errors of metabolism, for example, could lead to abnormal mental development and retardation. Nevertheless, hereditarian explanations in general met with hostility. ''In most cases of mental illness in a parent,'' Lawrence C. Kolb wrote, ''one can presuppose a prolonged period of maladjustment with difficulties and inconsistencies in personal relationships which preclude the existence of a home atmosphere conducive to healthy emotional growth and future mental health.'' Such pathology could be carried over to the next generation. Whatever the contribution of heredity, ''postnatal influences and growing-up experiences'' shaped personality structure and pattern.[67]

Psychodynamic psychiatry from the 1940s to the 1960s was virtually synonymous with the psychotherapies and, to a lesser extent, milieu and other psychosocial therapies. The function of practitioners was to interpret the inner meaning of symptoms, which, it was hoped, would lead to a resolution of the problem. Yet within the psychodynamic synthesis lay elements that would hasten its eventual decline. Whereas most medical specialties had committed themselves to biological and physiological research that would illuminate organ function and pathology, the eventual development of therapies that might be evaluated by randomized clinical trials, and an emerging hospital-based technology, psychiatrists all but abandoned research into brain pathology. As a consequence, the gap between psychiatry and medicine widened precipitously. Moreover, the psychotherapies were time consuming and labor intensive, and were often far removed from the needs of seriously and chronically mentally ill persons both within and without institutions. To design means of evaluating the efficacy of psychotherapy proved extraordinarily difficult, as the experiences of the Menninger Foundation's Psychotherapy Research Project—which lasted several decades—suggested.[68] That psychotherapy could not be defended in strictly medical terms meant that psychiatry was vulnerable to challenges from other professional groups, particularly clinical psychology. By the close of the decade of the 1960s, psychodynamic and psychoanalytic psychiatry had begun to lose the paramount position it had enjoyed since World War II. Its inability to maintain hegemony over psychological therapies, the development of an effective drug armamentarium, a growing disillusionment with purely environmental ideologies, and the improvement of technologies in the biological and neurosciences that promised fundamental break-throughs all combined to shift leadership and authority from psychodynamic to biologically oriented figures.

## Biological Psychiatry

Although the psychodynamic school shaped and dominated the evolution of psychiatry in the two decades following 1945, its hegemony had never been absolute. Admittedly relegated to a minority position, an older biological tradition persisted in one form or another. As early as 1946 a small group of somatically oriented psychiatrists, neurologists, and neurophysiologists had formed the Society of Biological Psychiatry. Its members were united by a faith in the neuronal basis of psychiatry, and insisted on the necessity of adding "substance to psychological concepts" by tracing "the anatomical structures which made these concepts possible."[69] Reiterating the older belief that the brain was the organ of the mind, its members tended to favor such therapies as electroshock and lobotomy in the late 1940s and the psychotropic drugs in succeeding decades. Indeed, the introduction of the psychotropic drugs during the 1950s had given a more biologically oriented psychiatry a foundation on which to build.

The application of clinical randomized trials tended to strengthen the claims of biological psychiatrists and weaken their psychodynamic colleagues. Two studies begun in 1961 suggested that the new drug therapies were particularly effective in dealing with severe mental disorders. The first was undertaken by the NIMH's Psychopharmacological Service Center. The double-blind study of the newer phenothiazines as well as chlorpromazine demonstrated the effectiveness of drug therapy with patients with schizophrenia.[70] In a classic randomized study with appropriate control groups extending over 6 years, Benjamin Pasamanick and his associates found that excellent results could be achieved with drug therapy in a home-care setting for schizophrenic patients. Yet psychiatrists were being trained in psychodynamic concepts that were "already obsolete, utilizing knowledge and skills which daily approach obsolescence."[71]

Although biological psychiatrists were united in their opposition to their psychodynamic and psychoanalytic colleagues, they did not necessarily constitute a single group. There was general agreement that mental disorders had a somatic foundation, even though differences persisted about the nature of the physiological mechanisms. Nor were environmental and interpersonal etiological elements excluded. Nevertheless, fundamental differences between the psychodynamic and biological traditions were self-evident. The development of a large psychotropic drug armamentarium elevated the importance of diagnosis and classification, if only because it was believed that medications were syndrome specific. By contrast, psychodynamic and psychoanalytic practitioners were preoccupied with the interpretation of symptomatology; nosology was of minor importance. Biological psychiatrists hastened the reemergence of neo-Kraepelinism, which was evident in the third and fourth editions of the American Psychiatric Association's *Diagnostic and Statistical Manual of Mental Disorders,* published in 1980 (issued in revised form in 1987) and 1994, respectively.[72]

Equally important, there was a resurgence of interest in the biological and physiological mechanisms that led to mental disorders. Time and space preclude even a small sampling of an immense literature published since the 1970s. In brief, psychiatric research took a variety of forms, but all involved a form of biological reductionism. The advantages were obvious. An emphasis on somatic mechanisms allied psychiatry with the biomedical sciences, whose prestige and funding had been growing at an exponential rate; it also seemed to eliminate the gap between psychiatry and other medical specialties. Some investigators pursued pedigree studies in an effort to identify the relationship between a given diagnosis and heredity. These included consanguinity research, co-twin control cases, and adoption comparisons. The most recent approach is the effort to identify a genetic marker for a specific diagnosis. Others pursued pharmacological studies designed to demonstrate that psychiatric illnesses were based on a biochemical foundation (generally involving neurotransmitters). Some undertook neuropsychological and neurophysiological research, and others attempted to identify specific biological markers for emotional disorders such as anxiety or depression.[73]

Like the leaders of earlier generations, many contemporary psychiatrists—as well as those associated with genetics and neuroscience—perceive of themselves as standing on the threshold of a new era. Laboratory findings will presumably shed light on the physiological and genetic etiological mechanisms that shape normal and abnormal behavior and thus set the stage for the development of effective curative interventions. The most extreme version of this dream are the claims about the potential inherent in the Human Genome Project. By mapping all of the genes on the twenty-three sets of chromosomes, the stage will be set for the identification of markers that govern disease and behavior. If God the Watchmaker cannot be comprehended by human beings, the mechanism and functioning of His creation will surely become crystal clear.

In his presidential address at the American Psychiatric Association in 1990 Herbert Pardes captured the enthusiasm and hope that differentiated the present from the past. In the past 10 years, he noted,

> We have been dazzled by advances in brain imaging, molecular genetics, psychopharmacology, and other fields of study that have revolutionized neuroscience. The most exciting aspect of these advances is what they promise for our capacity as physicians. Because of this panoply of research—encompassing high-tech findings about the brain, evaluation of the psychotherapies, linkage studies, epidemiology and the refinement of diagnostic criteria, the discovery of effective medications, and the development of effective rehabilitation—we can see and begin to deliver new kinds of relief for mental suffering. Few of us would have dreamed that these advances were possible in the days when we entered the field.[74]

If Pardes spoke for a generation preoccupied with biological and physiological mechanisms, his enthusiasm and confidence nevertheless did not differ in kind from his immediate or distant predecessors, each of whom also argued that their specialty stood on the threshold of wisdom and knowledge.

The rise of biological psychiatry had the inadvertent but beneficial result of shifting attention back to serious mental disorders. Psychodynamic and psychoanalytic psychiatrists had been less concerned with this category, if only because their psychotherapies often were inapplicable. Despite their affinity for a reductionist approach to mental illnesses, biological psychiatrists, by contrast, were far more receptive toward the use of psychotropic drugs. The rise of biological psychiatry, moreover, came at a time when the mental health scene was being transformed in fundamental ways. The creation of Medicaid in 1965 resulted in the transfer of aged patients to chronic nursing care facilities, thus permitting public mental hospitals to refocus their energies on treatment of severely and chronically mentally ill individuals. By the early 1970s other federal entitlement programs—Supplementary Security Income (SSI), Social Security Disability Insurance (SSDI), food stamps, and housing supplements—provided resources for mentally ill persons to live in the community. Moreover, Medicaid provided payments for medical treatment outside of hospitals, and quickly became one of the largest mental health programs in the nation. These programs, in conjunction with effective medications that controlled symptomatology and the development of coping strategies—helped launch what became known as deinstitutionalization.

To be sure, not all individuals benefited from the shift in policy away from long-term institutionalization. The major exception was a smaller subgroup of young adult chronic patients who had a dual diagnosis of a severe mental illness and substance abuse. Most were part of the baby boom that occurred between 1946 and 1960, when more than 59 million births were recorded. They were restless and mobile, and they were the first generation of psychiatric patients to reach adulthood in the community. Although their disorders were not fundamentally different than their predecessors, they behaved in quite different ways. They tended to emulate the behavior of their age peers who were often hostile toward conventions and authority. The young-adult mentally ill individual exhibited aggressiveness, volatility, and was noncompliant. They generally fell into the schizophrenic category, although affective disorders and borderline personalities were present. Above all, they lacked functional and adaptive skills. High rates of alcoholism and drug abuse only exacerbated their volatile behavior. Their mobility and lack of coping skills resulted in high rates of homelessness. Many traveled and lived together on the streets, thereby reinforcing each other's pathology. Socially isolated from their families, they aroused negative reactions from mental health professionals, if only because chronicity and substance abuse proved an intractable combination and contradicted the medical dream of cure. Despite the visibility of this group, we should not lose sight of the fact that a large proportion of severely and persistently mentally ill persons made a more or less successful transition to community life as a result of federal disability and entitlement programs, community support systems, and psychiatric medications.[75]

## Conclusion

The history of psychiatric thought and practice over two centuries offers a striking example of a cyclical pattern that has alternated between enthusiastic optimism and fatalistic pessimism. The former manifested itself in ubiquitous claims that the specialty stood on the threshold of fundamental breakthroughs that would revolutionize the ways in which mental disorders were treated. In the nineteenth century the instrument of change was the mental hospital. In the mid-twentieth century psychodynamic and psychoanalytic psychiatry became the vehicle by which the mysteries of normal and abnormal behavior would be revealed. At present the road to salvation is presumably through biological psychiatry, neuroscience, and genetics.

To call into question the exaggerated promises of future omniscience is not to deny the possibility of progress.[76] Science, medicine, and technology retain the possibility of generating knowledge that can be used for human betterment. Nevertheless, progress (i.e., change) usually creates new and unforeseen problems, as the history of disease illustrates so well. Through the close of the nineteenth century infectious diseases were the major causes of mortality, particularly among infants and children. Their decline (for reasons that still remain obscure) was surely a cause for celebration. Yet when these diseases ceased to kill the young, large numbers survived to old age, thus setting the stage for the emergence of chronic diseases as the major causes of mortality. This is not to suggest that medicine and science do not have important roles to play. It is merely to insist that solutions to one set of problems often create circumstances that give rise to new ones. Indeed, resistant organisms as well as viruses still retain the potential to become major causes of mortality among all age groups.

Nearly forty years ago Rene Dubos warned in eloquent terms about the illusory belief that "perfect health and happiness" were distinct possibilities or that the conquest of disease was a real possibility.[77] His warning is as applicable today as it was at that time. Given our mortal and therefore fallible nature, total mastery will probably always elude our grasp. To insist otherwise is to subscribe to the Faustian illusion of attaining total knowledge. I do not believe that a reductionist and mechanistic approach will achieve its stated ends. Recently Leon Eisenberg has argued that "major brain pathways are specified in the genome; [but that] detailed connections are fashioned by, and consequently reflect, socially mediated experience in the world."[78] If this assertion is correct, it will be futile to seek *final* answers from either genetics, molecular biology, or neuroscience.

Finally, let me return to a point made at the outset of this chapter—namely, that psychiatrists, like all human beings, reside in two different worlds. In the idealized one they provide seemingly logical and rational explanations of the nature and etiology of mental disorders as well as prescriptions for action. Offered with confidence, these explanations generally lack an empirical foundation, and represent an act of faith even though phrased in the language of science. More important, exaggerated claims of the impending conquest of disease raise expectations to unreasonable levels and detract attention from smaller, meliorative, and less spectacular steps that have substantial benefits.

In the everyday world, on the other hand, psychiatrists are compelled to deal with individuals whose pathologies are rarely clear-cut and certainly never simple. The founders of American psychiatry in the early nineteenth century were somewhat cognizant of their divided selves. They recognized that their hospitals provided care as well as treatment. They were also aware of the impact that persistent mental disorders had on individuals, families, and society, and that disability and dependency often followed. Hence their theoretical concerns were modified by their insistence that care and treatment, although conceptually distinct, were in fact inseparable and indispensable.

Like their predecessors, contemporary psychiatrists also deal with illnesses that are often chronic in nature. Severe and persistent mental disorders—such as cardiovascular, renal, and other chronic degenerative disorders—require a judicious mix of medical therapies and social support programs. Psychiatric therapies can alleviate symptoms and permit individuals to live in the community, but they do not lead to cures within the conventional meaning of that term. Yet there is persuasive evidence that programs that integrate mental health services, entitlements (SSI, SSDI, Medicaid, food stamps), housing, and social supports often minimize the need for prolonged hospitalization and foster a better quality of life. Many—but not all—contemporary psychiatrists recognize that chronic mental illnesses cannot be treated in isolation, and that care and management are as crucial as psychiatric therapies. It would be a disaster if preoccupation with the biological and genetic basis of mental disorders were to obscure the continuing need for supportive systems of care in community settings.

The evolution of psychiatry and mental health policy also holds important insights for those concerned with health policy. Unlike the general health care system, the mental health system has traditionally dealt with a population whose chronic illnesses created dependency. By definition, chronic diseases do not lend themselves to cure; they require a blend of care, management, and therapy. As the general health care system increasingly confronts chronic illnesses (as compared with the earlier preoccupation with acute infectious diseases), those involved in its reshaping may well benefit from the experiences of their psychiatric colleagues. It would be ironic if psychiatry—recently regarded as an anachronistic medical specialty—would become a model for the future reconstruction of America's health care system.[79] It would be an even greater irony if psychiatry, in its thrust to identify with pure neuroscience and genetics, would in effect abandon a clinical foundation that traditionally began with the human needs of severely and persistently mentally ill persons.

## Notes

1. Isaac Ray, *A Treatise on the Medical Jurisprudence of Insanity* (1838 ed.: Cambridge: Harvard University Press, 1962), 60. *See also* Gerald N. Grob, "Origins of DSM-I: A Study in Appearance and Reality," *American Journal of Psychiatry* 148 (1991):421–431.
2. Nancy Tomes has described in a sensitive and insightful manner Thomas S. Kirkbride's emphasis on the need to foster public confidence in mental hospitals. *See A Generous Confidence: Thomas Story Kirkbride and the Art of Asylum-Keeping, 1840–1883* (New York: Cambridge University Press, 1983).
3. For an analysis of nineteenth-century concepts of health and disease see Charles E. Rosenberg, *Explaining Epidemics and Other Studies in the History of Medicine* (New York: Cambridge University Press, 1992), and John H. Warner, *The Therapeutic Perspective: Medical Practice, Knowledge, and Identity in America, 1820–1885* (Cambridge, MA: Harvard University Press, 1986).
4. *See* Isaac Ray, *Mental Hygiene* (Boston: Ticknor and Fields, 1863), 1–2.
5. Amariah Brigham, *An Inquiry Concerning the Diseases and Functions of the Brain, the Spinal Cord, and the Nerves* (New York: George Adlard, 1840), 14.
6. Utica State Lunatic Asylum, *Annual Report* 1 (1843):36.
7. Worcester State Lunatic Hospital, *Annual Report,* (1839):72. *See also ibid.,* 9 (1841):40–41, 10 (1842):39, 13 (1845):50–51, and Samuel B. Woodward, "Observations on the Medical Treatment of Insanity," *American Journal of Insanity* 7 (1850):1–34.
8. Pliny Earle to Clark Bell (copy), April 16, 1886, Earle Papers, American Antiquarian Society, Worcester, MA.
9. For typical examples of etiological thinking, *see* Edward Jarvis's "On the Supposed Increase of Insanity," *American Journal of Insanity* 8 (1852):333–364, and "The Production of Vital Force," *Medical Communications of the Massachusetts Medical Society,* 1854, 2nd ser., IV:1–40. A general discussion can be found in Gerald N. Grob, *Mental Institutions in America: Social Policy to 1875* (New York: Free Press, 1973), 155–165.
10. Brigham, *An Inquiry Concerning the Diseases,* 288.
11. Ray, *Treatise on the Medical Jurisprudence of Insanity,* 109; Ray, *Mental Hygiene,* 285–286.
12. [Edward Jarvis] *Report on Insanity and Idiocy in Massachusetts by the Commission on Lunacy . . . 1854* (Massachusetts *House Document 144;* Boston: William White, 1855), 61–62.
13. Henry P. Stearns, *Insanity: Its Causes and Prevention* (New York: G. P. Putnam's Sons, 1883), 135–141.
14. William A. Hammond, *A Treatise on Insanity and Its Medical Relations* (New York: D. Appleton and Co., 1883), 77.
15. Grob, *Mental Institutions in America,* 243–255, and *Mental Illness and American Society,* 1875–1940 (Princeton, NJ: Princeton University Press, 1983), 38, 220–221; Peter McCandless, *Moonlight, Magnolias, & Madness: Insanity in South Carolina from the Colonial Period to the Progressive Era* (Chapel Hill: University of North Carolina Press, 1996), 151ff; John S. Hughes, "Labeling and Treating Black Mental Illness in Alabama, 1861–1910," *Journal of Southern History* 58 (1992):435–460.
16. *See* E. Y. Williams, "The Incidence of Mental Disease in the Negro," *Journal of Negro Education* 6 (1937):377–392; Philip S. Wagner, "A Comparative Study of Negro and White Admissions to

the Psychiatric Pavilion of the Cincinnati General Hospital,'' *American Journal of Psychiatry* 95 (1938):167–183; J. E. Greene, ''Analysis of Racial Differences Within Seven Clinical Categories of White and Negro Mental Patients in the Georgia State Hospital, 1923–32,'' *Social Forces* 17 (1938):201–211; and J. E. Greene and W. S. Phillips, ''Racial and Regional Differences Among White and Negro Mental Patients,'' *Human Biology* 11 (1939):514–528.

The rise of psychiatric epidemiology tended to create more interest in racial and ethnic categories. Benjamin Malzberg, a pioneer in the field, found higher rates of mental disorders among Blacks as compared with Whites. Malzberg speculated that higher rates of general paresis and alcoholic psychoses might be explained in historical terms. It was difficult, however, to account for dementia praecox (i.e., schizophrenia) and manic-depressive psychosis in terms of environment. He noted that Blacks had a ''more emotional makeup,'' as was evidenced by their music. ''Given such emotional instability,'' he added, ''it is likely that there is a fruitful ground for functional mental disorders.'' Malzberg, ''Mental Disease Among Negroes in New York State,'' Human Biology 7 (1935):471–513.

17. Ray, *Mental Hygiene,* 315–321.
18. For a forceful statement of this view see Amariah Brigham, *Observations on the Influence of Religion Upon the Health and Physical Welfare of Mankind* (Boston: Marsh, Capen and Lyon, 1835), 267–273.
19. Worcester State Lunatic Hospital, Annual Report 10 (1842):62. Evidence exists that such curability claims had at least some basis in fact. *See* Grob, *Mental Institutions in America,* 182–185.
20. Thomas S. Kirkbride, *On the Construction, Organization, and General Arrangements of Hospitals for the Insane,* 2d ed. (Philadelphia: J. B. Lippincott, 1880), 300.
21. Grob, *Mental Institutions in America,* chaps. V–VI; Tomes, *A Generous Confidence; Ellen Dwyer, Homes for the Mad: Life Inside Two Nineteenth-Century Asylums* (New Brunswick, NJ: Rutgers University Press, 1987).
22. Pliny Earle, *The Curability of Insanity: A Series of Studies* (Philadelphia: J. B. Lippincott, 1887).
23. For an elaboration of this theme, *see* Grob, *Mental Illness and American Society,* 89–92, 180–186, and ''Government and Mental Health Policy: A Structural Analysis,'' *Milbank Quarterly* 72 (1994):471–500.
24. For an outstanding study of this case, *see* Charles E. Rosenberg, *The Trial of the Assassin Guiteau: Psychiatry and Law in the Gilded Age* (Chicago: University of Chicago Press, 1968).
25. Gray's thinking is revealed in a number of his articles, including ''Insanity and Its Relation to Medicine,'' *American Journal of Insanity* 25 (1868):145–172; ''The Dependence of Insanity on Physical Disease,'' *id.* 27 (1871):377–408, ''Pathology of Insanity,'' *id.* 31 (1874):1–29, ''General View of Insanity,'' *id.* 31 (1875):443–465, and ''Heredity,'' *id.* 41 (1884):1–21. For secondary accounts *see* Rosenberg, *Trial of the Assassin Guiteau,* and Robert J. Waldinger, ''Sleep of Reason: John P. Gray and the Challenge of Moral Insanity,'' *Journal of the History of Medicine and Allied Sciences* 34 (1979):163–179.
26. *The United States vs. Charles J. Guiteau* (2 vols.: Washington, DC: n.p., 1882), II, 1593, 1629–1630 *et passim.*
27. Edward C. Spitzka, *Insanity: Its Classification, Diagnosis and Treatment* (New York: E. B. Treat, 1887), 81.
28. *See* Charles E. Rosenberg, ''Inward Vision and Outward Glance: The Shaping of the American Hospital, 1880–1914,'' *Bulletin of the History of Medicine* 53 (1979):346–391, and *The Care of Strangers: The Rise of America's Hospital System* (New York: Basic Books, 1987); Morris J. Vogel, *The Invention of the Modern Hospital: Boston, 1870–1930* (Chicago: University of Chicago Press, 1980); Kenneth M. Ludmerer, *Learning to Heal: The Development of American Medical Education* (New York: Basic Books, 1985); Rosemary Stevens, *In Sickness and in Wealth: American Hospitals in the Twentieth Century* (New York: Basic Books, 1989); and Joel Howell, *Technology in the Hospital: Transforming Patient Care in the Early Twentieth Century* (Baltimore: Johns Hopkins University Press, 1995).
29. *American Journal of Insanity* 10 (1853):85, 28 (1871):205–208, 212; American Medical

Association, *Transactions* 17 (1866):121ff., 18 (1867):399ff., 19 (1868):161ff., 22 (1871):101–109.

30. S. Weir Mitchell, ''Address Before the Fiftieth Annual Meeting of the American Medico-Psychological Association ... 1894,'' American Medico-Psychological Association, *Proceedings,* 1 (1894):101–121.
31. Edward Cowles, ''The Advancement of the Work of the Association, and the Advantages of a Better Organization,'' *American Journal of Insanity* 48 (1891):118–124; Cowles, ''The Advancement of Psychiatry in America,'' *id.* 52 (1896):364–386; Cowles, ''The Relation of Mental Diseases to General Medicine,'' *Boston Medical and Surgical Journal* 137 (1897):277–282.
32. Peter M. Wise, ''Presidential Address,'' *American Journal of Insanity* 58 (1901):79.
33. Henry J. Berkley, *A Treatise on Mental Diseases* (New York: D. Appleton, 1900), 53, 105–106.
34. Stewart Paton, *Psychiatry: A Text-Book for Students and Physicians* (Philadelphia: J. B. Lippincott, 1905), 179–181.
35. For an elaboration of the ''psychiatry of the normal'' *see* Elizabeth Lunbeck, *The Psychiatric Persuasion: Knowledge, Gender, and Power in Modern America* (Princeton, NJ: Princeton University Press, 1994).
36. Adolf Meyer, ''A Short Sketch of the Problems of Psychiatry,'' *American Journal of Insanity,* 53, 1896–1897, reprinted in *The Collected Papers of Adolf Meyer,* ed. Eunice Winters (4 vols.: Baltimore: Johns Hopkins University Press, 1950–1952), II, 273–282, and ''Objective Psychology or Psychobiology With Subordination of the Medically Useless Contrast of Mental and Physical,'' *Journal of the American Medical Association* 65 (1915):860–863.
37. Meyer, ''Organization of Eugenics Investigation, *Eugenical News* 2 (1917). In *Collected Papers,* IV, 304; Meyer, ''A Review of the Signs of Degeneration and of Methods of Registration,'' *American Journal of Insanity* 52 (1895–1896). In *Collected Papers,* II, 257, and ''A Short Sketch of the Problems of Psychiatry,'' *American Journal of Insanity* 53 (1897). In *Collected Papers,* II, 279.
38. Charles G. Wagner, ''Recent Trends in Psychiatry,'' *American Journal of Insanity* 74 (1917):14.
39. Flexner's observations were mentioned in a letter from E. E. Southard to Thomas W. Salmon, July 24, 1919, Salmon Boxes in American Foundation for Mental Hygiene Papers, Archives of Psychiatry, New York Hospital-Cornell Medical Center, New York, NY. *See also* Salmon to Southard, July 21, 1919, *id.*
40. Henry A. Cotton, *The Defective Delinquent and Insane: The Relation of Focal Infections to Their Causation, Treatment and Prevention* (Princeton, NJ: Princeton University Press, 1921).
41. Grob, *Mental Illness and American Society,* 126–143; Lunbeck, *Psychiatric Persuasion, passim.*
42. Thomas W. Salmon, ''Some New Fields in Neurology and Psychiatry,'' *Journal of Nervous and Mental Disease* 46 (1917):90–99.
43. (New York: Longmans, Green, 1908).
44. The history of the National Committee for Mental Hygiene can be followed in Grob, *Mental Illness and American Psychiatry,* 147–166, and Norman Dain, *Clifford W. Beers: Advocate for the Insane* (Pittsburgh, PA: University of Pittsburgh Press, 1980). *See also* Margo Horn, *Before It's Too Late: The Child Guidance Movement in the United States, 1922–1945* (Philadelphia: Temple University Press, 1989), and Theresa R. Richardson, *The Century of the Child: The Mental Hygiene Movement and Social Policy in the United States and Canada* (Albany: State University of New York Press, 1989).
45. Richard L. Dugdale, *The Jukes: A Story in Crime, Pauperism, Disease, and Heredity* (New York: G. P. Putnam's Sons, 1877). *See also* Charles E. Rosenberg, ''The Place of George M. Beard in Nineteenth-Century Psychiatry,'' *Bulletin of the History of Medicine* 36 (1962)::245–259, F. G. Gosling, *Before Freud: Neurasthenia and the American Medical Community, 1870–1910* (Urbana: University of Illinois Press, 1987); and George M. Beard, *A Practical Treatise on Nervous Exhaustion (Neurasthenia): Its Symptoms, Nature, Sequences, Treatment* (New York: William Wood, 1880); Beard, *American Nervousness: Its Causes and Consequences* (New York: G. P. Putnam's Sons, 1881).
46. The literature on eugenics is large, and includes Mark H. Haller, *Eugenics: Hereditarian Attitudes*

*in American Thought* (New Brunswick, NJ: Rutgers University Press, 1963); Donald K. Pickens, *Eugenics and the Progressives* (Nashville, TN: Vanderbilt University Press, 1968); Daniel J. Kevles, *In the Name of Eugenics: Genetics and the Uses of Human Heredity* (New York: Alfred A. Knopf, 1985); Philip P. Reilly, *The Surgical Solution: A History of Involuntary Sterilization in the United States* (Baltimore: Johns Hopkins University Press, 1991); and Edward J. Larson, *Sex, Race, and Science: Eugenics in the Deep South* (Baltimore: Johns Hopkins University Press, 1995).

47. G. Alder Blumer, ''Presidential Address,'' *American Journal of Insanity* 60 (1903):1–18.
48. Grob, *Mental Illness and American Society,* 166–178.
49. *See especially* Ian Dowbiggin, *Keeping American Sane: Psychiatry and Eugenics in the United States and Canada, 1880–1940* (Ithaca, NY: Cornell University Press, 1997).
50. William A. White, *The Mental Hygiene of Childhood* (Boston: Little, Brown, 1919), 10, 52; White, *Twentieth Century Psychiatry: Its Contribution to Man's Knowledge of Himself* (New York: W. W. Norton, 1936), 54–55.
51. Edward A. Strecker and Franklin G. Ebaugh, *Practical Clinical Psychiatry for Students and Practitioners,* 4th ed. (Philadelphia: P. Blakiston's Son, 1935), 35.
52. Abraham Myerson, *The Inheritance of Mental Diseases* (Baltimore: Williams and Wilkins, 1925), 25–26, 31–32. (Emphasis in original.)
53. *Id.,* 270, 278–282, 295–298, 309–311, 316–320.
54. National Research Council, Committee on Psychiatric Investigations, *The Problem of Mental Disorder* (New York: McGraw Hill, 1934).
55. Committee of the American Neurological Association for the Investigation of Eugenical Sterilization (Abraham Myerson, James B. Ayer, Tracy J. Putnam, Clyde E. Keeler, and Leo Alexander), *Eugenical Sterilization: A Reorientation of the Problem* (New York: Macmillan, 1936), *passim. See also* Myerson's ''A Critique of Proposed 'Ideal' Sterilization Legislation,'' *Archives of Neurology and Psychiatry* 33 (1935):453–466.
56. Franz J. Kallmann, *The Genetics of Schizophrenia* (New York: J. J. Augustin, 1938), and *Heredity in Health and Mental Disorder: Principles of Psychiatric Genetics in the Light of Comparative Twin Studies* (New York: W. W. Norton, 1953). A convenient summary of Kallmann's findings, along with the statistical results of comparable studies, can be found in David Rosenthal, *Genetic Theory and Abnormal Behavior* (New York: McGraw-Hill, 1970). In *The Genetics of Schizophrenia,* Kallmann indicated that he was not averse toward using the authority of the state to enforce eugenic prophylactic measures in certain cases.
57. Horatio M. Pollock, *Mental Disease and Social Welfare* (Utica: State Hospitals Press, 1941), 117–133. *See also* Pollock, Benjamin Malzberg, and Raymond G. Fuller, *Hereditary and Environmental Factors in the Causation of Manic-Depressive Psychoses and Dementia Praecox* (Utica: State Hospitals Press, 1939).
58. An overview of somatic therapies during the 1930s can be found in Grob, *Mental Illness and American Society.* Elliott S. Valenstein's *Great and Desperate Cures: The Rise and Decline of Psychosurgery and Other Radical Treatments for Mental Illness* (New York: Basic Books, 1986), provides a revealing portrait of Walter Freeman, but the author's hostility toward psychiatry vitiates some of his claims. Jack Pressman, *Last Resort: Psychosurgery and the Problem of Mental Disorder, 1935–1955* (New York: Cambridge University Press, 1998), is an outstanding analysis of how invasive somatic therapies were incorporated into psychiatric practice.
59. *Id.*
60. Morton Kramer, H. Goldstein, R. H. Israel, and N. A. Johnson, *A Historical Study of the Disposition of First Admissions to a State Mental Hospital: Experiences of the Warren State Hospital During the Period 1916–50.* U.S. Public Health Service Publication No. 445: Washington, DC: U.S. Government Printing Office, 1955).
61. Examples of this kind of thinking include John W. Appel and Gilbert M. Beebe, ''Preventive Psychiatry: An Epidemiologic Approach,'' *Journal of the American Medical Association* 131 (1946):1469–1475; John W. Appel, ''Incidence of Neuropsychiatric Disorders in the United States Army in World War II,'' *American Journal of Psychiatry* 102 (1945):433–436; Roy R.

Grinker and John P. Spiegel, *Men Under Stress* (Philadelphia: Blakiston, 1945), 427–460; Thomas A. C. Rennie and Luther E. Woodward, *Mental Health in Modern Society* (New York: Commonwealth Fund, 1948); and William C. Menninger, *Psychiatry in a Troubled World: Yesterday's War and Today's Challenge* (New York: Macmillan, 1948). A detailed history of post-World War II developments can be found in Grob, *From Asylum to Community: Mental Health Policy in Modern America* (Princeton, NJ: Princeton University Press, 1991).

62. *See especially* Lawrence Friedman, *Menninger: The Family and the Clinic* (New York: Alfred A. Knopf, 1990), and Nathan G. Hale, Jr., *The Rise and Crisis of Psychoanalysis in the United States: Freud and the Americans, 1917–1985* (New York: Oxford University Press, 1995).
63. Karl A. Menninger, Martin Mayman, and Paul Pruyser, *The Vital Valance: The Life Process in Mental Health and Illness* (New York: Viking Press, 1963), 5. *See also* Jack Ewalt, Edward A. Strecker, and Franklin G. Ebaugh, *Practical Clinical Psychiatry,* 8th ed. (New York: McGraw Hill, 1957).
64. Robert H. Felix and R.V. Bowers, ''Mental Hygiene and Socio-Environmental Factors,'' *Milbank Memorial Fund Quarterly* 26 (1948):125–147; Felix, ''Mental Public Health: A Blueprint.'' Presentation at St. Elizabeth's Hospital, Washington, DC, April 21, 1945, Felix Papers, National Library of Medicine, Bethesda, MD. *See also* Paul V. Lemkau, ''The Future Organization of Psychiatric Care,'' *Psychiatric Quarterly* 25 (1951):201–212.
65. Grob, ''The Origins of American Psychiatric Epidemiology,'' *American Journal of Public Health* 75 (1985):229–236.
66. Edward A. Strecker, *Basic Psychiatry* (New York: Random House, 1952), 16–17.
67. Lawrence C. Kolb, *Noyes' Modern Clinical Psychiatry,* 7th ed. (Philadelphia: W. B. Saunders, 1968), 117–118.
68. The Menninger Foundation's Psychotherapy Research Project generated numerous publications, many of which dealt with the barriers that impeded scientific measurement of psychotherapy. The most detailed was Robert Wallerstein's *Forty-Two Lives in Treatment: A Study of Psychoanalysis and Psychotherapy* (New York: Guilford Press, 1986). For a history of the project, *see* Friedman, *Menninger,* 287–289.
69. 'The Society of Biological Psychiatry,'' *American Journal of Psychiatry* 111 (1954):389–391.
70. NIMH Psychopharmacology Service Center Collaborative Study Group, ''Phenothiazine Treatment in Acute Schizophrenia,'' *Archives of General Psychiatry* 10 (1964):246–261.
71. Benjamin Pasamanick, Frank R. Scarpitti, and Simon Dinitz, *Schizophrenics in the Community: An Experimental Study in the Prevention of Hospitalization* (New York: Appleton-Century-Crofts, 1967), 271.
72. American Psychiatric Association, *Diagnostic and Statistical Manual of Mental Disorders: DSM-III [DSM-III-R] [DSM-IV]* (Washington, DC: Author, 1980, 1987, 1994).
73. For a convenient (although one-sided) review of the literature, *see* Colin A. Ross and Alvin Pam, *Pseudoscience in Biological Psychiatry: Blaming the Body* (New York: John Wiley and Sons, 1995), 7–84.
74. Herbert Pardes, ''Presidential Address: Defending Humanistic Values,'' *American Journal of Psychiatry* 147 (1990):1114. It is instructive to compare Pardes's address with those of his more psychodynamically oriented predecessors, which can be found in a collection of American Psychiatric Association presidential addresses titled *New Directions in American Psychiatry 1944–1968* (Washington, DC: American Psychiatric Association, 1969). The latter were equally proud of the achievements of their specialty and looked forward to an even brighter future.
75. A longer discussion of the mental health scene from the 1970s to the present can be found in Gerald N. Grob, *The Mad Among Us: A History of the Care of America's Mentally Ill* (New York: Free Press, 1994), 279–311.
76. In suggesting that the potential benefits of the Human Genome Project are unrealistic, I do not wish to imply that the pursuit of genetic knowledge is futile. Unfortunately, there is a prevailing belief that genetic findings will be both simple and deterministic. The most sophisticated work in this field, by contrast, suggests the very opposite. Single major gene diseases are the exception rather than the rule. More important, there is no simple correspondence between genotype and

phenotype; multiple genes interact with each other as well as the environment. The same genotype can give rise to a different phenotype; conversely, different genotypes can produce the same phenotype. In effect, the findings of modern genetics can only be stated in terms of probabilities, which hardly differs from the older concept of predisposition or diathesis.

77. Rene Dubos, *Mirage of Health: Utopias, Progress and Biological Change* (New York: Harper and Brothers, 1959).
78. Leon Eisenberg, "The Social Construction of the Human Brain," *American Journal of Psychiatry* 152 (1995):1563–1575. *See also* Eisenberg, "Seed or Soil: How Does Our Garden Grow?," *id.* 153 (1996):3–5, and "Medicine—Molecular, Monetary, or More than Both?," *JAMA* 274 (1995):331–334.
79. For a general discussion that relates the health care system to changing epidemiological patterns, *see* Daniel M. Fox, *Power and Illness: The Failure and Future of American Health Policy* (Berkeley: University of California Press, 1993).

# ELABORATION
## Criminal Determinism in Twentieth-Century America

Edward J. Larson

Social concepts of the causes of mental illness and criminal behavior have been closely intertwined over the past century. During the final quarter of the nineteenth century, an increasing number of American policy makers joined many of their European counterparts in embracing the view that heredity, rather than the environment, was the primary cause of criminal behavior. Support for this view came from the ongoing research of Italian criminologist Cesare Lombroso, who conducted extensive experiments on the physical anthropology of criminals. By the 1890s, Lombroso's theories became something of a fad among American criminologists and prison officials.

Criminal anthropology, as this science was called, undermined traditional American faith in the possibility of reforming degenerates. According to this view, criminals were born, not made, and their behavior could no more be changed than their supposedly atypical brains. As historian Mark Haller noted, "Criminal anthropology, in the name of science, would banish sentimental faith in reform and prescribe that those whose failures were inherited and incurrable should be forbidden to propagate."[1]

American criminologists readily accepted these grim conclusions because the groundwork for them had been laid by earlier developments in eugenics. The English gentleman scientist Francis Galton had coined the term *eugenics* in the 1880s to identify the theories that he had already begun developing from the evolutionary concepts of his cousin, Charles Darwin. Galton devised the term from the Greek for "born well," thus showing his fascination with the sources of natural ability rather than a preoccupation with the causes of human degeneration. He systematically studied the lineage of eminent persons, leading to his conclusion that talent and genius tended to run in families. Accordingly, he urged the government to encourage men and women of hereditary fitness to marry each other and to bear many children—proposals that became known as "positive eugenics."[2] Although he suggested that the unfit should be institutionalized to prevent their breeding, such efforts at "negative eugenics" did not become as much the focus of attention for Galton as they did for his followers.[3]

Pioneering the genealogy of degeneracy was Richard Dugdale. He became interested in the issue when, during an 1874 inspection of jail conditions in rural New York state, he found that six of the prisoners were related. Struck by this finding, Dugdale launched an investigation into the lineage of their family, which he called the Jukes, in an effort to uncover the causes of their criminal behavior. Using prison records, relief rolls, and court files, he traced the Jukes's family tree through five generations, back to six sisters, two of whom had married the sons of a Dutch colonist named Max. Dugdale found that, over the years, more than half the 709 people connected to this family by blood or marriage were criminals or prostitutes, or were on relief, and that this had cost the public a staggering $1.3 million.[4] The publication of Dugdale's findings created a sensation, and eventually led to a series of other family studies that claimed to reveal hereditary degeneracy.[5]

Fitting into the hereditarian implications of Galton's and Dugdale's work, Lombroso's ideas that atavism and morbid characteristics were major causes of crime became powerful forces in American criminological thought by the 1890s. "Everyone who has visited prisons and observed large numbers of prisoners together has undoubtedly been impressed from the appearance of prisoners alone, that a large portion of them were born to be criminals,'' Pennsylvania state penologist Henry M. Boies commented in 1893. ''The laws of biology, that like begets like . . . [and] that certain inherited defects and deficiencies induce criminality . . . are well-known, and generally accepted to be as invariable and immutable as the law of gravitation.'' Accordingly, Boies urged that criminals be sexually segregated or castrated.[6] Picking up on this suggestion in 1901, a speaker for the newly forming Southern Medical Association argued that, because criminality is hereditary, ''It is essentially a state function'' to restrain ''the pro-creative powers'' of the unfit. ''Emasculation is the simplest and most perfect plan that can be adopted to secure the perfection of the race,'' he maintained, and added that it offered a preferable alternative to lynching for the ''animalized negroes'' who raped White women.[7]

These early hereditary determinists appreciated the efficiency of surgically preventing reproduction by criminals suffering from hereditary defects, but they lacked an acceptable means to carry out this remedy. Both euthanasia and castration were suggested, but neither gained much popular support. At most, the emergence of hereditarian explanations for criminal behavior revived interest in castrating rapist and other violent criminals, though the primary argument for such a penalty still focused on its direct deterrent effects rather than its indirect hereditary benefits. The physical and psychological impact of castration on its victims, however, made its use on anyone other that ''animalized'' criminals seem unacceptably inhumane to the general public. For example, a popular outcry ended an experiment with eugenic castration at a Kansas youth institution in the 1890s after 44 boys and 14 girls were mutilated.[8]

Fortuitously for these determinists, new surgical techniques for sexual sterilization were developed around the turn of the century. Vasectomy offered a safe and simple means of terminating a man's ability to reproduce, without inhibiting his sexual desire or pleasure. Salpingectomy provided a similarly effective but somewhat more dangerous sterilization procedure for women. Proponents of a deterministic view of criminal behavior immediately recognized the significance of these developments for their cause. Famed Chicago surgeon A. J. Ochsner announced his pioneering research in vasectomy in an 1899 article aptly titled ''Surgical Treatment of Habitual Criminals,'' which appeared in the prestigious *Journal of the American Medical Association.* In this article, Ochsner never mentioned the possible use of vasectomy as a means of voluntary birth control, but instead stated that the procedure ''seems to me to be of especial interest in pointing out a reasonable plan for the surgical treatment of habitual criminals of the male sex.'' He went on to explain,

> It has been demonstrated beyond a doubt that a very large proportion of all criminals, degenerates, and perverts have come from parents similarly afflicted. It has also been shown, especially by Lombroso, that there are certain inherited anatomic defects which characterize criminals, so that there are undoubtedly born criminals. . . . Taking these two facts into consideration it would seem that if it were possible to eliminate all habitual criminals from the possibility of having children, there would soon be a very marked decrease in this class.[9]

American determinists needed little encouragement. An Indiana reformatory promptly launched a sexual sterilization effort that began in 1899 as a voluntary experiment with prisoners, but rapidly expanded until 1907, when the state enacted America's first compulsory sterilization statute. ''Vasectomy,'' the reformatory's physician reported, ''is indeed very simple and easy to perform; I do it without administering an anaesthetic, either general or local. It requires about three minutes' time to perform the operation and the subject returns to his work immediately, suffers no inconvenience, and is in no way impaired for his pursuit of life, liberty and happiness.''[10] This last point was meant to counter constitutional objections to the law, but betrayed an utter disregard for any liberty interest in or joy coming from having a child, whether ''defective'' or not. The Indiana law mandated the sterilization of ''confirmed criminals, idiots, rapists and imbeciles'' whose condition was pronounced incurable by a committee of three physicians.[11]

Fifteen other states followed suit over the next decade, including such progressive trendsetters as California, Iowa, Michigan, New Jersey, New York, Oregon, and Wisconsin. Except for the one in Michigan, all of these laws applied to certain criminals. Typically the subject class was identified as ''confirmed criminals,'' as in the Indiana and New Jersey laws, or ''habitual criminals,'' as in the New York and Oregon laws, or persons ''twice convicted for sexual offenses or three times for other crimes,'' as in the California and Iowa laws.[12] Most of these compulsory sterilization laws also applied to persons in state institutions for the mentally ill or retarded, but some, such as the laws in Washington state and Nevada, applied only to criminals. A surge in forced sterilization practices followed the enactment of these statutes, with California leading the way by sterilizing hundreds of persons each year, mostly at its mental health facilities.

Despite the confident assurance of proponents that sterilization was a harmless procedure that did not deprive the hereditarily unfit of anything worth having, constitutional challenges dogged the early sterilization laws, especially as they were applied to criminals. Seven of the first 16 statutes were struck down by state or federal courts for violating individual rights of equal protection, due process, or freedom from cruel or unusual punishment. These decisions so hobbled the enforcement of all the early sterilization laws that only the California program, which was repeatedly revised and reenacted, lived up to proponents' expectations. Even Indiana's pioneering program was shut down, first by a hostile governor in 1913 and finally by court order in 1920. A landmark decision to uphold a Virginia sterilization statute by the U.S. Supreme Court in 1927 did not resolve questions about the legality of imposing the procedure on criminals. That test case dealt solely with compulsory sterilization for persons diagnosed with hereditary forms of mental retardation, for which the evidence of inheritability then appeared stronger than for criminal behavior.[13]

Nevertheless, this Supreme Court decision fueled a resurgence of law making that led to the enactment of new compulsory sterilization statutes in 16 American states, with most of these laws applying to criminals incarcerated in state prisons as well as to patients confined to state mental health institutions. More critically, confidence in their general constitutionality contributed to the increased use of such laws. The annual average number of operations performed under compulsory sterilization statutes in the United States jumped ten fold, from 230 during the second decade of the century to 2273 during the fourth. Although the number of operations gradually declined thereafter, it did not become insignificant until the 1960s, by which time the total number of individuals sterilized had passed 60,000.[14] Widespread popular support for these procedures was reflected in a 1937 public opinion survey, which found that more than 60% of the Americans surveyed favored the forced sterilization of habitual criminals.[15]

During the inter-war period, a broad coalition of geneticists, psychologists, physicians, sociologists, and criminologists supported the view that hereditary factors could predispose persons toward criminal behavior. Chicago jurist Harry Olson, who hired prominent eugenics researcher H. H. Laughlin to conduct an official study of criminal determinism, played a leading role in advocating a eugenic solution to crime. In a typical speech from the 1920s, Olson confidently proclaimed, ''Formerly many believed that environment was everything, but science has repudiated that view today; heredity is of paramount importance, and every day more and more scientists are coming to believe this. Surely the criminal is not an environmental product.'' His remedy for crime involved testing all schoolchildren, and then sexually segregating or sterilizing those found to carry an ''inborn tendency'' to commit crime. ''The medical man, rather than the legal profession, are the men who will curb crime in the future,'' Olson predicted.[16]

Deterministic explanations for criminal behavior did not necessarily lead to advocacy of forced sterilization, however. During the 1920s, for example, Clarence Darrow made headlines across the nation with his use of deterministic arguments to save accused murderers from the death penalty, culminating in the sensational 1924 Loeb–Leopold case, in which he saved two wealthy and brilliant teenagers from execution for their cold-blooded murder of an unpopular schoolmate. Yet Darrow never clearly differentiated between environmental and hereditary determinism, and adopted whichever better explained the case at hand. In the Loeb–Leopold case, for example, he blamed the defendants' genetic makeup, indulgent upbringing, and their reading of the work of Friedrich Nietzsche. Summarizing his arguments in this and other cases in his 1932 autobiography, Darrow concluded that the cause of crime by a person ''is so obscured that no man can trace or solve it. It may go back to his youth. . . . Or, perhaps, it may reach back to remote ancestors and have come to him through subterranean caverns affecting the brain or nerves or other parts of the structure.'' The materialistic attorney was certain about one thing, however: People are not responsible for their actions.[17] Sorting through Darrow's many writings on the subject, biographer John C. Livingston simply concluded that Darrow believed that ''crime occurred when an unhealthy environment acted upon a defective or weak heredity.''[18] Yet Darrow so feared the abuse of governmental power that he utterly rejected compulsory eugenic remedies for crime and denounced compulsory sterilization practices.

Persistant challenges to the legality of sterilizing criminals from the defense bar shifted the primary locus of such practices out of the prisons and into mental health institutions. But this did not necessarily exempt criminals. A common strategy for sterilizing prisoners believed to be mentally retarded or mentally ill involved temporarily transferring them to state mental health institutions—sometimes for just a few hours—when their sterilization would be performed.[19] The procedure could then be categorized as therapeutic rather than punitive. The result was the same for the prisoner, however. In other instances, the operations continued to be performed within the prison, but only on criminals diagnosed as mentally ill or retarded.

Some states continued to enforce their compulsory sterilization statutes strictly with respect to confirmed, habitual, or repeat criminals. In 1942, Oklahoma's persistence in this respect finally produced a definitive Supreme Court test of the practice in *Skinner v. Oklahoma.* The Oklahoma statute mandated the sterilization of persons convicted three times for crimes ''amounting to felonies involving moral turpitude,'' but the legislation exempted ''offenses arising out of the violation of the prohibitory laws, revenue acts, embezzlement, or political offenses.''[20] In short, white-collar criminals (including crooked politicians) were likely escape the treatment.

The U.S. Supreme Court easily disposed of this law in a appeal challenging a court-ordered vasectomy for an armed robber, one of whose three convictions involved stealing chickens. ''Sterilization of those who have thrice committed grand larceny with immunity for those who are embezzlers is a clear, pointed, unmistakable discrimination,'' the high Court wrote in *Skinner.* ''We have not the slightest basis for inferring that that line has any significance in eugenics nor that the inheritability of criminal traits follows the neat legal distinctions which the law has marked between those two offenses.'' Therefore, the Oklahoma law was held to violate the constitutional guarantee of equal protection under law even as the Court reaffirmed its approval of the eugenic sterilization of the mentally retarded.

"This Court has sustained such an experiment with respect to an imbecile," one justice wrote, but "there are limits to the extent to which a legislatively represented majority may conduct biological experiments at the expense of the dignity and personality and natural powers of a minority—even those who have been guilty of what the majority define as crimes." Another justice questioned the notion that "criminal tendencies of any class of habitual offenders are universally or even generally inheritable."[21]

With this ruling, the American experiment with the forced sterilization of criminals ended. Thereafter, for another two decades, sexual sterilization was imposed only on persons committed to state mental health institutions. This might include the criminally insane, but the justification no longer rested solely on evidence of past criminal behavior, though such antisocial actions could still be considered in an evaluation of the criminal's mental health.

By this time, however, hereditary explanations for human behavior had been eclipsed. Various factors contributed to this development. *Skinner,* of course, was handed down during World War II, at a time when the extreme and violent excesses of Nazi eugenics had discredited hereditarian ideas in general. Indeed, it was the specter of Nazi practices that finally emboldened the genetic community to repudiate eugenics at the 1939 International Eugenics Congress and, a year later, led the Carnegie Institution of Washington, DC, to close its Eugenics Record Office, from which H. H. Laughlin had championed the enactment of state sterilization laws.[22] By the 1950s, the unrepentant leader of eugenics in California, Paul Popenoe, insisted that "the major factor" in America's rejection of hereditarian explanations for human behavior "was undoubtedly Hitlerism."[23]

Historian Carl N. Degler has demonstrated, however, that, during the first several decades of the twentieth century, a new school of American social scientific thought driven by an ideological commitment to human freedom, equality, and fellowship, and led by the likes of Franz Boas, had already laid the foundation for reviving cultural explanations for human behavior.[24] Although these nonhereditarian theories still dominate social scientific thought, Degler posited a return to hereditarian explanations, and quoted a 1985 book by Harvard social scientists James Q. Wilson and Richard Herrnstein titled, *Crime and Human Nature.* Seeking to account for the persistent disparities in the crime rates of men and women, for example, Wilson and Herrnstein asserted that "their roots go . . . deep into the biological substratum."[25] Based on this and other examples of sociobiological thinking, Degler concluded, "The return of biology to the social sciences is still in its infancy, as some of its uses make clear. The process, however, is likely to continue if only because the biological sciences continue to throw fresh light on the nature of human beings."[26] At least that is how it appears to an increasing number of Americans as the nation enters the twenty-first century.

## Notes

1. M. H. Haller, *Eugenics: Hereditarian Attitudes in American Thought* (New Brunswick, NJ: Rutgers University Press, 1933).
2. Francis Galton, *Hereditary Genius: An Inquiry into Its Causes and Consequences* (London: Macmillan, 1869).
3. ———, "Hereditary Improvements," *Fraser's Magazine* 87 (1873):125–128.
4. R. L. Dugdale, *The Jukes: A Study in Crime, Pauperism, Disease and Heredity,* 5th ed. (New York: Putnam, 1895).
5. E. J. Larson, *Sex, Race, and Science: Eugenics in the Deep South* (Baltimore: Johns Hopkins University Press, 1995).
6. H. M. Boies quoted in P. R. Reilly, *The Surgical Solution: A History of Involuntary Sterilization in the United States* (Baltimore: Johns Hopkins University Press, 1991).

7. J. E. Purdon, ''Social Selection: The Extirpation of Criminality and Hereditary Disease,'' *Transactions of the Medical Association of the State of Alabama* 4 (1901):457–467.
8. Reilly, *The Surgical Solution.*
9. A. J. Ochsner, ''Surgical Treatment of Habitual Criminals,'' *Journal of the American Medical Association* 32 (1899):867–868.
10. D. J. Kevles, *In the Name of Eugenics: Genetics and the Uses of Human Heredity* (New York: Knopf, 1985).
11. Indiana Acts, ch. 215 (1907).
12. J. H. Landman, *Human Sterilization: The History of the Sexual Sterilization Movement* (New York: Macmillian, 1932; B. Van Wagenen, ''Preliminary Report of the Committee of the Eugenic Section.'' In *Problems in Eugenics: First International Eugenics Congress* (London: Knight, 1912), 465–479.
13. Buck v. Bell, 274 U.S. 200, 207 (1927).
14. J. Robitscher, ed., *Eugenic Sterilization* (Springfield, IL.: Thomas, 1973).
15. Reilly, *The Surgical Solution,* 125.
16. ''Doctors, Not Lawyers, Due to Stop Crime, Judge Olson Declares, Blaming Heredity,'' *Times Picayune* (New Orleans), April 29, 1926, 1.
17. C. Darrow, *The Story of My Life* (New York: Grosset, 1932).
18. J. C. Livingston, *Clarence Darrow: The Mind of a Sentimental Rebel* (New York: Garland, 1988).
19. For a relatively late example of such a transfer, *see* letter from William S. Hull to South Carolina State Board of Health, October 8, 1959, South Carolina State Hospital Archives (Sterilization File) regarding the transfer of a prisoner from the state prison to the state mental health hospital by order of then governor Ernest F. Hollings.
20. Oklahoma Stat., tit. 57, §§ 173, 195 (1935).
21. Skinner v. Oklahoma, 316 U.S. 535, 541–542, 544, 546 (1942).
22. K. M. Ludmerer, *Genetics and American Society: A Historical Appraisal* (Baltimore: Johns Hopkins University Press, 1972), 121–129; F. Hassenchal, 1970, ''Harry H. Laughlin, 'Expert Eugenics Agent' for the House Committee on Immigration and Naturalization, 1921 to 1931,'' PhD diss., Case Western Reserve University.
23. P. Popenoe, quoted in D. K. Pickens, *Eugenics and the Progressives* (Nashville, TN: Vanderbilt University Press, 1968).
24. C. N. Degler, *In Search of Human Nature: The Decline and Revival of Darwinism in American Social Thought* (New York: Oxford University Press, 1990).
25. James Q. Wilson and R. J. Herrnstein, *Crime and Human Nature* (New York: Simon and Schuster, 1985).
26. Degler, *In Search of Human Nature.*

# Chapter 2
# "BIG IDEAS, IMAGES AND DISTORTED FACTS": The Insanity Defense, Genetics, and the "Political World"

Michael L. Perlin

It was impossible to escape the Menendez trial in the winter of 1995. The case had all the high cards of any media circus: an affluent victim, an Oedipal plot, sex abuse, high-power counsel, superstar experts, and physically attractive defendants.

And the basis of the Menendez brothers' defense—that they suffered from Battered Child Syndrome (BCS)—became a staple of television talk shows, radio call-in programs, letters to the editor, and cocktail party chat. "So, Michael," friends would ask me, "What *is* it with this BCS stuff anyway? I mean, this is all a lawyer's invention, isn't it? There's no scientific *proof* of any of this, is there?"

And that question, of course, would immediately create a dilemma. No matter what I said, I knew what the response would be. If I were to say, "Yeah, you're right, it's all garbage," then the retort would be, "Just another example of a shifty lawyer trying to con a jury with a bogus insanity defense, right?" And that would require me to go into my set response number 1—that the insanity defense is *not* overpled, is rarely successful, exposes the defendant to periods of greater institutionalization, and so on.[1]

If I were to say, "Who knows?" then the response would be, "But with enough money you can get an expert who'd testify that *any* excuse is a mental illness, right? Remember the twinkie defense?"[2] And that would require me to go into set response number 2—that there is agreement as to criminal responsibility in nearly 90% of all insanity defense trials,[3] and that, in a frequently relied on study, there was agreement that 138 of 141 insanity acquittees were seriously mentally ill.[4] And that the defendant in the "twinkie" case (Dan White) wasn't even pleading insanity; like the Menendez brothers, he was seeking to use his diminished responsibility as a reductive element to reduce a murder charge to manslaughter.

On the other hand, if I were to say, "No really, that's *not* right. We're really learning an awful lot about the interplay between genetics and the environment, and why *some* people do inexplicably 'crazy' things (though others similarly situated do not),"[5] then the answer would likely be, "But so what? Who *cares* if there really *was* a genetic reason for what they did? These guys were cold-blooded killers. Fry 'em." And then I have to stop a second. Because there really is no set response to this position (one that demurs to the possible scientific basis of an insanity defense, nihilistically substituting a punitive retributivism).[6] I cannot respond to this logically or rationally—and it is this point to which I will return in a minute.

But first let me go back briefly to the Menendez case. I wanted to see the extent to which arguments about genetics permeated that trial, so I ran a simple Boolean search (GENETIC! or BIOLOG!) in WESTLAW'S MENENDEZ-TRANS database. I found, to my surprise, that there were 52 separate entries (an "entry" might be direct or cross-examination of an individual witness, a motion to exclude evidence, an opening or summation, or a judge's remarks to the jury). Again and again, this question resurfaced in all aspects of the trial.

Title of chapter taken from Bob Dylan, "Idiot Wind" (1974) and Dylan, "Political World" (1989).

A few examples should suffice. At the mistrial motion, defense witness Dr. William Vicary testified,

> I tried my best, as you do in every case, to get more information about the history [of the defendants] because you want to know why. Why is this person sick? Is this something genetic? . . . Is this because of some kind of situational stress? Is this because of some kind of traumatic abusive experience in the person's past?[7]

Vicary had previously testified, "Some of this is just genetics and just biology."[8] On the other hand, prosecution expert witness Park Dietz testified that "the whole theory of posttraumatics stress disorder is not genetic or inherited, or in that sense biological, but [is, rather,] caused by trauma and stress," noting further that "individuals may differ biologically in their vulnerability in acquiring that disorder."[9]

In her summation, defense counsel Leslie Abramson told the jury:

> It is understood that there's a biological component with respect to lots of emotions, but particularly the fear response. It is understood that built into us, going way back in the evolutionary scale, and repeated again every time one of us is born, this wiring system that is keyed to our survival. That's why it's there. People who have been traumatized, who develop these anxiety disorders, okay, are operating—as Dr. Wilson called it—their idle screw is turned up. They're operating at a much higher level in preparation for threat and the response to threat. They're aroused ordinarily. But certainly, once they're in a situation that poses for them, given the history of their trauma, the potential of threat, they are up there sort of ready biologically.
>
> We're not talking mentally thinking, cognitively. We're talking total unconsciously and beyond their control, at this level where they are hyper-aroused.[10]

On the other hand, in *his* summation, Deputy District Attorney David Conn struck back:

> And it was suggested to you that perhaps there's a biological change in response to abuse, and Dr. Dietz testified that there is really no evidence of that. He said there is no biological evidence that posttraumatic stress disorder impairs brain function; no biological evidence that it impairs the ability to control impulses, and no biological evidence that posttraumatic stress impairs your behavior or predicts your behavior in any necessary way.[11]

These excerpts should not be a surprise. The public has always been hostile to the insanity defense (or to the use of behavioral testimony that supports any excuse reducing an individual's degree of guilt or responsibility), and that hostility shows no sign of abating. Its enmity toward the defense has always been informed by multiple factors—among them, mental illness's invisibility and a lack of consensus as to the "cause" of mental illness.[12] If we could pinpoint a "visible" gene or chromosome or *whatever* (and I say "whatever" advisedly) that we could authoritatively say "causes" otherwise inexplicable aberrant behavior, at least some of the opposition to the insanity defense might be remediated. And thus the players in the Menendez trial attempted to "unpack" the underlying issues by searching for a genetic key that would unlock the doors of understanding and shed some light on the brothers' behavior.

Or perhaps not. Because although the participants in the Menendez trial appeared to believe that genetic evidence would do this, it is not at all clear that this was a correct assumption. This lack of clarity is critical to an understanding of the issues to be discussed in this volume. The questions that appear to be posed—is there a genetic basis for an insanity defense (or for other defenses reducing degrees of culpability based on mental status), and, if there is, what can contemporary research teach us about that basis?—are important ones that are certainly worthy of addressing. But they are not the most important questions for us to consider. Those questions are these: Does it *matter* to legislators, judges, and jurors (the public at large) if there *is* some sort of genetic "key" (provable through valid and reliable scientific methodology) to understanding why certain criminal defendants commit otherwise inexplicable criminal acts (or, in some cases, vide Menendez, all *too* explicable), and if it *doesn't* matter, why doesn't it?

Anything we consider in this context must be assessed constantly in the light of these follow-up questions. Assuming that we are all convinced—beyond a reasonable moral, legal, or scientific doubt—that, say, there is a genetic explanation for certain kinds of bizarre criminal behavior and that science has (or will soon have) the tools for determining whether an individual possesses that genetic makeup that "slots" that individual into such a category—will that explanation have any real-life impact on the language used to define a substantive insanity defense test, on the procedures used at insanity defense trials, on the mechanisms in place when a successful insanity acquittee is retained in a psychiatric facility following the jury verdict, or, most important, on the unshakable myths that the public retains about the use and abuse of the insanity defense?[13]

If it has no such impact, why *doesn't* it? Why are we so willing to blind ourselves wilfully to science (just as we are willing to blind ourselves wilfully to both behavioral and empirical reality) in the way we construct the insanity defense? By *construction* I refer to both the words that are used in statute and case law and—more important—in the folkways that dominate our insanity defense practice. If we fail to look at these issues, any illumination that comes from talking about new research, new discoveries, new science, will shed little more than false light on the question at hand.

In this chapter I will present a brief overview of the insanity defense—how it developed, the roots of the hostile attitudes, the substantive tests, and the significance of changes in the post-Hinckley universe. Next I will consider briefly the historic role of physiology and genetics in insanity defense jurisprudence and how genetic issues have been read by courts in physiological disorder cases, XYY chromosome cases, and so forth. Then I will look at some of the new interest in genetic issues, both through the seeming expansion of what I refer to as *syndrome cases* and through reports of new scientific discoveries purporting to explain the links between genetics and criminal behavior. Finally, I will conclude by talking about what I call "the missing link"—why none of this will matter much if we do not explore the roots of the underlying dilemma of why-do-we-feel-the-way-we-do-about-these-people. I do this in part by exploring the roots of "sanism" and "pretextuality," the two factors that I think best help explain our insanity defense jurisprudence.

The first line of this chapter title comes from Bob Dylan's brilliant song *Idiot Wind.* Interpretations of this song abound, and all focus on Dylan's image of the world as a *trompe l'oeil* creation: Is anything quite the way it appears at first blush? The line here completes Dylan's depiction of "people" who "can't remember how to act," and whose "minds are filled with big ideas, images and distorted facts."

When I started writing this chapter I instantly thought of this line, because it was my intuitive hunch that this aptly describes the public's approach to precisely this question. And I then thought about Dylan's more recent song, *Political World,* where "wisdom is thrown in jail/ It rots in a cell, is misguided as hell/ Leaving no one to pick up a trail." In this line, he captures the entire insanity defense debate.

## The Insanity Defense: An Overview

Our insanity defense[14] jurisprudence is incoherent. It reflects the public's episodic outrage at apparently inexplicable exculpations of obviously "guilty" acts, the legislatures' pandering, prereflective responses to constituency cries, and the judiciary's desperate ambivalence about having to decide hard cases involving mentally disabled criminal defendants.[15] We are beginning to come to grips with some of the scientific, biological, neurological, and psychological reasons that play a role in the commission of some otherwise inexplicable crimes, but we simultaneously narrow and limit the substance of the insanity defense and the procedures used in such cases (and in postacquittal commitment hearings).[16] We do this narrowing ostensibly both to lessen the possibility of a "moral mistake" (i.e., the entry of an insanity acquittal in cases in which we cannot be "sure" of the defendant's nonresponsibility,[17] and to make the choice of an insanity defense—never a high card in any criminal defense lawyer's hand—an even less attractive option.

The jurisprudence's incoherence is important. It is important because of the full scope of its social impact. First, through a series of legislative "reform" measures, it sanctions the criminal punishment of a significant number of individuals who—by any substantive standard—are not "responsible" for their "criminal acts." In addition to its evident punitive and damaging impact on these defendants, this outcome also makes prisons more chaotic and dangerous places for other inmates and for correctional staff. Second, it allows us—perhaps *forces* us—to deplete our intellectual and emotional resources and our creative energies by debating endlessly issues that are fundamentally irrelevant to the real-life impact of the defense (e.g., whether there should be a volitional as well as a cognitive standard employed), and that lead to, at the best, illusory change. At the same time, it allows us—perhaps, *encourages* us—to ignore empirical evidence, scientific study, and moral reasoning that seek to shed light on the underlying issues.

Third, it leads us to spend money in counterproductive ways. Recent reforms will lead to more individuals being institutionalized for longer periods of time in more punitive facilities, at precisely the same time that community resources are becoming scarcer.[18] If the insanity defense is successful only in a fraction of 1% of all cases, why do we devote such time and capital to this question, and why do we dramatically and egregiously exaggerate the impact that these cases have on the operation of our criminal justice system?

Fourth, it leads us to avoid consideration of the single most important issue in mental disability law (one that is magnified many times in insanity defense analysis): Why do we feel the way we do about "these people," and how do these feelings control our legislative, judicial, and administrative policies?[19] I believe that it is the answer to this question that is the wild card, and it is essential that we see its role in the incoherence of the policies I am discussing.

No aspect of the criminal justice system is more controversial than is the insanity defense. Nowhere else does the successful employment of a defense regularly bring about cries for its abolition; no other aspect of the criminal law inspires position papers from trade associations spanning the full range of professions and political entities. When the defense is successful in a high-level publicity case (especially when it involves a defendant whose "factual guilt" is clear), the acquittal triggers public outrage and serves vividly as a screen on which each relevant interest group can project its fears and concerns. It symbolizes "the most profound issues in social and criminal justice."[20]

Although, on one hand, the defense is a reflection of the "fundamental moral principles of our criminal law,"[21] and serves as a bulwark of the law's "moorings of condemnation for moral failure,"[22] we remain fixated on it as a symbol of all that is "wrong" with the criminal justice system and as a source of social and political anger. It is thus attacked by a former U.S.

attorney general as a major stumbling block in the restoration of the ''effectiveness of Federal law enforcement'' and as tilting the balance ''between the forces of law and the forces of lawlessness.''[23]

Our fixation on the insanity defense has evolved into a familiar story. The insanity defense, so common wisdom goes, encourages the factually (and morally) guilty to seek refuge in an excuse premised on pseudoscience, shaky rehabilitation theory, and faintly duplicitous legal *légèrdemain.* The defense, allegedly, is used frequently (mostly in abusive ways), is generally successful, and often results in brief slap-on-the-wrist periods of confinements in loosely-supervised settings; because it is basically a no-risk maneuver, the story continues, even when it fails the defendant will suffer no harm. Purportedly, the defense is used disproportionately in death penalty cases (often involving garish multiple homicides), and inevitably results in trials in which high-priced experts do battle in front of befuddled jurors who are inevitably unable to make sense of contradictory, highly abstract, and speculative testimony. Finally, the defense is seen as being subject to the worst sort of malingering or feigning, and it is assumed that, through this gambit, clever defendants can con gullible, soft experts into accepting a fraudulent defense.[24]

The largely unseen counterworlds of empirical reality, behavioral advance, scientific discovery, and philosophical inquiry paint quite a different picture. Empirically, the insanity defense is rarely used, is less frequently successful, and in general results in lengthy stays in maximum security facilities (often far more restrictive than many prisons or reformatories) for far longer periods of time than the defendants would have been subject to had they been sentenced criminally.[25]

It is also a risky plea; where it fails, penal terms are generally significantly longer than in like cases where the defense was not raised. The defense is most frequently pled in cases *not* involving a victim's death, and is often raised in cases involving minor property crimes. The vast majority of cases are so-called ''walk-throughs''—that is, where both state and defense experts agree both about the severity of the defendant's mental illness and his or her lack of responsibility. Feigned insanity is rare; successfully feigned insanity even rarer. It is far more likely for a jury to convict in a case in which the defendant meets the relevant substantive insanity criteria than to acquit where the defendant does not.[26]

As a result of these myths, we demand legislative ''reform.'' This reform leads to a variety of changes in insanity defense statutes—in substantive standards, in burdens of proof, in standards of proof, in the creation of hybrid verdicts such as ''guilty but mentally ill,'' even, in a few instances, in supposed abolition of the defense itself. No matter what their final reform, these reforms stem from one primary source: ''the public's overwhelming fear of the future acts of [released insanity] acquittees.''[27]

From a behavioral standpoint researchers are beginning to develop sophisticated assessment tools that can translate insanity concepts into quantifiable variables that appear to easily meet the traditional legal standard of ''reasonable scientific certainty.''[28] From a scientific standpoint the development of ''hard science'' diagnostic tools (such as CT scanning or Magnetic Resonance Imaging) has helped determine the presence and severity of certain neurological illnesses that may be causally related to some forms of criminal behavior.[29] From a moral standpoint philosophers are increasingly trying—with some measure of success—to clarify such difficult underlying issues as the contextual meaning of terms such as *causation, responsibility,* and *rationality.*[30]

At the same time, the U.S. Supreme Court's decision in *Daubert v. Merrill Dow Pharmaceuticals*[31]—in overruling the so-called *Frye* standard that had required that novel scientific evidence be ''generally accepted''[32] (as a predicate for admissibility of expert testimony), and in substituting the broader standard of ''relevant and reliable'' scientific evidence[33]—will focus even greater attention on scientific studies that explore the insanity–genetics link.

Yet these discoveries and developments have had virtually no impact on the basic debate. They are ignored, trivialized, denied, and distinguished. And the gap between myth and reality is a vast one that widens exponentially with the passage of time. Although the gap is acknowledged by virtually every empirical researcher who has studied any of these issues,[34] it continues to grow. We continue to honor and reify the myths through legislative action and judicial decisions, and in public forums: The public continues to endorse a substantive standard for insanity that approximates the ''wild beast'' test of 1724,[35] and legislators look to the potential abolition of the insanity defense as a palliative for rampant crime problems in spite of incontrovertible statistics that show that the defense is raised in a fraction of 1% of felony prosecutions (and is successful only about a quarter of the time).[36] Our response to the most celebrated insanity acquittal of the twentieth century—that of John W. Hinckley, who shot President Ronald Reagan in 1981—was to shrink the insanity defense in federal jurisdictions to a more narrow and restrictive version of an 1843 test that was seen as biologically, scientifically, and morally outdated at the very time of its creation.[37]

Why? What is it about the insanity defense that allows for (and perhaps encourages) such a discontinuity between firmly held belief and statistical reality? Why is our insanity defense jurisprudence so irrational? Why do we continue to obsess about questions that are fundamentally irrelevant to the core jurisprudential inquiry of who should be exculpated because of lack of mental responsibility? Why do we allow ourselves to be immobilized by an irresoluble debate? Why does our willful blindness allow us (lead us?) to ignore scientific and empirical developments and, instead, force us to waste time, energy, and passion on a series of fruitless inquiries that will have negligible impact on any of the underlying social problems? Most important, why do we continue to ignore the most fundamental and core question: Why do we feel the way we do about ''these people,'' and why, when we engage in our endless debates and incessant retinkering with insanity defense doctrine, do we not seriously consider our answer to this question?[38]

I believe that our insanity defense jurisprudence is a prisoner of a combination of these empirical myths and related social meta-myths. Born of a medievalist and fundamentalist religious vision of the roots of mental illness and the relationships between mental illness, crime, and punishment, the myths continue to dominate the landscape in spite of (and utterly independently of) the impressive scientific and behavioral evidence to the contrary.[39]

The legal system is a prisoner of these myths and of the concomitant powerful symbols that permeate any criminal trial (especially any highly visible criminal trial) at which a nonresponsibility defense is raised. It rejects psychodynamic explanations of human motivation and behavior, and remains intensely suspicious of concepts of mental health and disability, of mental health professionals, and of the ability of such professionals to assess or ameliorate mental disability.[40] As a result, it remains most comfortable with all-or-nothing tests of mental illness,[41] and demands that nonresponsible defendants match visual images of ''deranged madmen'' who, indisputably, ''look crazy.''[42] Again, it does this in utter disregard of the past 150 years of scientific and behavioral learning.

Our efforts to understand the insanity defense and insanity-pleading defendants, in short, are doomed to eternal intellectual, political, and moral gridlock unless we are willing to take a fresh look at the underlying doctrine through a series of filters: empirical research, scientific discovery, moral philosophy, cognitive and moral psychology, sociology, communications theory, and political science. Only in this manner can we attempt to articulate the sort of coherent and integrated perspective that is necessary if we are to unpack the myths from the defense's facade and reconstruct a meaningful insanity-defense jurisprudence.

## The Development of Insanity Defense Doctrine

I will now turn briefly to an explanation of the different insanity defense tests in an effort to provide a better sense of the ways that the tests have developed, the ways that myths have similarly dominated these developments, and the level of rigidity that has shaped the creation of our insanity defense jurisprudence.[43]

### *Pre-*M'Naghten *History*

The development of the insanity defense prior to the mid-nineteenth century tracked both the prevailing scientific and popular concepts of mental illness, "craziness," responsibility, and blameworthiness.[44] Prior to the *M'Naghten*[45] decision in 1843, the substantive insanity defense went through three significant stages: the "good and evil" test, the "wild beast" test, and the "right and wrong" test.

The "good and evil" test—which apparently first appeared in a 1313 case involving the capacity of an infant under the age of 7—reflected the "moral dogmata reflected in [the medieval] theological literature."[46] The insane, like children, were incapable of "sin[ning] against [their] will," because man's freedom "is restrained in children, in fools, and in the witless who do not have reason whereby they can choose the good from the evil."[47]

During the fourteenth through sixteenth centuries, this test—the source of which was most likely biblical—remained constant in English law, and, by the end of that time, insane persons who met this test were treated as "nonpersons"—not fit subjects for punishment—"since they did not comprehend the moral implications of their harmful acts." It was thus no surprise when this test was transfigured in 1724 to the "wild beast test."

Under the "wild beast" formulation, in *Rex v. Arnold,* a case in which the defendant had shot and wounded a British Lord in a homicide attempt, Judge Tracy instructed the jury that it should acquit by reason of insanity in the case of "a mad man . . . must be a man that is totally deprived of his understanding and memory, and doth not know what he is doing, no more than *a brute, or a wild beast,* such a one is never the object of punishment."[48]

In short, the emphasis was on lack of *intellectual ability,* rather than the violently wild, ravenous beast image that the phrase calls to mind; the test continued to be used until at least 1840.[49]

The following step—the "right and wrong" test (the true forerunner of *M'Naghten*)—emerged in two 1812 cases, and was finally expanded on in 1840 in *Regina v. Oxford,* in which Lord Denman charged the jury that it must determine whether the defendant, "from the effect of a diseased mind," knew that the act was wrong, and that the question that must thus be answered was whether "he was quite unaware of the nature, character, and consequences of the act he was committing."[50]

Even with these rigid tests in place, the public's perceptions of abuse of the insanity defense differed little from its reactions in the aftermath of the Hinckley acquittal nearly a century and a half later. The public's representatives demanded an "all or nothing" sort of insanity, a conceptualization that has been "peculiarly foreign" to psychiatry since at least the middle of the nineteenth century.[51] Similarly, the "demonological" concept of mental illness retained its power centuries after it became clear that such a view was never supported by scientific data. Psychiatry can thus be seen as having always "inextricably tied" to the values of the prevailing culture.[52]

## M'Naghten

In 1843, the "most significant case in the history of the insanity defense in England" arose out of the shooting by Daniel M'Naghten of Edward Drummond, the secretary of the man he mistook for his intended victim: Prime Minister Robert Peel.[53] After *nine* medical witnesses testified that M'Naghten was insane, and after the jury was informed that an insanity acquittal would lead to the defendant's commitment to a psychiatric hospital, M'Naghten was found not guilty by reason of insanity (NGRI).[54]

Enraged by the verdict, Queen Victoria questioned why the law was of "no avail," because everybody is morally convinced that "[the] malefactor . . . [was] perfectly conscious and aware of what he did," and demanded that the legislature "lay down the rule" so as to protect the public "from the wrath of madmen who they feared could now kill with impunity."[55] In response, the House of Lords asked the Supreme Court of Judicature to answer five questions regarding the insanity law, and the judges' answers to two of these five became the *M'Naghten* test:

> [T]he jurors ought to be told in all cases that every man is presumed to be sane, and to possess a sufficient degree of reason to be responsible for his crimes, until the contrary be proved to their satisfaction; and that to establish a defence on the ground of insanity, it must be clearly proved that, at the time of the committing of the act, the party accused was labouring under such a defect of reason, from disease of the mind, as not to know the nature and quality of the act he was doing; or, if he did know it, that he did not know he was doing what was wrong.[56]

This rigid, cognitive-only responsibility test, established under royal pressure, reflected "the prevailing intellectual and scientific ideas of the times," and stemmed from an "immutable philosophical and moral concept which assumes an inherent capacity in man to distinguish right from wrong and to make necessary moral decisions."[57]

The M'Naghten Rules reflected a theory of responsibility that was outmoded far prior to its adoption, and which bore little resemblance to what was known about the human mind, even at the time of their promulgation. Nonetheless, with almost no exceptions, they were held as sacrosanct by American courts that eagerly embraced this formulation, and codified it as the standard test "with little modification" in virtually all jurisdictions until the middle of the twentieth century.[58]

### *Post-M'Naghten Developments*

Although there was some interest in the post *M'Naghten* years in the so-called irresistible impulse test—allowing for the acquittal of a defendant if his mental disorder caused him to experience an irresistible and uncontrollable impulse to commit the offense"[59]—the first important theoretical alternative to *M'Naghten*[60] emerged in the District of Columbia in the 1954 case of *Durham v. United States.*[61]

### Durham

Writing for the court in *Durham,* Judge David Bazelon rejected both *M'Naghten* and the irresistible impulse tests on the theory that the mind of a human was a functional unit, and that a far broader test would be appropriate.[62] *Durham* thus held that an accused would not be criminally responsible if his or her "unlawful act was the product of mental disease or mental defect."[63] This test would provide for the broadest range of psychiatric expert testimony, "unbound by narrow or psychologically inapposite legal questions."[64] Further, it reiterated the jury's function in such a case:

> Juries will continue to make moral judgments, still operating under the fundamental precept that "Our collective conscience does not allow punishment where it cannot impose blame." But in making such judgments, they will be guided by wider horizons of knowledge concerning mental life. The question will be simply whether the accused acted because of a mental disorder, and not whether he displayed particular symptoms which medical science has long recognized do not necessarily or even typically, accompany even the most serious disorder.[65]

*Durham* was the first modern, major break from the *M'Naghten* approach, and created a "feeling of ferment" as the District of Columbia "became a veritable laboratory for consideration of all aspects of insanity (medical *and* legal), in its fullest substantive and procedural ramifications."[66] In other words, both the question of the substantive test—how nonresponsibility should be defined—and the procedural aspects—for example, allocation of burdens of proof; admissibility of expert testimony—were being debated vigorously anew in the aftermath of *Durham.* Within a few years, however, *Durham* was judicially criticized, modified,[67] and ultimately dismantled by the D.C. circuit (the court that had decided *Durham*),[68] its burial being completed by the 1972 decision in *United States v. Brawner*[69] to adopt the Model Penal Code/American Law Institute test.[70]

## United States v. Brawner

*Brawner* discarded *Durham's* "product" test, but added a volitional question to *M'Naghten's* cognitive inquiry. Under this test, "A defendant would not be responsible for his criminal conduct if, as a result of mental disease or defect, he 'lack[ed] substantial capacity either to appreciate the criminality of his conduct or to conform his conduct to the requirements of law.' "[71]

Although the test was rooted in *M'Naghten,* there were several significant differences. First, its use of the word "substantial" was meant to respond to caselaw developments that had required "a showing of total impairment for exculpation from criminal responsibility." Second, the substitution of the word "appreciate" for the word "know" showed that "a sane offender must be emotionally as well as intellectually aware of the significance of his conduct," and "mere intellectual awareness that conduct is wrongful when divorced from an appreciation or understanding of the moral or legal import of behavior, can have little significance." Third, by using broader language of mental impairment than had *M'Naghten,* the test "capture[d] both the cognitive and affective aspects of impaired mental understanding." Fourth, its substitution in the final proposed official draft of the word "wrongfulness" for "criminality" reflected the position that the insanity defense dealt with "an impaired moral sense rather than an impaired sense of legal wrong."[72]

Although it was incorrectly assumed that the spreading adoption of *Brawner* would augur the death of *M'Naghten,*[73] *Brawner* did serve as the final burial for the *Durham* experiment.[74]

## The Insanity Defense Reform Act

The entire insanity debate was radically shifted in March 1981, after John Hinckley shot President Reagan. Hinckley's acquittal galvanized the American public in a way that led directly to the reversal of 150 years of study and understanding of the complexities of psychological behavior and the relationship between mental illness and certain violent acts. The public's outrage over a legal system that could allow a defendant who shot an American president on national television to plead "not guilty" (for *any* reason) became a "river of fury" after the jury's verdict was announced.[75]

Members of Congress responded quickly to the public's outpouring of outrage by introducing 26 separate pieces of legislation designed to limit, modify, severely shrink, or abolish the insanity defense in federal trials.[76] The Reagan administration originally had called loudly for the abolition of the insanity defense.[77] However, in the face of a nearly unified front presented by most of the relevant professional organizations and trade associations, it eventually quietly dropped its public call for abolition and supported the Insanity Defense Reform Act (IDRA) as a "reform compromise."[78] The legislation ultimately enacted by Congress—legislation that closely comported with the public's moral feelings—returned the insanity defense to "*status quo ante* 1843: the year of . . . *M'Naghten.*"[79]

Besides changing the burden of proof in insanity trials from the government to defendants,[80] establishing strict procedures for the hospitalization and release of defendants found not guilty by reason of insanity,[81] and severely limiting the scope of expert testimony in federal insanity cases,[82] the IDRA discarded the Model Penal Code test (applied in almost all federal judicial districts),[83] and adopted a more restrictive version of *M'Naghten* by specifying that the level of mental disease or defect that must be shown to qualify be "severe."[84]

Two thirds of the states quickly followed the lead of the federal government by reevaluating their laws, but not, in all cases, with the same results. Twelve states adopted the guilty but mentally ill (GBMI) test,[85] 7 narrowed the substantive test, 16 shifted the burden of proof, and 25 tightened release provisions in the cases of those defendants found to be NGRI. Three states adopted legislation that purported to abolish the defense, but actually retained a *mens rea* exception—that is, if the defendant were so *severely* impaired that he, for instance, thought he were squeezing a lemon rather than strangling a person, he would be exculpated.[86] All of this had the ultimate effect of returning the law to "do its punitive worst," that had "the rigidity of an army cot and the flexibility of a Procrustean bed," that retained the flavor "of the celebrated concepts of Hale and Coke of the 17th century," and that was, simply, "bad psychiatry and bad law."[87]

## The Insanity Defense and Physiological Disorders

With this lengthy backdrop, I will now turn to subsets of cases that bear more directly on our specific interests here and on the relationship between the insanity defense and cases involving defendants with physiological disorders[88] and those with genetic abnormalities. As I discuss these, keep in mind my discussions of both insanity defense myths and of the ways that the public constructs the insanity defense.

A series of cases has considered the availability of the insanity defense to defendants with such physiological and developmental disorders[89] as minimal brain dysfunction,[90] psychomotor epilepsy,[91] Huntington's Disease,[92] fetal alcohol syndrome,[93] cocaine-induced psychosis,[94] neurosarcoidosis,[95] Tourette syndrome,[96] pedophilia,[97] obsessive–

compulsive disorder,[98] and hypoglycemia.[99] Professor Fox has suggested that a consideration of those physiological conditions that can cause "impaired consciousness"[100] will also include, inter alia, "hypopituitarianism ... liver disease, head trauma, and delirium."[101]

Recent scientific developments have spurred more interest in this field.[102] Pointing out that a "defect in the brain" can now be determined through a variety of diagnostic tests—such as CAT scan, PET scan, NMR test, and EEG readings[103]—Jeffrey and White have surveyed the physiological disorders in question.[104] They suggest that such "biological brain disorders do not constitute insanity or mental illness,"[105] and recommend the establishment of special institutions to treat such individuals,[106] and the creation of a new "scientific biological definition of mental illness":[107] "a physical condition of the brain demonstrated by physical evidence which ties the behavior in question to neurological and biochemical processes within the brain and the nervous system."[108]

Such a definition would, according to its proponents, "tie the criminal law directly to science and psychiatry, and allow the lawyer to make the best possible use of modern psychiatry."[109] They concluded, "[This] psychobiological view ... is holistic and physicalistic in its concept of man [, offering] great hope in the future for the early diagnosis, prevention and treatment of behavioral disorders."[110]

### *Defendants With Genetic Abnormalities*

Beginning in the early 1960s, geneticists and criminologists began to explore the connection between genetic abnormality and criminal behavior,[111] an exploration initially spurred by the discovery of the so-called XYY chromosome.[112] Research into the XYY condition has since gone through three phases: the initial period (1961–1965), characterized by the discovery of XYY cases, often in conjunction with physical abnormality and mental retardation;[113] the second period (1965–early 1970s), centering around a "classic study"[114] in a British maximum security institution, which identified what appeared to be a statistically significant percentage of XYY males among the institution population as compared to the population at large;[115] and the third period (early 1970s–present), consisting of "more sophisticated studies"[116] of this population and of exploratory studies of individuals with other chromosomal abnormalities, including the XXY and XXYY chromosomes.[117]

This most recent generation of studies cannot be relied on to support sufficiently the thesis that "a positive link exists between antisocial behavior, aggression, and a genetic abnormality in men."[118] On the other hand, because the data appears to present fairly uniform evidence that there are "disproportionate numbers of XYY, XXY, and XXYY males in security and mental penal institutions,"[119] it becomes necessary to inquire about the possible linkage between such abnormality and the insanity defense.[120]

One additional complication is added as a result of the U.S. Supreme Court's recent decision in *Daubert v. Merrill Dow Pharmaceuticals,*[121] which I discussed previously. By establishing a new, broader standard of "relevant and reliable" scientific evidence[122] in cases involving expert testimony, *Daubert* may lead to new courtroom inquiries into the scope of a potential insanity–genetics link.

Although the current case law is scanty, there is at least one Australian case reported in which an NGRI verdict (in a *M'Naghten* jurisdiction) was reached, in which psychiatric evidence was introduced on the question of the link between the XYY defect and the defendant's responsibility.[123] On the other hand, efforts at employing the defense in the American jurisdictions have been unsuccessful so far, with reported decisions concluding that: (a) scientific evidence was sufficiently inconclusive so as to deny a defendant's motion to withdraw his guilty plea and substitute a plea of NGRI,[124] (b) the evidence presented was not sufficient to allow a jury to consider a defendant's NGRI plea because it was not based on "reasonable medical certainty,"[125] and (c) it was not improper for a trial court to deny a defendant's motion for the appointment of a cytogeneticist to perform a chromosome test,[126] or for an additional psychiatric examination to determine whether he was an individual with the XYY chromosome.[127] In one of these cases, however, a New York state trial court set out a rule to govern the future introduction of genetic evidence, which may be read to indicate at least some receptivity to the admissibility of such evidence in the future:

> Thus, in New York an insanity defense based on chromosome abnormality should be possible only if one establishes with a high degree of medical certainty an etiological relationship between the defendant's mental capacity and the genetic syndrome. Further, the genetic imbalance must have so affected the thought processes as to interfere substantially with the defendant's cognitive capacity or with his ability to understand the basic moral code of his society.[128]

Although there are clear ethical issues raised any time the insanity defense is employed,[129] the use of genetic evidence involves specific additional ethical considerations as well as the usual trial strategy decisions.[130] Because of the conflict in scientific evidence, public fear of "genetically abnormal" individuals, and the general empirical evidence—showing that: (a) individuals found NGRI are institutionalized for longer periods than individuals found guilty of the same charge,[131] and (b) individuals unsuccessful in the assertion of an insanity defense, a universe that includes all defendants in American jurisdictions who have offered such a defense based on genetic abnormality, serve significantly longer sentences than defendants convicted of similar charges who do not assert the defense[132]—it is clear that the "ante is raised" whenever a genetic abnormality insanity defense is offered.[133]

This analysis should not be read as a recommendation that further exploration in this area be curtailed, but as a reflection that courts will "walk slowly and carefully in these early . . . days and until more universal and reliable scientific acceptance has been achieved."[134]

## Syndromes and Recent Research

On the other hand, there has been far greater recent public attention focused on cases (and academic writing) suggesting new and additional cultural, medical, and behavioral bases for the insanity defense—in the context of what is frequently referred to as "syndrome evidence."[135]

### *On Syndromes*

The developments that I have just discussed—the relationship between the insanity defense and both physiological disorders and genetic aberration—actually pose fewer doctrinal and "real life" problems than the topic to which I will now turn briefly: the availability of "syndromes" as a basis for an insanity defense or other mental status defense seeking to reduce levels of culpability.[136]

For it is here that the public's most profound ambivalence about the defense surfaces for a variety of reasons: the sense that syndromes do not "really" exist; the sense that—even if they *do* exist—they can be overcome (e.g., "*I* went to Vietnam and *I* didn't come back and start shooting up bars!"; "*I* had a baby, and sure I was sad, but *I* didn't stuff *my* baby's head into a toilet!"), and the sense that the acceptance of any syndrome is the first step on a slippery slope that might ultimately be an excuse to exculpate anyone charged with serious crime.

Most academic attention has focused on posttraumatic stress disorder-related syndromes (especially Vietnam Veterans' Stress Syndrome),[137] on battered spouse syndrome,[138] premenstrual stress syndrome,[139] and postpartum depression,[140] but evidence has been proferred in other syndrome cases as well: rape trauma syndrome,[141] captivity syndrome,[142] Russian emigre syndrome,[143] adoptive child syndrome,[144] and battered child syndrome.[145]

It is the latter syndrome—that of the battered child—that served as the impetus for a search for a genetic explanation of the behavior of the Menendez brothers. Although the defense team sought to use the defense to reduce the crime of murder to that of manslaughter (rather than as a pure insanity defense),[146] the same constellation of issues that erupts in insanity defense cases burst forth there as well.

The next question to turn to, then, is the ongoing search for the most contemporary research exploring the relationship between genetics and criminal behavior.

## *Recent Genetics Research*

Almost a century ago, Cesare Lombroso described a ''criminal type'' as one with ''a low slanting forehead, long ear lobes (or none at all), a large jaw with no chin, heavy ridges above the eye socket, and either excessive body hair or an abnormally small amount of body hair.''[147] Although we allegedly decry this sort of physiological reductionism now—I say ''allegedly'' because law enforcement officers still work on a daily basis from a foundation of ''informed hunches'' about who ''looks guilty,'' and almost everyone of *us* has a picture in *our* minds of who ''looks like a criminal''—we still yearn for an explanation, a category, a set of variables that might help us answer the perennial question: What ''causes'' criminal behavior?

Scientists and researchers have spent much of the twentieth century trying to answer this question. And the answers, of course, are ''lots of things'' and ''it depends.'' Studies tell us variously that there is a significant relationship between testosterone levels and adult criminal behavior,[148] that individuals with ''Negative Emotionality'' (the tendency to experience aversive affective states) are more likely to commit antisocial acts,[149] that alcohol consumption is positively related to propensity to violence,[150] that a constellation of social factors centering on the ''macho personality'' type correlates with violence,[151] that the prevalence of EEG abnormalities appears to be highest among the more violent habitual offenders,[152] that Attention Deficit Disorder and Minimal Brain Dysfunction are significantly correlated to juvenile delinquency,[153] and that head injury, physical growth and development, and a history of child abuse and neglect are all significantly linked to crime.[154] In short, both biological and environmental factors predict crime and violence.[155]

The Philadelphia Biosocial Study—extensively and expertly analyzed by Professor Deborah Denno—maintains that crime is directly related to family instability, a lack of behavioral control, and with neurological and central nervous system disorders;[156] correlates of this complex include central nervous system trauma and neurodevelopmental delay.[157]

I expect that none of this will surprise readers. On the other hand, it leads to a series of other questions that are relevant to our inquiries. First, can we find one Rosetta Stone, one factor, that predominates and unlocks the underlying mysteries? Second, should we spend our intellectual capital on such an inquiry? Third, assuming that we can and should, so what?

I am not being nihilistic with this inquiry, but am pointing out the need to make two additional inquiries: First, if we do find a genetic factor, does it ''fit'' within our current insanity defense formulations? Second, if it does, do decision makers—judges, jurors, legislators—care?

The meta-research appears to be fairly consistent on the first two questions. Virtually every modern expert agrees that both environmental and social forces play ''essential roles'' in shaping conduct and attitudes.[158] There are multiple factors that affect criminality, but none of the researchers can identify a single genetic marker, handicapping condition, or physiological attribute that serves as a legally sufficient ''cause.'' Although comprehensive models of biological and environmental variables have predicted as much as 25% of future adult criminality among males, and 19% among females, this still leaves three fourths to four fifths of such behavior unexplained.[159] Yet ''evidence of 'genetic factors in crime' cannot be ignored.''[160]

But beyond this lack-of-proof-as-to-causation is another issue. In an important and thoughtful article, Stephen Morse framed the issue this way:

> The ''fundamental psycholegal error'' is the mistaken belief that if science or common sense identifies a cause for human action, including mental or physical disorders, then the conduct is necessarily excused. But causation is neither an excuse per se nor the equivalent of compulsion, which is an excusing condition.[161]

''Causes of behavior are not excuses per se,'' Morse concluded.[162] Denno's research similarly demonstrated to her that there is ''no strong evidence [that] suggests that [biological deficiency defenses] *totally* impair free will or directly 'causes' the crime.''[163]

Yet we remain fascinated with genetic explanations of deviant behavior. Dreyfuss and Nelkin characterized this absorption as a ''preoccupation with biological determinism,'' a phenomenon they define as ''genetic essentialism'' (one that posits that personal traits are predictable and permanent, determined at conception, hardwired into the human constitution.''[164] When carried to its logical extreme, it minimizes the importance of social context and threatens to obscure the importance of ''life experiences'' in determining behavior.[165]

## Sanism, Pretextuality, the Menendezes, Genetics, and the Insanity Defense

What sense are we to make of all of this? On one hand, we are inordinately suspicious of psychodynamic explanations of criminal behavior, and we become hyperskeptical when a defendant proffers such an excuse to allow him or her to avoid criminal responsibility. On the other, we resolutely search for a genetic factor that would illuminate criminal behavior, while, at the same time, showing absolutely no sign that we wish to use such knowledge in the criminal justice system in a way that would take such genetic factors into account in such a way that would affect either responsibility determinations or ultimate punishment schemes.

### *Some Alternative Approaches*

In an effort to search for an explanation of this conundrum, I decided to focus on several overlapping constructs—drawn from cognitive psychology, from law, from sociology, from philosophy, and from my own invention. [166] I will first explain what I mean by each of these concepts, and then turn to their relationship to the way we feel about the insanity defense and, eventually, on the role of genetics in insanity defense decision making.

Let me start with heuristics. *Heuristics* is a cognitive psychology construct that refers to the implicit thinking devices that individuals use to simplify complex, information-processing tasks. The use of heuristics leads to distorted and systematically erroneous decisions, and causes decision makers to "ignore or misuse items of rationally useful information." One single vivid, memorable case overwhelms mountains of abstract, colorless data on which rational choices should be made.[167]

Thus, through the "availability" heuristic, we judge the probability or frequency of an event based on the ease with which we recall it.[168] Through the "typification" heuristic, we characterize a current experience via reference to past stereotypic behavior; through the "attribution" heuristic, we interpret a wide variety of additional information to reinforce preexisting stereotypes.[169]

Next is *ordinary common sense* (OCS). The positions frequently taken by Chief Justice Rehnquist and Justice Thomas in criminal procedure cases best highlight the power of OCS as an unconscious animator of legal decision making. Such positions frequently demonstrate a total lack of awareness of the underlying psychological issues and focus on such superficial issues as whether a putatively mentally disabled criminal defendant bears a "normal appearance."[170]

These are not the first jurists to exhibit this sort of close-mindedness. Trial judges will typically say, "[The defendant] doesn't look sick to me," or, even more revealing, "He is as healthy as you or me."[171] In short, in cases in which defendants do not conform to "popular images of 'craziness,'" the notion of a handicapping mental disability condition is flatly, and unthinkingly, rejected.[172] Views such as these reflect a false kind of OCS.[173] In criminal procedure, OCS presupposes two "self-evident" truths: "First, everyone knows how to assess an individual's behavior. Second, everyone knows when to blame someone for doing wrong."[174]

I next want to turn to what I call "sanism." *Sanism* is an irrational prejudice of the same quality and character of other irrational prejudices that cause (and are reflected in) prevailing social attitudes of racism, sexism, homophobia, and ethnic bigotry. It infects both our jurisprudence and our lawyering practices. Sanism is largely invisible, largely socially acceptable, and is based predominantly on stereotype, myth, superstition, and deindividualization.[175]

Judges, legislators, attorneys, and laypersons all exhibit sanist traits and profess sanist attitudes. It is no surprise that jurors reflect and project the conventional morality of the community, and judicial decisions in all areas of civil and criminal mental disability law continue to reflect and perpetuate sanist stereotypes.[176]

The concept of sanism must be considered hand-in-glove with that of *pretextuality*. Sanist attitudes often lead to pretextual decisions. By this I mean simply that fact finders accept (either implicitly or explicitly) testimonial dishonesty and engage similarly in dishonest (frequently meretricious) decision making, specifically where witnesses, especially *expert* witnesses, show a "high propensity to purposely distort their testimony in order to achieve desired ends." The pretexts of the forensic mental health system are reflected both in the testimony of forensic experts and in the decisions of legislators and fact finders. Experts frequently testify in accordance with their own self-referential concepts of "morality" and openly subvert statutory and case-law criteria that impose rigorous behavioral standards as predicates for commitment or that articulate functional standards as prerequisites for an incompetency-to-stand-trial finding.[177]

Finally, we need to consider *teleology*. The legal system selectively—teleologically—either accepts or rejects social science evidence, depending on whether or not the use of that data meets the *a priori* needs of the legal system. In cases in which fact finders are hostile to social science teachings, such data thus often meets with tremendous judicial resistance, and by the courts express their skepticism about, suspicions of, and hostilities toward such evidence.[178]

Courts are often threatened by the use of such data. Social science's "complexities [may] shake the judge's confidence in imposed solutions."[179] Courts' general dislike of social science is reflected in the self-articulated claims that judges are unable to understand the data and unable to apply it properly to a particular case.[180] Thus social science literature and studies that enable courts to meet predetermined sanist ends are often privileged, and data that would require judges to question such ends are frequently rejected. Judges often select certain proferred data that adheres to their preexisting social and political attitudes, and use heuristic reasoning in rationalizing such decisions. Social science data is used pretextually in such cases and is ignored in other cases to rationalize otherwise baseless judicial decisions.

How do these concepts play out in insanity defense cases? At the outset, consider that the insanity defense is a textbook example of the power of heuristic reasoning. Insanity defense defenders attempt to use statistics (to rebut empirical myths), scientific studies (to demonstrate that "responsibility" is a valid, externally verifiable term, and that certain insanity pleading defendants are, simply, "different") and principles of moral philosophy (to "prove" that responsibility and causation questions are legitimate ones for moral and legal inquiry). On the other hand, the vivid anecdote or the self-affirming attribution overwhelm all attempts at rational discourse. Insanity defense decision making is a uniquely fertile field in which the distortive "vividness" effect can operate and in which the legal system's poor mechanisms of coping with "systematic errors in intuitive judgment" made by heuristic "information processors" become especially troubling. The chasm between perception and reality on the question of the frequency of use of the insanity defense, its success rate, and the "appropriateness" of its success rate all reflect this effect.

Also, reliance on OCS is one of the keys to an understanding of why and how our insanity defense jurisprudence has developed. Not only is it "prereflexive" and "self-evident," it is susceptible to precisely the type of idiosyncratic, reactive decision making that has traditionally typified insanity defense legislation and litigation. Paradoxically, the insanity defense is necessary precisely because it rebuts "common-sense everyday inferences about the meaning of conduct."[181]

Empirical investigations corroborate the inappropriate application of OCS to insanity defense decision making. Judges "unconsciously express public feelings . . . reflect[ing] community attitudes and biases because they are 'close' to the community."[182] Virtually no members of the public can actually articulate what the substantive insanity defense test is. The public is seriously misinformed about both the "extensiveness and consequences" of an insanity defense plea."[183] And the public explicitly and consistently rejects any such defense substantively broader than the "wild beast" test.[184]

What about sanism? Insanity defense decision making is often irrational. It rejects empiricism, science, psychology, and philosophy, and substitutes myth, stereotype, bias, and distortion. In short, our insanity defense jurisprudence is the jurisprudence of sanism.

Like the rest of the criminal trial process, the insanity defense process is riddled by sanist stereotypes and myths. Think of these:

- reliance on a fixed vision of popular, concrete, visual images of "craziness";
- an obsessive fear of feigned mental states;
- a presumed absolute linkage between mental illness and dangerousness;
- sanctioning of the death penalty in the case of mentally retarded defendants, some defendants who are "substantially mentally impaired," or defendants who have been found guilty but mentally ill (GBMI);
- the incessant confusion and conflation of substantive mental status tests; and

- the regularity of sanist appeals by prosecutors in insanity defense summations, arguing that insanity defenses are easily faked, that insanity acquittees are often immediately released, and that expert witnesses are readily duped.

Sanism, in short, regularly and relentlessly infects the courts in the same ways that it infects the public discourse. It synthesizes all of the irrational thinking about the insanity defense, and helps create an environment in which groundless myths can shape the jurisprudence. As much as any other factor, it explains why we feel the way we do about "these people." As I will discuss next, it also provides a basis for courts to engage in pretextual reasoning in deciding insanity defense cases.

Further, pretextual decision making riddles the entire insanity defense decision-making process; it pervades decisions by forensic hospital administrators, law enforcement officers, expert witnesses, and judges. Hospital decision making is a good example. An NIMH task force convened in the wake of the Hinckley acquittal underscored this in its final report: "From the perspective of the Hospital, *in controversial cases such as Hinckley,* the U.S. Attorney's Office can be counted upon to oppose *any* conditional release recommendation."[185] As John Parry has explained, "Hospitals have been pressured by public outrage to bend over backwards to make sure that no insanity acquittee is released too soon, *even* if such pressure is contrary to the intent and spirit of being found not guilty by reason of insanity."[186]

Expert witnesses are often similarly pretextual. In one case, a testifying doctor conceded that he may have "hedged" in earlier testimony (as to whether an insanity acquittee could be released) "because he did not want to be criticized should [the defendant] be released and then commit a criminal act."[187] Most important, all aspects of the judicial decision-making process embody pretextuality. To a significant extent, this fear that defendants will "fake" the insanity defense to escape punishment continues to paralyze the legal system in spite of an impressive array of empirical evidence that reveals (a) the minuscule number of such cases, (b) the ease with which trained clinicians are usually able to "catch" malingering in such cases, (c) the inverse greater likelihood that defendants—even at grave peril to their life—will more likely try to convince examiners that they're "not crazy," (d) the high risk in pleading the insanity defense (leading to statistically significant greater prison terms meted out to *unsuccessful* insanity pleaders), and (e) the far greater length in stay that most successful insanity pleaders (a minute universe to begin with) remain in maximum security facilities than they would have served had they been convicted on the underlying criminal indictment. In short, pretextuality dominates insanity defense decision making. The inability of judges to disregard public opinion and inquire into whether defendants have had fair trials is both the root and the cause of pretextuality in insanity defense jurisprudence.

Finally, little attention has been paid in general to the role of social science data in insanity defense decision making. The law's suspicion of the psychological sciences is well-documented. Also, the issues before the courts in insanity defense cases raise such troubling issues for decision makers that the courts' inherent suspicion of the social sciences will be further heightened. This, though, should not surprise us. Traditionally, social science has played less of a role in the establishment of legal policy in areas "dominated by clear ideological division" or "political debate." The more that social science contradicts "sentiments essential to other legal institutions," the less likely it will influence legal policy.[188]

I believe that much of the incoherence of insanity defense jurisprudence can be explained by these phenomena. Stereotyped thinking leads to sanist behavior. Sanist decisions are rationalized by pretextuality on the part of judges, legislators, and lawyers, and are buttressed by the teleological use of social science evidence and empirical data. This combination of sanism and pretextuality ''fits'' with traditional ways of thinking about (and acting toward) mentally disabled persons; it reifies centuries of myths and superstitions, and is consonant with both the way we use heuristic cognitive devices and our own faux, nonreflective ''ordinary common sense.''

### *Seeking a Connective Tissue*

How, if at all, does this link back to the Menendez trial? Recall that the Menendezes proferred a BCS defense as an excuse to explain the killings of their parents. Think about how many emotionally freighted issues are conflated and confounded: a mental status defense, a syndrome, a ''flashy'' syndrome that serves as a projective test for a nation frustrated with a seemingly irreversibly flawed criminal justice system, and a syndrome that may or may not have a genetic or biological component.[189]

Consider also how our debate on the insanity defense is, fundamentally, an *irrational* one; how it reflects our society's sanist attitudes toward mental disability and mentally disabled persons and our courts' pretextual attitudes toward cases involving mentally disabled criminal defendants; how it reflects our desire to punish, our search for an all-or-nothing, good-or-evil dyadic universe; how it reflects our rejection of empiricism and of philosophy, substituting an episodic and reactive jurisprudence built on the heuristic of the most recent, awful, vivid case.

Recent literature on the relationship between genetics and the criminal law is, when we come down to it, a gigantic tease. The recent studies suggest relationships between genetic factors and criminality, but virtually always in the context of a host of qualifiers (both environmental qualifiers and empirical qualifiers).[190] And, moreover, criminal law scholars regularly counsel caution in playing the ''causation'' card in this context.

So, this is *not* a case in which I will say, ''Aha! There *is* a smoking gun! There *is* a genetic factor that shows us why *A* commits an otherwise inexplicable criminal act and why *B*—who seems to be a lot like *A*—doesn't.'' Because I do not think that we have that evidence before us. And frankly, I do not think we ever will.

But that does *not* say that we should put our heads comfortably back in the policy sands and give implicit aid and comfort to the deans and doyennes of the television afternoon talk shows (our true town squares as we near the new century). It is vitally important that we continue to study genetics in an insanity defense context, that we continue to think about these relationships, and, most important, the extent to which any scientific discoveries/research/breakthroughs are likely to alter the paradigms of the insanity defense that we continue to fruitlessly debate.

The Menendez case serves as a wonderful projective test for all of this that I have been discussing and as a reflection of the difficulties we face in trying to bring some measure of doctrinal coherence to this subject matter. Because the Menendezes do not ''match up'' with our picture of the typical insanity defendant—one exhibiting ''wild eyes and trembling''[191]—we become even more frustrated in trying to ''slot'' the case into our rigid mental status–behavior categories. And that frustration causes a certain measure of cognitive dissonance[192] that dominates conversations about this entire topic of law.

## Conclusion

The development of the insanity defense has tracked the tension between psychodynamics and punishment, and reflects our most profound ambivalence about both. On one hand, we are especially punitive toward the mentally disabled, "the most despised and feared group in society"; on the other, we recognize that in some narrow and carefully circumscribed circumstances, exculpation is—and historically has been—proper and necessary. This ambivalence infects a host of criminal justice policy issues that involve mentally disabled criminal defendants beyond insanity defense decision making: on issues of expert testimony, mental disability as a mitigating (or aggravating) factor at sentencing and in death penalty cases, and the creation of a "compromise" GBMI verdict.

The post-*Hinckley* debate revealed the fragility of our insanity defense policies, and demonstrated that there was simply not enough "tensile strength" in the criminal justice system to withstand the public's dysfunctionally heightened arousal that followed the jury verdict. In spite of doctrinal changes and judicial glosses, the public remains wedded to the "wild beast" test of 1724, a reflection of how we *truly* feel about "those people."[193] It should thus be no surprise that when Congress chose to replace the ALI/Model Penal Code insanity test with a stricter version of *M'Naghten,* that decision was seen as a victory by insanity defense supporters.

These dissonances, tensions, and ambivalences—again, rooted in medieval thought—continue to control the public's psyche. They reflect the extent of the gap between academic discourse and social values, and the "deeply rooted moral and religious tension" that surrounds responsibility decision making.[194] They lead to sanism, to pretextuality, and to teleological decision making. They seek confirmation in "ordinary common sense" and in the use of heuristic cognitive devices. Ours is a culture of punishment, a culture that grows out of our authoritarian spirit. Only when we acknowledge these psychic and physical realities—and the anthropology of insanity defense attitudes—can we expect to make sense of the underlying jurisprudence.

Let me conclude by returning to my title. The concepts to be discussed in this volume are, I am confident, "big ideas." But before we can incorporate these big ideas into our jurisprudence, we need to consider the remainder of the first line I chose to quote: the public "images" in this area of the law have totally overwhelmed its reality. Beyond that, the discourse is premised—inevitably—on "distorted facts." And finally, perhaps most important, we are living in a "political world." It is this final reality that will never go away.

## Notes

1. *See generally* Michael L. Perlin, *The Jurisprudence of the Insanity Defense* (Durham, NC: Carolina Academic Press, 1994), 105–114.
2. *See* People v. White, 172 Cal. Rptr. 612, 614–615 (App. 1981), the case of Dan White, convicted on voluntary manslaughter based on a diminished capacity defense in the Moscone–Milk killings in San Francisco in 1984. *See also* Charles Hobson, "Reforming California's Homicide Law," *Pepperdine Law Review* 23 (1990):495–563.
3. Perlin, *The Jurisprudence of the Insanity Defense,* 113.
4. *Id.*, 111, discussing results reported in Joseph Rodriguez, L. M. LeWinn, & Michael L. Perlin, "The Insanity Defense Under Siege: Legislative Assaults and Legal Rejoinders," *Rutgers Law Journal* 14 (1983):397–430.
5. For an excellent article cautioning about the ascription of too much causality in this context, *see* Stephen Morse, "Brain and Blame," *Georgetown Law Journal* 84 (1996):527–549.
6. For two expressions of this position, *see, e.g.,* Louisiana v. Perry, 610 So.2d 746, 781 (La. 1992)

(Cole, J., dissenting) (''Society has the right to protect itself from those who would commit murder and seek to avoid their legitimate punishment by a subsequently contracted, or feigned, insanity''); Gilbert Geis and Robert F. Meier, ''Abolition of the Insanity Plea in Idaho: A Case Study,'' *Annals* 477 (1985):72, irrelevant to Idaho residents whether defendant's reliance on insanity defense was real or feigned.

7. People v. Menendez, Docket No. BA068880 (April 10, 1996), 1996 WL 167755 (Cal. Super. Trans. 1996), at *23.
8. *Id.* (April 4, 1996), 1996 WL 155281, at *3.
9. *Id.* (Feb. 9, 1996), 1996 WL 56037, at **6–7.
10. *Id.* (Feb. 27, 1996), 1996 WL 83110, at *24.
11. *Id.* (Feb. 23, 1996), 1996 WL 77762, at *35.
12. Perlin, *The Jurisprudence of the Insanity Defense,* 232–233.
13. I explore this extensively in *id.*, 73–142.
14. This section is adapted from Michael L. Perlin, ''The Current Status of the Insanity Defense.'' In *13 Innovations in Clinical Practice: A Source Book,* Leon VandeCreek, Samuel Knapp, and Thomas L. Jackson, eds. (Sarasota, FL: Professional Resource Exchange, 1994).
15. Jodi English, ''The Light Between Twilight and Dusk: Federal Criminal Law and the Volitional Insanity Defense,'' *Hastings Law Journal* 40 (1988):1–52; Michael L. Perlin, ''On Sanism,'' *Southern Methodist University Law Review* 46 (1992):373–407.
16. *See* Lisa Callahan, Connie Mayer, & Henry J. Steadman, ''Insanity Defense Reform in the United States—Post-Hinckley,'' *Mental and Physical Disability Law Report* (1987):54–59.
17. *See, e.g.,* Richard Bonnie, ''Morality, Equality, and Expertise: Renegotiating the Relationship Between Psychiatry and Law,'' *Bulletin of the American Academy of Psychiatry and Law* 12 (1984):5–20.
18. *See, e.g.,* Michael L. Perlin, '' 'The Borderline Which Separated You From Me': The Insanity Defense, The Authoritarian Spirit, the Fear of Faking, and the Culture of Punishment,'' *Iowa Law Review* 82 (1997):1375–1426.
19. *See generally* Perlin, *The Jurisprudence of the Insanity Defense.*
20. Ingo Keilitz, ''Researching and Reforming the Insanity Defense,'' *Rutgers Law Review* 39 (1987):289–322.
21. United States v. Lyons, 739 F.2d 994, 994 (4th Cir. 1984) (Rubin, J., dissenting).
22. Joseph Livermore and Paul Meehl, ''The Virtues of M'Naghten,'' *Minnesota Law Review* 51 (1967):7.
23. U.S. Congress, Senate Committee on the Judiciary, 97th Cong., 2d Sess., *Insanity Defense Hearings* 27 (1982).
24. *See generally* Perlin, *The Jurisprudence of the Insanity Defense.*
25. *See generally* Michael L. Perlin, ''Unpacking the Myths: The Symbolism Mythology of Insanity Defense Jurisprudence,'' *Case Western Reserve Law Review* 40 (1989–90):599–731; Michael L. Perlin, *Mental Disability Law: Civil and Criminal* § 15.37 (Charlottesville, VA: Lexis Law, 1989).
26. Perlin, *The Jurisprudence of the Insanity Defense,* 171–186.
27. Henry J. Steadman, Margaret A. McGreevey, Joseph P. Morrissey, Lisa A. Callahan, Pamela C. Robbins, & Carmen Cirincione, *Before and After Hinckley: Evaluating Insanity Defense Reform.* (New York: Guilford Press, 1993).
28. Rodriguez et al., ''The Insanity Defense Under Seige.''
29. H. J. Garber, ''Use of Magnetic Resonance Imaging in Psychiatry,'' *American Journal of Psychiatry* 145 (1988):164–168.
30. *See, e.g.,* Michael Moore, *Law and Psychiatry: Rethinking the Relationship* (Cambridge: Cambridge University Press, 1984).
31. 509 U.S. 579 (1993).
32. Frye v. United States, 293 F.1013, 1014 (D.D.C. 1923).
33. *Daubert,* 113 S. Ct. at 2795. *See generally* Bert Black, F. J. Ayalo, & C. Saffron-Brinks, ''Science and the Law in the Wake of *Daubert:* A New Search for Scientific Knowledge,'' *Texas Law*

*Review* 72 (1994):715–802; Andrew Taslitz, *Daubert*'s "Guide to the Federal Rules of Evidence: A Not-So-Plain-Meaning Jurisprudence," *Harvard Journal of Legislation* 32 (1995):3–77.

34. *See, e.g.*, Lisa Callahan et al., "The Volume and Characteristics of Insanity Defense Pleas: An Eight-State Study," *Bulletin of the American Academy of Psychiatry and Law* 19 (1991):331–338; Steadman, *Before and After Hinckley.*
35. Caton Roberts, Stephen L. Golding, & Frank D. Fincham, "Implicit Theories of Criminal Responsibility Decision Making and the Insanity Defense," *Law and Human Behavior* 11 (1987):207–232, discussing *Rex v. Arnold,* 16 How. St. Tr. 695 (1724).
36. Callahan et al., "The Volume and Characteristics of Insanity Defense Pleas," 334–336.
37. *See* Perlin, *Mental Disability Law,* § 15.04, 292–294.
38. *See generally* Michael L. Perlin, "Myths, Realities, and the Political World: The Anthropology of Insanity Defense Attitudes," *Bulletin of the American Academy of Psychiatry and Law* 25 (1996):5–26.
39. Richard Rogers, W. Seman, & C. R. Clark, "Assessment of Criminal Responsibility: Initial Validation of the R-CRAS with the M'Naghten and GBMI Standards," *International Journal of Law and Psychiatry* 9 (1986):67–75.
40. Michael L. Perlin, "The Supreme Court, the Mentally Disabled Criminal Defendant, and Symbolic Values: Random Decisions, Hidden Rationales, or 'Doctrinal Abyss'?, *Arizona Law Review* 29 (1987):1–98; Perlin, *Symbolic Values*; Michael L. Perlin, "Psychodynamics and the Insanity Defense: 'Ordinary Common Sense' and Heuristic Reasoning," *Nebraska Law Review* 69 (1990):3–70; Michael L. Perlin and Deborah A. Dorfman, "Sanism, Social Science, and the Development of Mental Disability Law Jurisprudence," *Behavior Science and Law* (1993):47–66.
41. Johnson v. State, 439 A.2d 542, 552 (Md. 1982) ("For the purposes of guilt determination, an offender is either wholly sane or wholly insane").
42. People v. Battalino, 199 P.2d 897, 901 (Colo. 1948), finding no evidence of defendant exhibiting "paleness, wild eyes and trembling."
43. This section is generally adapted from Michael L. Perlin, "The Insanity Defense: Deconstructing the Myths and Reconstructing the Jurisprudence." In *Law, Mental Health, and Mental Disorder,* Bruce D. Sales and Daniel W. Shuman, eds. (Pacific Grove, CA: Brooks/Cole, 1996).
44. Perlin, *Mental Disability Law,* §§ 15.02–15.03, 279–286.
45. M'Naghten's Case, 8 Eng. Rep. 718 (H.L. 1843).
46. Anthony Platt and Bernard Diamond, "The Origins of the 'Right and Wrong' Test of Criminal Responsibility and Its Subsequent Development in the United States: An Historical Survey," *California Law Review* 54 (1966):1227, 1231–1233.
47. *Id.*, 1233.
48. Rex v. Arnold, 16 How. St. Tr. 695 (1724), *reprinted in A Complete Collection of State Trials,* T. B. Howell, ed. (London, 1812).
49. Platt and Diamond, "The Origins of the 'Right and Wrong' Test," 1236.
50. 9 Carr & P. 525 (1840).
51. Bernard Diamond, "Criminal Responsibility of the Mentally Ill," *Stanford Law Review* 14 (1961):59.
52. Lola Romanucci-Ross and Lawrence Tancredi, "Psychiatry, the Law and Cultural Determinants of Behavior," *International Journal of Law and Psychiatry* 9 (1986):265–293.
53. *See* Perlin, *Mental Disability Law,* § 15.04.
54. *See generally* Donald Herrmann and Yvonne Sor, "Convicting or Confining? Alternative Directions in Insanity Law Reform: Guilty But Mentally Ill Versus New Rules for Release of Insanity Acquittees," *Brigham Young University Law Review* (1983):499–638.
55. *See e.g.*, Julian Eule, "The Presumption of Sanity: Bursting the Bubble," *University of California Los Angeles Law Review* 25 (1978):637–644; Richard Moran, *Knowing Right From Wrong: The Insanity Defense of Daniel McNaughtan* (New York: Free Press, 1981); Hermann and Sor, "Convicting or Confining?" 510.

56. M'Naghten's Case, 8 Eng. Rep. 718, 722 (1843).
57. Herbert Hovenkamp, "Insanity and Responsibility in Progressive America," *North Dakota Law Review* 57 (1981):541–575.
58. Barbara Weiner, "Not Guilty by Reason of Insanity—A Sane Approach," *Chicago-Kent Law Review* 56 (1980):1057–1085; *see generally* Callahan et al., "Insanity Defense Reform in the United States."
59. *See, e.g.*, Parsons v. State, 2 So. 854, 866–867 (Ala. 1886).
60. Although there was some interest in the post-*M'Naghten* years in the so-called irresistible impulse exception—allowing for the acquittal of a defendant if his mental disorder caused him to experience an "irresistible and uncontrollable impulse to commit the offense, even if he remained able to understand the nature of the offense and its wrongfulness"—this formulation was not more than a transitory detour in the development of an insanity jurisprudence. *See* Perlin, *Mental Disability Law,* § 15.05; George Dix, "Criminal Responsibility and Mental Impairment in American Criminal Law: Responses to the Hinckley Acquittal in Historical Perspective." In *Law and Mental Health: International Perspectives,* David Weisstub, ed. (New York: Pergamon Press, 1986).
61. 214 F.2d 862 (D.C. Cir. 1954), *overruled in* United States v. Brawner, 471 F.2d 969 (D.C. Cir. 1972). *Also, see generally* Perlin, *Mental Disability Law,* § 15.06.
62. 214 F.2d 862, 870–874.
63. *Id.,* 874–875.
64. Barbara Weiner, "Mental Disability and Criminal Law." In *The Mentally Disabled and the Law,* 3d ed., Samuel Brakel, John Parry, and Barbara A. Weiner, eds. (Buffalo, NY: William Stein, 1985).
65. *Durham,* 214 F.2d at 876.
66. Abraham Goldstein, *The Insanity Defense* (New Haven, CT: Yale University Press, 1967).
67. Dean Acheson, "*McDonald v. United States:* The *Durham* Rule Refined," *Georgetown Law Journal* 51 (1963):580.
68. *See, e.g.*, Frigillana v. United States, 307 F.2d 665 (D.C. Cir. 1962); McDonald v. United States, 312 F.2d 847 (D.C. Cir. 1962).
69. 471 F.2d 969 (D.C. Cir. 1972), *overruling* Durham v. United States, 214 F.2d 862 (D.C. Cir. 1954).
70. American Law Institute, Model Penal Code § 4.01 (New York: 1956).
71. *Brawner,* 471 F.2d at 973.
72. Goldstein, *The Insanity Defense,* 87.
73. Bernard Diamond, "From *M'Naghten* to *Currens* and Beyond," *California Law Review* 50 (1962):189.
74. Bernard Diamond, "From *Durham* to *Brawner:* A Futile Journey," *Washington University Law Quarterly* (1973):109.
75. Perlin, *The Jurisprudence of the Insanity Defense,* 13.
76. Perlin, "Unpacking the Myths"; Michael L. Perlin, "'The Things We Do For Love': John Hinckley's Trial and the Future of the Insanity Defense in the Federal Courts (Book Review of L. Caplan, *The Insanity Defense and the Trial of John W. Hinckley, Jr.* (1984))," 30 *New York Law School Law Review* 30 (1985):857–868.
77. Perlin, "The Things We Do for Love," 860 n. 9.
78. Perlin, *Mental Disability Law,* § 15.39, 398–399.
79. Perlin, "The Things We Do for Love," at 862.
80. 18 U.S.C. § 17.
81. 18 U.S.C. § 4243.
82. Fed. R. Evid. 704(b).
83. Perlin, *Mental Disability Law,* § 15.07.
84. 18 U.S.C. § 17(a).
85. Perhaps the most important development in substantive insanity defense formulations in the twenty years post-*Brawner* has been the adoption in more than a dozen jurisdictions of the hybrid "guilty but mentally ill" (GBMI) verdict. Bradley McGraw, D. Farthing-Capouich, & D. Keilitz,

''The 'Guilty But Mentally Ill' Plea and Verdict: Current State of the Knowledge,'' *Villanova Law Review* (1985): 117. It received its initial recent impetus in 1975 in Michigan as a reflection of legislative dissatisfaction with and public outcry over a state Supreme Court decision that prohibited automatic commitment of insanity acquittees. Ira Mickenberg, ''A Pleasant Surprise: The Guilty But Mentally Ill Verdict Has Succeeded on Its Own Right and Successfully Preserved the Insanity Defense,'' *University of Cincinnati Law Review* 55 (1987):943–946, discussing People v. McQuillan, 211 N.W. 2d 569 (Mich. 1974).

The rationale for the passage of GBMI legislation was that the implementation of such a verdict would decrease the number of persons acquitted by reason of insanity (as it would give the jury an alternative verdict that would acknowledge the defendant's mental illness but allow the jury to not enter an NGRI verdict), and would ensure treatment of those who were GBMI within a correctional setting. People v. Smith, 465 N.E.2d 101, 106 (Ill. App. 1984). A GBMI defendant would purportedly be evaluated on entry to the correctional system and be provided appropriate mental health services either on an in-patient basis as part of a definite prison term or, in specific cases, as a parolee or as an element of probation. Weiner, ''Not Guilty by Reason of Insanity,'' 715.

Practice under GBMI statutes reveals that the verdict does little or nothing to ensure effective treatment for mentally disabled offenders. As most statutes vest discretion in the director of the state correctional or mental health facility to provide a GBMI prisoner with such treatment as she ''determines necessary,'' the GBMI prisoner is not ensured treatment ''beyond that available to other offenders.'' Chris Slobogin, ''The Guilty But Mentally Ill Verdict: An Idea Whose Time Should Not Have Come,'' *George Washington Law Review* 53 (1985):494–527. One comprehensive study of the operation of the GBMI verdict in Georgia revealed that only 3 of the 150 defendants who were found GBMI during the period in question were being treated in hospitals. Steadman, *Before and After Hinckley,* 195.

86. Steadman, *Before and After Hinckley;* Callahan et al., ''Insanity Defense Reform in the United States,'' Callahan, ''The Volume and Characteristic of Insanity Defense Pleas,'' 3 Perlin, *Mental Disability Law,* § 15.41.
87. Robert Sadoff, *Insanity: Evolution of a Medicolegal Concept.* A paper presented at College Night, College of Physicians of Philadelphia, September 1986; English, *The Light Between Twilight and Dusk,* 47.
88. *See generally* Perlin, *Mental Disability Law,* § 15.27 and *id.* (1998 Cum. Supp.).
89. On the ancient questions of somnambulism (sleepwalking) and automatism, *see, e.g.,* Richard Bonnie and Chris Slobogin, ''The Role of the Mental Health Professional in the Criminal Process: The Case for Informed Speculation,'' *Virginia Law Review* 66 (1980):427–495. The classic case is Fain v. Commonwealth, 78 Ky. 183 (1879). *See also* People v. Rothrock, 68 P.2d 364 (Cal. App. 1937); People v. Grant, 377 N.E.2d 4 (Ill. 1978) (discussing the relationship between the insanity defense and ''involuntary conduct'' as it applies to epileptic seizures); Bratty v. Attorney-General for Northern Ireland, 1963 A.C. 386, 3 All E.R. 523 (1961); Fulcher v. State, 633 P.2d 142 (Wyo. 1981); Wise v. State, 580 So.2d 329 (Fla. Dist. Ct. App. 1991).

On the related question of dissociative disorder, *see, e.g,* Elyn Saks, ''Multiple Personality Disorder and Criminal Responsibility,'' *University of California Davis Law Review* 25 (1992):383–461; Schiele, ''Dissociation and Unconsciousness as a Defense in the Criminal Case,'' *American Journal of Forensic Psychiatry* 12 (1991):35; Marlene Steinberg, B. Rounsaville, and D. Cicchetti, ''Detection of Dissociative Disorders in Psychiatric Patients by a Screening Instrument and a Structured Diagnostic Interview,'' *American Journal of Psychiatry* 148 (1991):1050–1054; Marlene Steinberg, J. Bancroft, and J. Buchanan, ''Multiple Personality Disorder in Criminal Law,'' *Bulletin of the American Academy of Psychiatry & Law* 21 (1993):345; Felicia Rubenstein, ''Committing Crimes While Experiencing a True Dissociative State: The Multiple Personality Defense and Appropriate Criminal Responsibility,'' *Wayne Law Review* 38 (1991):353–381; Elyn Saks, ''Does Multiple Personality Disorder Exist? The Beliefs, the Data, and the Law,'' *International Journal of Law and Psychiatry* 17 (1994):43–78; A. Piper, ''Multiple Personality Disorder,'' *British Journal of Psychiatry* 164 (1994):600–612. For a recent thoughtful case, *see* United States v. Denny-Shaffer, 2 F.3d 999 (10th Cir. 1993) (reversing conviction trial judge refused to submit instructions to jury in multiple personality disorder case), discussed in Mark Hindley, ''*United States v. Denny-Shaffer* and Multiple Personality Disorder: 'Who Stole the Cookie from the Cookie Jar?,''' *Utah Law Review* (1994):961–997.

90. People v. Wright, 648 P.2d 665, 668-669 (Colo. 1982) (expert testimony supporting insanity defense claim appropriately admitted into evidence). *See also* Henderschott v. People, 653 P.2d 385 (Colo. 1982) (effect of minimum brain dysfunction on reduction of gravity of specific-intent crime); State v. Hollis, 731 P.2d 260 (Kan. 1987) (testimony as to organic brain syndrome); Commonwealth v. Brennan, 504 N.E.2d 612 (Mass. 1987) (same).
91. Clark v. State, 436 N.E.2d 779, 780-781 (Ind. 1982) (trial court was within its authority to order neurological evaluation of defendant who claimed insanity defense, alleging he committed act ''in the grip of a blackout brought on by psychomotor epilepsy''). *See also* Matter of Torsney, 420 N.Y.S.2d 192 (1979). *But cf.* Government of Virgin Islands v. Smith, 278 F.2d 169, 175 (3d Cir. 1960) (epilepsy not defense to involuntary manslaughter).
92. Roach v. Martin, 757 F.2d 1463, 1473–1475 (4th Cir. 1985) (affirming denial of *habeas corpus* petition sought on grounds that contemporary technological advances in neuroscience research could presymptomatically diagnose Huntington's Disease; even assuming defendant had genetic disorder, Eighth Amendment would not preclude imposing death sentence). *See generally* Cheryl Becker, ''Legal Implications of the G-8 Huntington's Disease Genetic Marker,'' *Case Western Reserve Law Review* 39 (1988–89):273–305.
93. *See, e.g.*, David Davis, ''A New Insanity—Fetal Alcohol Syndrome,'' *Florida Bar Journal* 66 (Dec. 1992):53–57; Claire Dineen, ''Fetal Alcohol Syndrome: The Legal and Social Responses to Its Impact on Native Americans,'' *North Dakota Law Review* 70 (1994):1–66.
94. *See, e.g.*, Gregory Bunt, ''New Perspectives in the Legal Psychiatry of Cocaine-Related Crimes,'' *Journal of Forensic Science* 37 (1992):894.
95. United States v. Haga, 740 F. Supp. 1493, 1499 (D. Colo. 1990). Neurosarcoidosis is a rare neurological disease that, if untreated, causes mental disturbance. *See id.* n.6 (listing sources).
96. United States v. Rigatuso, 719 F. Supp. 409 (D. Md. 1989). Tourette syndrome is a neuropsychiatric disorder characterized by the childhood onset of involuntary motions and vocal tics and compulsive thoughts and actions. *See* J. Sverd, A. D. Curley, L. Jandorf, and L. Volkersz, ''Behavior Disorder and Attention Deficits in Boys with Tourette Syndrome,'' *Journal of the American Academy of Child and Adolescent Psychiatry* 27 (1988):413–417.
97. People v. Mawhinney, 163 Misc. 2d 329, 622 N.Y.S.2d 182 (Sup. 1994), discussed in Matthew Goldstein, ''Mental Illness Defense Denied to Molester,'' *New York Law Journal* (Jan. 10, 1995): 1.
98. Merrill Rotter and Wayne Goodman, ''The Relationship Between Insight and Control in Obsessive–Compulsive Disorder: Implications for the Insanity Defense,'' *Bulletin of the American Academy of Psychiatry and Law* 21 (1993):245.
99. *See, e.g.*, R. v. Quick, 1973 All E.R. 347. *See also* People v. Morton, 473 N.Y.S.2d 66, 67–68 (App. Div. 1984) (evidence as to defendant's hypoglycemia condition required submitting manslaughter, as well as murder, charges to jury). *Cf.* State v. Parker, 416 So.2d 545 (La. 1982) (affirming trial judge's decision to deny new trial motion following jury's rejection of defendant's evidence of hypoglycemia in support of proffered insanity defense).

    Perhaps the most famous and controversial case involving the use of hypoglycemia as a mitigation of charge was that of Dan White, who murdered the mayor of San Francisco and a city council representative, and pled (unsuccessfully) insanity resulting from hypoglycemia. White was eventually convicted of voluntary manslaughter, a ''lenient verdict which led to rioting in the streets.'' *See* Larry Bodine, ''Rx Sought for Plea of Insanity,'' *National Law Journal* (July 23, 1979):1, 13. *See generally* Thomas Szasz, ''J'Accuse: How Dan White Got Away with Murder,'' *Inquiry*( Aug. 6, 20, 1979):17; *see also* People v. White, 172 Cal Rptr. 612.
100. Sanford Fox, ''Physical Disorder, Consciousness, and Criminal Liability,'' *Columbia Law Review* 63 (1963):645.
101. *Id.* at 650. *See* Rose, ''Criminal Responsibility and Competency as Influenced by Organic Disease,'' *Missouri Law Review* 35 (1970):326, ''The organic diseases to be considered are (1) metabolic and endocrine disturbances; (2) infections; (3) neoplasia (cancer); (4) senility; (5)

genetic abnormalities; (6) alcohol, delirium tremens, and drugs; and (7) organic brain lesions, rage reactions, and seizures.''

For other more recent inquiries into this subject matter area, *see, e.g.*, Deborah Denno, ''Human Biology and Criminal Responsibility: Free Will or Free Ride?,'' *University of Pennsylvania Law Review* 137 (1988):615; J. Arturo Silva, G. B. Leong, and R. Weinstock, ''The Dangerousness of Persons With Misidentification Syndromes,'' *Bulletin of the American Academy of Psychiatry and Law* 20 (1992):615–671; Maureen Coffey, ''The Genetic Defense: Excuse of Explanation?,'' *William and Mary Law Review* 35 (1993):353–399; Jeremy Horder, ''Pleading Involuntary Lack of Capacity,'' *Cambridge Law Journal* 52 (1993):298–318; Richard Restak, ''The Neurological Defense of Violent Crime: 'Insanity Defense' Retooled,'' *Archives of Neurology* 58 (1993):869–870; Waltes, ''A Meta-Analysis of the Gene-Crime Analysis,'' *Criminology* 30 (1992):595–596.

102. *See, e.g.*, A. Raine, Clarence Ray Jeffrey, James D. White, and Rolando Del Carmen, ''Relationships Between Central and Autonomic Measures of Arousal at Age 15 Years and Criminality at Age 24 Years,'' *Archives of General Psychiatry* 47 (1990):1003 (psychophysiological factors alone cannot fully account for criminal behavior, and do not negate role of social variables as predictors of such behavior).

103. C. R. Jeffrey and James White, ''Criminal Law and the Medical Model.'' In *Attacks on the Insanity Defense,* C. R. Jeffrey, ed. (Springfield, IL: Thomas, 1985). *See generally* Barbara Ward, ''Competency for Execution: Problems in Law and Psychiatry,'' *Florida State University Law Review* 14 (1986):35–107, 50% of death row inmates ''intermittently insane''; Dorothy Lewis, et al., ''Psychiatric, Neurological, and Psychoeducational Characteristics of 15 Death Row Inmates in the United States,'' *American Journal of Psychiatry* 143 (1986):838–845; Lincoln Caplan, *The Insanity Defense and the Trial of John W. Hinckley, Jr.* (Boston: David R. Godine, 1984), significance of evidence of CAT scan in shaping insanity determination.

*Compare* State v. Zimmerman, 802 P.2d 1024 (Ariz. 1990), *rev. den.* (1991) (not abuse of discretion to exclude evidence of Brain Electrical Activity Mapping study that would have been proffered in support of insanity defense); United States v. McQuiston, 998 F.2d 627 (8th Cir. 1993) (no abuse of discretion where district court failed to order MRI testing); *see also* People v. Weinstein, 591 N.Y.S.2d 715 (Sup. 1992) (PET scan results admissible), discussed in Anderson, ''Brain Scan Deemed Admissible at Trial,'' *New York Law Journal* (Oct. 20, 1992):1; Rojas-Burke, ''PET Scans Advance as Tool in Insanity Defense,'' *Journal of Nuclear Medicine* 34 (1988):34.

104. *See* Jeffrey and White, ''Criminal Law and the Medical Model,'' 87–89, discussing ''automatism and episodic cerebral dysfunction,'' a category that authors use to subsume cerebral tumors, hypoglycemia, arteriosclerosis, epilepsy, shock, trauma, sleepwalking, and influence of certain drugs, and 92–93, discussing hypoglycemia. The authors also discuss other stress disorders—*e.g.*, premenstrual stress syndrome, *id.*, 89–92 (*see* Perlin, *Mental Disability Law* § 15.30) and posttraumatic stress disorder, *id.*, 94–96 (Perlin, *Mental Disability Law,* § 15.31), as well as psychopathy, *id.*, 97–98, and genetic defects such as XYY syndrome, *id.*, 98–101.

105. Jeffrey and White, ''Criminal Law and the Medical Model,'' 104.

106. *Id.* On the somewhat related question of the propriety of special status and institutions for sex offenders, *see* Perlin, *Unpacking the Myths,* § 16.18.

In the past few years, at least 40 states have enacted new legislation targeting violent sex offenders. Such bills—mostly emulating New Jersey's Megan's Law (named after the victim of one such attack)—provide for (a) the registration of sex offenders and the creation of a central registry; (b) community notification; (c) notification procedures for the release of certain offenders; (d) extended terms of incarceration for sexually violent predators; (e) the consideration of murder of a child under 14 as an aggravating factor in death penalty proceedings; (f) involuntary civil commitment of dangerous criminals; (g) lifetime community supervision; (h) the collection of a DNA sample from sex offenders for the creation of a DNA database and data bank; and (i) no ''good time'' credits for sex offenders who refuse treatment. *See e.g.*, N.J. STAT. ANN. § 2C:43-7; 2C:11-3c(4)(k), 2C:43-6.4, 53:1-20.20, 2C:47-8, 30:4-82.4. For a comprehensive discussion, *see, e.g.*, Elga Goodman, ''Megan's Law: The New Jersey Supreme Court Navigates Uncharted Waters,'' *Seton Hall Law Review* 26 (1996):764–802.

107. Jeffrey and White, ''Criminal Law and the Medical Model,'' 105.
108. *Id.*, 107.
109. *Id.* The authors add, however, that the adoption of such an approach should go hand-in-glove with the preservation of the defendant's rights through a ''therapeutic bill of rights'' governing the right to treatment and right to refuse treatment. On the question of the current right of prisoners to mental health treatment, *see* Perlin, *Mental Disability Law,* § 16.22.
110. Jeffrey and White, ''Criminal Law and the Medical Model,'' 117.

    Approaches such as the one suggested by Jeffrey are criticized in Richard Delgado, ''Organically Induced Behavior Change in Correction Institutions: Release Decisions and the 'New Man' Phenomenon,'' *Southern California Law Review* 50 (1977):215, and Michael Shapiro, ''Legislating the Control of Behavior Control: Autonomy in the Coercive Use of Organic Therapies,'' *Southern California Law Review* 47 (1974):237. For a recent debate on a similar issue, *compare* R. L. Bonn and A. B. Smith, ''The Case Against Using Biological Indicators in Judicial Decision Making,'' *Criminal Justice and Ethics* 9 (Winter/Spring 1988) 3, with '''Crime and Human Nature' Revisited: A Response to Bonn & Smith,'' *Criminal Justice and Ethics* 9 (Winter/Spring 1988):10. For a recent overview, *see* Andrew Lelling, ''Eliminative Materialism, Neuroscience and the Criminal Law,'' *University of Pennsylvania Law Review* 141 (1993):141. For recent important perspectives, *see* David Faust, ''Forensic Neuropsychology: The Art of Practicing a Science That Does Not Yet Exist,'' *Neuropsychology Review* 2 (1991):205; D. Wedding, ''Clinical Judgment in Forensic Neuropsychology: A Comment on the Risks of Claiming More Than Can Be Delivered,'' *Neuropsychology Review* 2 (1991):233; J. T. Barth, T. V. Ryan, and G. L. Hawk, ''Forensic Neuropsychology: A Reply to the Method Skeptics,'' *Neuropsychology Review* 2 (1991):251–266.

111. *See generally* Perlin, *Mental Disability Law,* § 15.28.
112. The structure of the XYY chromosome is explained in David Skeen, ''The Genetically Defective Offender,'' *William Mitchell Law Review* 9 (1983):217–265; for an overview of genetics in this context, *see* Gordon Valentine, *The Chromosome Disorders: An Introduction for Clinicians* (Philadelphia: Lippincott, 1975). For a strong statement in support of a genetic theory of criminal behavior, *see* Lawrence Taylor, *Born to Crime: The Genetic Causes of Criminal Behavior* (Westport CT: Greenwood, 1984). *But see* ''Book Review,'' *Michigan Law Review* 83 (1985):1218–1228.

    Early studies are discussed in W. M. Brown, ''Men With an XYY Sex Chromosome Complement,'' *Journal of Medical Genetics* 5 (1968):341, Patricia Jacobs, M. Brunton, M. M. Melville, R. P. Brittain, and W. F. McClemont, ''Aggressive Behavior, Mental Subnormality and the XYY Male,'' *Nature* 208 (1965):1351; W. H. Price, J. A. Strong, P. B. Whatmore, and W. F. McClemont, ''Criminal Patients With XYY Sex-Chromosome Complement,'' *Lancet* 1 (1966):565. The first law review articles on this topic were apparently John Money et al., ''Impulse, Aggression, and Sexuality in the XYY Syndrome,'' *Saint John's Law Review* 44 (1969):220, and Peter Farrell, ''The XYY Syndrome in Criminal Law: An Introduction,'' *Saint John's Law Review* 44 (1969):217. *See also* ''Note, The XYY Syndrome: A Challenge to Our System of Criminal Responsibility,'' *New York Law Review* 16 (1970):232.

113. Skeen, ''The Genetically Defective Offender,'' 219. *See* Brown, ''Men with an XYY Sex Chromosome Complement,'' 341.
114. Skeen, ''The Genetically Defective Offender,'' 219.
115. Jacobs et al., ''Aggressive Behavior,'' 1351. Another study of the same population found that XYY men were also unusually tall (with a mean height of more than 6 ft.). *Id.,* 1351–1352.
116. Skeen, ''The Genetically Defective Offender,'' 230.
117. *Id.*, 220 n. 6. *See, e.g.*, M. B. Garry, ''Two Cases of 48, XXYY: With Discussion on the Behaviour of Prepubertal and Postpubertal Patients,'' *New Zealand Medical Journal* 92 (1980):49.
118. Skeen, ''The Genetically Defective Offender,'' 220. *See also* Fred E. Inbau, James R. Thompson, and Andre P. Mocrosens, *Cases and Comments on Criminal Law,* 3d ed. (Mineola, NY:

Foundation Press, 1983). The thesis that such a link exists is propounded in, *inter alia,* Taylor, *Born to Crime.*

Thus, more recent studies have been far more equivocal about the chromosomal link with violence. An exhaustive Danish study published in 1976 (for a discussion of the methodology employed, *see* Alice Theilgaard, ''Aggression and the XYY Personality,'' *International Journal of Law and Psychiatry* 6, 14 (1983):413–421, concluded that XYY males were *not* more likely to commit crimes than genetically normal men (*see* H. A. Witkin, S. A. Medrick, F. Schulinger, E. Bakkestrom, K. O. Christiansen, D. R. Goodenough, K. Hirschorn, J. Lundsteen, D. R. Owen, J. Philip, D. B. Rubin, and M. Stocking, ''Criminality in XYY and XXY Men,'' *Science* 193 (1976):547–555, but a Danish follow-up found that XYY males had engaged in criminal behavior at a far higher rate than a control group. *See* J. Nielsen, S. G. Johnson, and K. Sorensen, ''Follow-Up Ten Years Later of 35 Klinefelter Males With Karyotype 47, XXY, and 16 Hypogonadal Males With Karotype 46,XY,'' *Psychological Medicine* 10 (1980):345–352. On the other hand, after considering all recent literature, Theilgaard has concluded that the ''weight of the evidence . . . does not support the notion that XYY men are particularly violent or aggressive'' (Theilgaard, at 420, and *see id.*, 421, listing references and that the critical difference between XYY men and others is in their ''ways of exhibiting aggression.'' *Id.*, 420. Still, Skeen is probably accurate when he suggests that research ''paints a confusing picture.'' *Id.*, 222, and *see id.*, n.19 (listing articles).

119. Skeen, ''The Genetically Defective Offender,'' 227 (discussing, *inter alia,* the conclusions of Ernest Hook, ''Behavioral Implications of the Human XYY Genotype,'' *Science* 179 (1973):139–150, XYY White males *eighteen* times more likely to be found in mental–penal settings than genetically normal white males).
120. Skeen, ''The Genetically Defective Offender,'' 227. *See also* Jeffrey and White, ''Criminal Law and the Medical Model,'' 100 (a link between XYY and crime ''exists [and] must be further investigated'').
121. 113 S. Ct. 2786 (1993).
122. *Id.* at 2795.
123. Skeen, ''The Genetically Defective Offender,'' 250–251, discussing *Regina v. Hannell,* case study reported in ''An English XYY Murder Trial,'' *British Medical Journal* (1969):201, and in Bartholomew and Sutherland, ''A Defence of Insanity and the Extra Y Chromosome: *R. v. Hannell,*'' *Australia and New Zealand Journal of Criminology* (1968):29. In other foreign cases, evidence of the presence of the XYY chromosome apparently resulted, in one case, in the imposition of a less severe sentence—*see* Skeen, 248–249, discussing the French trial of Daniel Hugon (*see id.*, 248 n. 145, citing *New York Times,* April 21, 1968, 1), and *see* Skeen, 245 nn. 73–74 (citing *Medical Tribune,* November 4, 1968, 1, 26)—and, in another, in the imposition of the maximum sentence—*see* Skeen, 250, discussing the German trial of Ernest Dieter Beck, *see* R. G. Fox, ''The XYY Offender: A Modern Myth?,'' *Journal of Criminal Law, Criminology and Political Science* 62 (1971):59–73.
124. People v. Tanner, 91 Cal. Rptr. 656, 657 (App. 1970). *Tanner* is distinguished in People v. Wright, 648 P.2d 665, 668 (Colo. 1982) (minimal brain dysfunction case).
125. Millard v. State, 261 A.2d 227, 230 (Md. App. 1970). On the other hand, the *Millard* court noted,

> We do not intend to hold, as a matter of law, that a defense of insanity based upon the so-called XYY genetic defect is beyond the pale of proof. . . . We only conclude that on the record before us the trial judge properly declined to permit the case to go to the jury.

*Id.*, 231–232.

126. People v. Yukl, 372 N.Y.S.2d 313, 320 (Sup. Ct. 1975). For an earlier, unofficially reported case in which the introduction of XYY evidence was unsuccessful in attempts to establish an insanity defense, *see* People v. Farley, Ind. No. 1827 (N.Y. Sup. Ct., Queens Cty., Apr. 30, 1969), *discussed in* Skeen, ''The Genetially Defective Offender,'' 246–247, and Farrell, ''The XYY Syndrome in Criminal Law,'' 218. Skeen concluded that the introduction of genetic evidence ''seemed to have little or no effect on the sentencing decision of the judge.'' Skeen, 251.

127. State v. Roberts, 544 P.2d 754, 758 (Wash. App. 1976) (''the behavioral impact of this chromosome defect ha[s] not been precisely determined'').
128. *Yukl,* 372 N.Y.S.2d at 319. *See also Millard,* 261 A.2d at 231–232.
129. *See generally* Michael L. Perlin and Robert L. Sadoff, ''Ethical Issues in the Representation of Individuals in the Commitment Process,'' *Law and Contemporary Problems* 45 (Summer 1982):161.
130. Skeen, ''The Genetically Defective Offender,'' 238.
131. Rodriguez et al., ''The Insanity Defense Under Seige,'' 403–04.
132. *Id.*, 401–402 and n.32. A possible reason for this disparity is suggested in Perlin, *Symbolic Values.*
133. It is not unreasonable to expect that few of these institutions will be staffed by professionals and therapists skilled in dealing with individuals with chromosomal disorders. *Cf.* C. Wiedeking, J. Money, and P. Walker, ''Follow-up of 11 XYY Males with Impulse and/or Sex-Offending Behavior,'' *Psychological Medicine* 9 (1979):287–292. *See also* Jeffrey and White, ''Criminal Law and the Medical Model,'' 101 (the ''most useful suggestion'' for treating XYY defendants is ''special behavior therapy and remedial education'').
134. State v. Smith, 362 A.2d 578, 581 (N.J. App. Div. 1976) (referring to polygraph testimony; *see generally,* Rodriguez et al., ''The Insanity Defense Under Seige,'' 421 n.151).
135. Perlin, *Mental Disability Law* § 15.29, 369–371.
136. *See* Perlin, *Mental Disability Law,* §§ 15.29–15.32.
137. *See e.g.*, Adele Beckerman and Leonard Fontana, ''Vietnam Veterans and the Criminal Justice System: A Selected Review,'' *Criminal Justice and Behavior* 16 (1989):412–428; Anthony Higgins, ''Post-Traumatic Stress Disorder and Its Role in the Defense of Vietnam Veterans,'' *Law and Psychology Review* 15 (1991):259–276. For recent cases, *see e.g.*, State v. Jensen, 735 P.2d 781 (Ariz. 1987); Silagy v. Peters, 905 F.2d 986 (7th Cir. 1990), *cert. denied,* 111 S. Ct. 1024 (1991); United States v. Whitehead, 896 F.2d 432 (9th Cir. 1990); United States v. Long Crow, 37 F.3d 1319 (8th Cir. 1994), *cert. denied,* 115 S. Ct. 1167 (1995).
138. *Cf.* Charles Ewing, ''Psychological Self-Defense: A Proposed Justification for Battered Women Who Kill,'' *Law and Human Behavior* 114 (1990):579, *to* Stephen Morse, ''The Misbegotten Marriage of Soft Psychology and Bad Law: Psychological Self-Defense as a Justification for Homicide,'' *Law and Human Behavior* 14 (1990):595–618. For recent cases, *see, e.g.*, State v. Hennum, 441 N.W.2d 793 (Minn. 1989); State v. Williams, 787 S.W.2d 308 (Mo. App. 1990); State v. Clark, 377 S.E.2d 54 (N.C. 1989).
139. *See* Perlin, *Mental Disability Law,* § 15.30 at 371 n. 555 (listing scholarly articles).
140. *See e.g.*, Kimberley Waldron, ''Postpartum Psychosis as an Insanity Defense: Underneath a Controversial Defense Lies a Garden Variety Insanity Defense Complicated by Unique Circumstances for Recognizing Culpability in Causation,'' *Rutgers Law Journal* 21 (1990):669–697.
141. *See* Perlin, *Mental Disability Law,* § 15.29, 370 n.550, listing scholarly articles.
142. *See e.g.*, United States v. Kozminski, 771 F.2d 125 (6th Cir. 1985).
143. *See e.g.*, People v. Rylkin, 541 N.Y.S.2d (App. Div.), *appeal denied,* 545 N.Y.S.2d 121 (1989).
144. *See e.g.*, Kirschner, ''Understanding Adoptees Who Kill: Dissociation, Patricide, and the Psychodynamics of Adoption,'' *International Journal of Offender Therapy and Comparative Criminology* 36 (1993):323–333.
145. *See e.g.*, Lauren Goldman, ''Nonconfrontational Killings and the Appropriate Use of Battered Child Syndrome Testimony,'' *Case Western Reserve Law Review* 45 (1994):185–249.
146. *See* Ann O'Neill, ''Menendez 'Abuse Excuse' Is Attacked by Noted Psychiatrist,'' *Los Angeles Times,* Feb. 9, 1996, B1.
147. Susan Horan, ''The XYY Supermale and the Criminal Justice System: A Square Peg in a Round Hole,'' *Loyola Law Review* 25 (1992):1343–1376, citing Cesare Lombroso, *Crime: Its Causes and Remedies* (1911).
148. Alan Booth and D. Wayne Osgood, ''The Influence of Testosterone on Deviance in Adulthood: Assessing and Explaining the Relationship,'' *Criminology* 31 (1993):93. Denno, ''Human Biology and Criminal Responsibility,'' 627.

149. Avshalom Caspi et al., ''Are Some People Crime Prone? Replications of the Personality–Crime Relationship Across Countries, Genders, Races and Methods,'' *Criminology* 32 (1994):163.
150. Jo Borrill and David Stevens, ''Understanding Human Violence: The Implications of Social Structure, Gender, Social Perception, and Alcohol,'' *Criminal Behavior and Mental Health* 3 (1993):129–141.
151. Matt Zaitchik and David Mosher, ''Criminal Justice Implications of the Macho Personality Constellation,'' *Criminal Justice and Behavior* 20 (1993):227.
152. Denno, *Human Biology and Criminal Responsibility,* 634–635.
153. *Id.*, 643–644.
154. *Id.*, 645.
155. *See generally* Zaitchik and Mosher, ''Criminal Justice Implications of the Macho Personality Constellation.''
156. Denno, ''Human Biology and Criminal Responsibility,]'' 658–659; Deborah Denno, ''Gender, Crime and the Criminal Law Defenses,'' *Journal of Criminology Law and Criminology* 85 (1994):80–180.
157. Denno, ''Human Biology and Criminal Responsibility,'' 658-659.
158. Maureen Coffey, ''The Genetic Defense: Excuse or Explanation,'' *William and Mary Law Review* 35 (1993):353–399.
159. Denno, ''Human Biology and Criminal Responsibility,'' 660–661.
160. Coffey, ''The Genetic Defense,'' 399.
161. Morse, ''Brain and Blame,'' 531.
162. *Id.*, 549.
163. Denno, ''Human Biology and Criminal Responsibility,'' 667.
164. Rochelle Dreyfuss and Dorothy Nelkin, ''The Jurisprudence of Genetics,'' *Vanderbilt Law Review* 45 (1992):313–348.
165. *Id.*, 321.
166. This section is generally adapted from Perlin, ''Myths, Realities, and the Political World.''
167. *See generally* Perlin, ''Psychodynamics and the Insanity Defense.''
168. Michael Saks and Robert Kidd, ''Human Information Processing and Adjudication: Trial by Heuristics,'' *Law and Society Review* 15 (1980–81):123.
169. Mark Snyder, Elizabeth D. Tanke, and Ellen Berscheid, ''Social Perception and Interpersonal Behaviors: On the Self-Fulfilling Nature of Social Stereotypes,'' *Journal of Personal and Social Psychology* 35 (1977):656–666
170. *See, e.g.*, State Farm Fire & Casualty Ltd. v. Wicka, 474 N.W.2d 324, 327 (Minn. 1991) (both law and society always more skeptical about putatively mentally ill person who has a ''normal appearance'' or ''doesn't look sick'').
171. Michael L. Perlin, ''Psychiatric Testimony in a Criminal Setting,'' *Bulletin of the American Academy of Psychiatry and Law* 3 (1975):143.
172. Harold Lasswell, ''Foreword.'' In Richard Arens, ed., *The Insanity Defense* (New York: Philosophical Library, 1974).
173. Richard Sherwin, ''Dialects and Dominance: A Study of Rhetorical Fields in the Law of Confessions,'' *University of Pennsylvania Law Review* 136 (1988):729–849.
174. *Id.*, 738.
175. *See, e.g.*, Perlin, ''On Sanism.''
176. *See, e.g.*, Michael L. Perlin, ''Therapeutic Jurisprudence: Understanding the Sanist and Pretextual Bases of Mental Disability Law,'' *New England Journal of Criminal and Civil Confinement* 20 (1994):243–383.
177. *See, e.g.*, Michael L. Perlin, ''Pretexts and Mental Disability Law: The Case of Competency,'' *University of Miami Law Review* 47 (1993):625–688.
178. Paul Appelbaum, ''The Empirical Jurisprudence of the United States Supreme Court,'' *American Journal of Law and Medicine* 13 (1987):335–349; J. Alexander Tanford, ''The Limits of a Scientific Jurisprudence: The Supreme Court and Psychology,'' *Indiana Law Journal* 66 (1990):137–173; David Faigman, '''Normative Constitutional Fact-Finding': Exploring the

Empirical Component of Constitutional Interpretation,'' *University of Pennsylvania Law Review* 139 (1991):541–613.

179. Anne Woolhandler, ''Rethinking the Judicial Reception of Legislative Facts,'' *Vanderbilt Law Review* 41 (1988):111, quoting Donald L. Horowitz, *The Courts and Social Policy* (Washington, DC: Brookings Institution, 1977).
180. *See, e.g.*, Michael L. Perlin, ''Are Courts Competent to Decide Questions of Competency?: Stripping the Facade From *United States v. Charters*,'' *University of Kansas Law Review* 38, (1990):957–1001, discussing decision in *United States v. Charters*, 863 F.2d 302 (4th Cir. 1988) (en banc), *cert. denied*, 494 U.S. 1016 (1990) (limiting right of pretrial detainees to refuse medication).
181. Benjamin Sendor, ''Crime and Communication: An Interpretive Theory of Insanity the Defense and the Mental Elements of Crime,'' *Georgetown Law Journal* 74 (1986):1371–1434.
182. Richard Arens and Jackwell Susman, ''Judges, Jury Charges, and Insanity,'' *Howard Law Journal* 12 (1966):1.
183. Valerie Hans, ''An Analysis of Public Attitudes Toward the Insanity Defense,'' *Criminology* 24 (1986):393.
184. Caton F. Roberts, Stephen L. Golding, & Frank D. Fincham, ''Implicit Theories of Criminal Responsibility Decision Making and the Insanity Defense,'' *Law and Human Behavior* 11 (1987):207–232.
185. ''Final Report of the National Institute of Mental Health (NIMH) Ad Hoc Forensic Advisory Panel,'' *Mental and Physical Disability Law Reporter* 12 (1988):77–109 (emphasis added).
186. John Parry, ''The Civil–Criminal Dichotomy in Insanity Commitment and Release Proceedings: Hinckley and Other Matters,'' *Mental and Physical Disability Law Reporter* 11 (1987):218–223.
187. Francois v. Henderson, 850 F.2d 231, 234 (5th Cir. 1988).
188. *See* Woolhandler, *Rethinking the Judicial Reception of Legislative Facts*, 119; Richard Berk, ''The Role of Subjectivity in Criminal Justice Classification and Prediction Methods,'' *Criminal Justice and Ethics* 7 (1988):35.
189. For a flavor of the academic commentary on the Menendez case, *see, e.g.*, Steven Jupiter, ''Constitution Notwithstanding: The Political Illegitimacy of the Death Penalty on American Democracy,'' *Fordham Urban Law Journal* 23 (1996):437–481, discussing predictable responses of a ''horrified and angered'' public to Menendez murders; John Robinson, ''Crime, Culpability, and Excuses,'' *Notre Dame Journal of Law, Ethics and Public Policy* 10 (1996):1–10, asking whether Menendezes should have ''escaped conviction'' for their parents' murders; Catherine S. Ryan, ''Battered Children Who Kill: Developing an Appropriate Legal Response,'' *Notre Dame Journal of Law, Ethics and Public Policy* 10 (1996):301–339, discussing Alan Dershowitz's critique of the battered child defense, using Menendez case as example.
190. *See* text accompanying note 156, discussing research reported in Denno, ''Human Biology and Criminal Responsibility,'' 661 (¾ to ⅘ of criminal behavior unexplained by genetic factors).
191. People v. Battolino, 199 P.2d 897, 901 (Colo. 1948).
192. *See* Michael L. Perlin, ''Morality and Pretextuality, Psychiatry and Law: Of 'Ordinary Common Sense,' Heuristic Reasoning, and Cognitive Dissonance,'' *Bulletin of the American Academy of Psychiatry and Law* 19 (1991):131–150.
193. Roberts et al., ''Implicit Theories of Criminal Responsibility Decision Making and the Insanity Defense,'' 226.
194. Stephen Golding, ''Mental Health Professionals and the Courts: The Ethics of Expertise,'' *International Journal of Law and Psychiatry* 13 (1990):281–307.

# Chapter 3
# THE GENETICS OF BEHAVIOR AND CONCEPTS OF FREE WILL AND DETERMINISM

Dan W. Brock and Allen E. Buchanan

James Watson, codiscoverer of the structure of DNA and the first director of the Human Genome Project, has been quoted as proclaiming that "we used to think that our fate was in the stars. Now we know, in large measure, our fate is in our genes." If, as Watson has suggested, our destiny is our molecular biology, there seems to be no room for human freedom. Instead of agents choosing the lives we live, and so in control of and responsible for our actions and the kinds of persons we become, we are the unwitting effects of unseen and unconscious biochemical entities whose activities we are only beginning to understand, molecular programs written by evolution. At least in the Western philosophical tradition, a central defining characteristic of persons is their freedom or agency. Philosophers as different as Aristotle, Aquinas, Kant, and Marx have thought that what distinguishes humans from other inhabitants of the world is that we are capable of free choice—of being self-determining or autonomous. We become unique individuals, with fully developed and unique identities, through the exercise of choice and through an awareness of ourselves as agents who are distinct from and in control over our own inclinations and impulses and of the natural and social environment in which we choose and act.

The literature of sociology and social psychology frequently speaks of individual identity in a self-identification sense, according to which one's identity is to a large extent determined by how one identifies one's self or the aspects of the self with which one identifies. For example, for a particular individual, being a Serb or a Croat or a father or a Catholic might be an important part of his identity in this self-identification sense of the term. The characteristics that an individual includes in his or her identity in its self-identification sense, and the order in which they are included ("I am first and foremost a Muslim, then an Iranian, etc.) reveal the individual's values and commitments and the priorities among them. To say that most persons believe that a significant sphere of agency or free choice is part of their identity is to imply that they value themselves as free individuals and that they think of their freedom as constitutive not only of who they are but of what they value in themselves and who they aspire to be. To this extent, an individual's sense of his or her worth or value depends on maintaining the belief that he or she does indeed possess the characteristics in question. Thus, at least for those of us who think of freedom as being a part of our identity in its self-identification sense, any understanding that calls our agency into question thereby assaults our identity and thereby our sense of our worth or value.

How might the value that we place on human freedom be thought to be imperiled by anticipated advances in genetic knowledge? Some have expressed a concern—one understandably evoked by Watson's equation of genes with fate—that the whole notion of freedom will become problematic once we fully understand how our genes work. The thought can be elaborated in the following, very oversimplified argument:

1. Free will involves the capacity to choose how one will act and to have acted otherwise than one in fact did.
2. Human action, like all other parts of the natural world, is causally determined.

3. If human action is causally determined, then given relevant initial conditions and causal laws, no one could ever act otherwise than he or she in fact does.
4. It follows from 1, 2, and 3 that no one could ever act otherwise than he or she in fact does, and so free will is an illusion.
5. If no one could ever act otherwise than he or she in fact does, then no one can ever justifiably be held morally responsible for his or her actions.

According to this argument, causal determinism is true regarding human action and it is incompatible with free will and moral responsibility; this position is typically called Incompatibilist Determinism. According to Compatibilist Determinism, human action is causally determined, but it is not incompatible with free will or moral responsibility for one's actions; defenders of Compatibilist Determinism typically reject number 3 and argue that the relevant sense of "could have done otherwise" necessary for free will and moral responsibility is compatible with determinism. A third position is that determinism regarding human action is incompatible with free will and moral responsibility but that determinism is not true of human action; on this position, number 2 is false. There are centuries of complex and unresolved philosophical debate over these issues whose details we cannot pursue, nor shall we defend any version of one of the three positions distinguished previously. Instead, the issue is what import the fruits of the HGP might have for these issues.

At this point, one further distinction is crucial. There are two quite different concerns that sometimes get lumped promiscuously together. On the one hand, there is the contention that an ever-increasing wealth of information about the functioning of genes will somehow show that human freedom is an illusion, because molecular biology will prove the philosophical thesis of Incompatibilist Determinism—namely, that human action is determined and that determinism excludes freedom. On the other hand, there is the suggestion that regardless of whether our growing knowledge of the functioning of genes proves Incompatibilist Determinism, the *psychological effect* of this knowledge will be to convince more people that Incompatibilist Determinism is true and that hence we will suffer a sense of loss of control or responsibility, and of diminution of ourselves, to the extent that we have thought of our identity as human beings as including the ability to act freely.

The first thesis is false, insofar as Incompatibilist Determinism (that human action is determined and this excludes any scope for freedom) is a *metaphysical* view rather than an empirical hypothesis.[1] Hence no increase in data, genetic or otherwise, can prove or confirm it. Conversely, the most plausible and forceful objection to the Incompatibilist Determinist position is not that we do not yet have sufficient data to support an empirical generalization that everything is determined, but rather an argument or arguments to the effect that even if everything is determined it does not follow that there is no freedom.[2]

The second interpretation of the claim that increased knowledge of the functioning of genes somehow threatens human freedom, which as we have seen is better described as a view about how this knowledge will affect our *belief* that we are free, cannot be dismissed so easily. Even if a vast increase in genetic knowledge cannot empirically confirm the Incompatibilist Determinist Thesis (because it is a metaphysical thesis, not an empirical generalization), it may increase our propensity to believe it. As our knowledge expands not only of the functioning of genes but also of the interaction of genes and environments in producing outcomes, it may become more difficult to think of ourselves or others as free agents. And to the extent that we think it is essential to us as human beings that we are free, this change in belief may be very traumatic. It will be an assault both on the concept of human identity (as including freedom) and on our "identity" in the self-identification sense, assuming that the latter includes the belief that we are free and that our freedom is one of our most valued characteristics.

Alternatively, if Hume was right, such a growing mountain of instances of detailed causal explanations of our character and conduct may have little effect on most people, simply because, even if they intellectually assent to the thesis of Incompatibilist Determinism, most will neither act nor feel that they are unfree—except perhaps fleetingly and episodically when they wax philosophical. In any case, it is worth emphasizing that whatever expansion of our fund of causal explanations of human characteristics and conduct genetic science produces will only be one more instance of a phenomenon that goes back to the beginning of the so-called Scientific Revolution of the sixteenth and seventeenth centuries. Every advance in our knowledge of causal interactions, from Newtonian mechanics to evolutionary biology, has caused additional anxieties about whether freedom is illusory.

There are, then, two distinct worries about the implications of accumulating genetic knowledge for our views about human freedom and responsibility that need our attention. The first concerns whether the sheer accumulation of new knowledge of genetic causal contributions to behavior is likely, as a psychological matter, to undermine people's *beliefs* in human freedom and responsibility, and more specifically the conception of ourselves as responsible agents that underlies many of our moral and legal beliefs and practices, including important aspects of the criminal law. The second concerns whether the accumulating knowledge from the HGP is likely to undermine the *justification* for considering ourselves and others as free, responsible agents, and to undermine the justification for the moral and legal beliefs and practices that embody these conceptions. The first concern is about the likely actual effect on beliefs and practices of freedom and responsibility; the second is about the effect on the justification of those beliefs and practices. The two concerns are not unrelated, of course, because what undermines the justification for our beliefs and practices of freedom and responsibility should undermine our beliefs in them as well, and will do so, to the extent that we are rational. But the two concerns are distinct, and we shall suggest that one impact of the new genetic knowledge from the HGP will be not to create a new problem for our beliefs and practices of freedom and responsibility but to make an old concern more pressing to a much broader public. What then is the problem the new genetics raises for freedom and responsibility?

Human beings in most modern societies conceive of themselves as responsible agents, and more specifically as morally, and as well legally, responsible for their actions, for the lives they live, and for the kinds of persons they become. This is not to deny, of course, that the social and natural environment, together with our biological nature, does not place substantial limits on what any individual can become or do with his or her life; but within these limits people exercise their choice or free will and construct themselves and their lives. The conception we have of ourselves as responsible agents is reflected in the value given in common moral beliefs and in important social and legal institutions and practices to individual self-determination or autonomy.[3] We use the term *self-determination* to mean people's interest in forming, revising over time, and pursuing their own conception of a good life. Exercising our self-determination on particular occasions involves making significant decisions about our lives for ourselves and according to our own aims and values. This conception of ourselves as responsible, self-reflective agents is embodied as well in our practice of holding ourselves and others morally and legally responsible for the actions we perform.

To illustrate how social and legal practices presuppose a conception of persons as responsible agents, consider the criminal law. H. L. A. Hart has been perhaps the most prominent proponent of the view of the criminal law as what he has called a "choosing system."[4] The law publicly announces that specified acts, such as homicide, assault, and rape, are not to be done and holds people responsible for conforming their behavior to these requirements of the law. Thus the criminal law in effect commands people to "conform your behavior to the criminal law or else suffer the punishment for your transgression." For this to be an effective means of controlling behavior and preventing the actions prohibited by the criminal law, as well as to be seen to be fair, those to whom it is addressed must be capable of understanding the command and have a reasonable opportunity to choose whether to conform to it. Only human beings, or more specifically persons, are believed to have the capacities required for the use of this means of social control. Dangerous animals, unlike dangerous people, must be controlled in other ways, such as by the preventive use of force to render them no longer dangerous or harmful. The preventive detention employed with people who are dangerous by reason of mental illness also reflects assumptions about their lack of capacity for controlling and taking responsibility for their behavior.[5]

A part of our complex practices of holding people morally responsible for their behavior, as well as of holding them legally responsible for conforming to the criminal law, is a set of important attitudes that we have toward both ourselves and others as responsible agents, which we will call *responsibility attitudes.* The first-person responsibility attitudes we have toward ourselves and our own behavior include, among the most prominent, guilt, shame, and pride.[6] Each of these is a quite complex attitude and we shall not attempt to analyze any of them fully. However, we will use the example of guilt to show how these attitudes presuppose that those who are justifiably the object of the attitude be responsible agents.

Rational guilt requires the belief that one has acted in a way that one knew, or should have known, was wrong and that it was reasonably within one's power not to have so acted. If a person experiences feelings of guilt for behavior that he does not believe to be wrong, for example, behavior that he was taught was wrong by his parents while a very young child, is behavior for which he should no longer feel guilty; the guilt is irrational because he does not believe that he has done anything wrong. If a person feels guilty about having injured another person in an accident that was not avoidable and for which she was in no way negligent or culpable, it is appropriate to feel regret about the accident and injury, but not guilt because she could not have reasonably avoided causing the injury. Experiencing guilt, or feeling guilty, is the way we acknowledge that we could and should have conformed to a legal or moral requirement when we have, in fact, failed to do so. Guilt involves more than merely having certain feelings, as the common expression "feeling guilty" might suggest, and more also than having the relevant beliefs. The experience of guilt also involves having complex dispositions to behave in a variety of ways, such as the dispositions to acknowledge our wrongdoing, to seek the forgiveness of those we have wronged, and to resolve to act differently on similar occasions in the future. Thus responsibility attitudes are complex, structured, and rational phenomena that include, besides feelings, beliefs—both moral and empirical or factual, and dispositions to behave in a variety of ways in various sorts of contexts. These different kinds of components of responsibility attitudes, as well as the scope of the various behavioral dispositions, play a deep part in everyday social life, as well as in more formal legal practices.

When directed toward others, the practices of moral and legal responsibility involve such attitudes or feelings as praise, blame, indignation, and resentment, and these too exhibit complexities similar to those of the first-person responsibility attitudes. For example, we resent and are properly indignant about an injury or insult another person does us that we believe we do not deserve. We do not resent or feel indignant, on the other hand, toward a physical object, an animal, or even a very young child who causes us injury, although we may equally be angry about and regret the injury. Resentment and indignation are properly reserved for responsible agents. The practices of moral and legal responsibility, with their component feelings, attitudes, beliefs, and dispositions for behavior are deeply embedded not only in institutions like the criminal law but also in ordinary human life and social interaction. They are in at least apparent conflict, however, with a different way of understanding and responding to others' and our own behavior to which our expanding genetic knowledge contributes.

We will call the alternative view of people and their actions the *scientific view of human behavior.* Unlike the view of people as responsible agents whose behavior is morally evaluated, the scientific view takes human behavior as a natural phenomenon to be scientifically understood and explained.[7] In many circumstances, and especially in practical contexts, a relatively surface explanation will be sufficient. Sometimes an explanation in the concepts and framework of "folk psychology," such as the agent's beliefs and desires, will satisfy our desire to understand the behavior. If we want to know why Michael broke into the building and stole electronic equipment, then learning of his desire to obtain money to buy liquor together with his beliefs that he would find items that he could sell and that he was not likely to be caught will usually be sufficient. Sometimes we may want to push the explanation deeper to understand the physical and environmental factors that caused him to become an alcoholic when other apparently similar people develop no such dependency. This may be from practical concerns—how can we treat his alcoholism?—or from theoretical concerns for scientific understanding.

From a theoretical scientific perspective on human behavior, as in science in general, the scientific view seeks the deeper causes of human behavior and is usually not satisfied with merely surface explanations of those causes. Instead, as deep and fundamental an explanation as is possible is ultimately sought. In physics, for example, this quest for a deeper understanding takes the form of a search for, and understanding of, the most basic constituents of matter. In the quest for a deeper understanding of human behavior, whether its motivation is practical or theoretical, we can expect in the future to increasingly find genetic factors among the causes of the behavior. In the science of human behavior in the age of modern genetics, these quests often take the form of a search for the genetic bases of human behavior and of people's character traits and dispositions to behave in particular ways. Although there may be special concern in the HGP with the genetic bases of human disease and with pathological conditions and behavior, there is no reason to believe that our genes have any less importance in the causation of normal than of pathological conditions and behavior. But if pathological behavior is understood in parallel to pathological physical conditions, that is as behavior that deviates from species-typical behavior, then the genetic underpinnings of pathological behavior are themselves likely to be deviations from the species-typical genotype, a particular form of bad luck in the genetic lottery.

The scientific view of human behavior, of course, is not committed to the view that all human behavior is fully explained by genetic factors alone. There is no reason to believe in genetic determinism, either in such a simplistic or more sophisticated form. It is the interaction over time of a particular human being and his or her specific genetic inheritance with a specific environment that will explain that person's behavior. Nevertheless, we believe there is good reason to expect that the HGP will enable us eventually to understand important genetic influences on human motivational and character traits, and thereby to fill in an important part of the causal explanation of those traits.

What is the connection of this potential new knowledge to the view of people as responsible agents and to the various moral, social, and legal practices that embody that view? The responsible agent's action is considered within his or her control to a reasonable extent; this is what justifies holding that agent morally or legally responsible for conforming his or her behavior to moral or legal requirements. When we ascribe the action to the agent's character, that is just to ascribe it to him or her. The new genetic knowledge is likely to be seen by many people as threatening this view. Setting aside the possibilities of changing a particular individual's genetic inheritance, one's specific and unique genetic structure is a paradigm of what is viewed as beyond one's control and for which one cannot be held responsible. But it will be that same genetic inheritance that increasingly will be seen to be an important component of the explanation of behavior for which we will want to continue to hold the individual responsible.

If one's genetic structure is a principal cause of one's behavior and one's genetic structure is completely beyond one's control, then how can it be justified to hold an individual responsible for the resultant behavior? To reward some such behavior will seem unfairly to favor those individuals genetically predisposed to that behavior, whereas to punish other such behavior will seem unfairly to burden or even to stigmatize those other individuals genetically predisposed to that behavior. Moreover, the fact that one's behavior is also shaped or caused by one's environment, or by the interaction between one's genetic structure and one's environment, will seem of little help in justifying holding the person responsible, because one's environment is often largely beyond one's control as well. The problem is seen in the old adage, "To understand all is to forgive all." Once we understand the full and deepest causes of human behavior, we see that those causes seem to be largely beyond an individual's control. This will be seen by many people to throw into question the responsibility framework for viewing human action and behavior. Is it fair to hold people responsible and to blame and punish them for their behavior once we understand that behavior as having a substantial genetic basis?

It is important to emphasize that a diminution in attributions of responsibility in the face of advancing genetic knowledge need not be total or consistent. As many of the eugenicists of decades ago did, we might come to believe it unfair to blame individuals for their undesirable "genetically based" traits, while at the same time blaming their parents for not having availed themselves of interventions that would have prevented the traits from being transmitted.

We noted previously that the view of ourselves as responsible agents is embodied in a wide range of attitudes toward both ourselves and others, as well as in important social and legal practices. To the extent that this underlying view comes under pressure from the new knowledge the HGP may yield, those attitudes and practices will be seen as resting on untenable foundations. Thinking of ourselves as responsible agents could hardly be more important to our conceptions of our identity as human beings. The philosopher Peter Strawson has suggested that these attitudes and practices are so deep-seated as to be largely ineradicable despite the pressure on them from the scientific view of behavior.[8] But even if this is so, to come to believe that these attitudes and practices have lost their foundations, even if we are unable to give up believing in and using them, will place them in an intellectually incoherent and unstable position whose full consequences are likely to be profound, however difficult they may now be to predict.

We emphasize that to the extent the problem raised by the HGP for beliefs and practices of freedom and responsibility is just whether those beliefs and practices are compatible with our actions having causes that make them subject to causal explanation, then the problem is hardly new to that project. There is a long tradition going back many centuries of philosophical discussion and debate of the problem of free will and determinism together with its implications for moral and legal responsibility. Even before we began to understand the specific genetic causal influences on human behavior in the way the HGP will eventually likely permit, it was commonly assumed that there were causal antecedents, both properties of people and properties of their environment, for all human behavior. Thus the problem we sketched earlier in the five-step argument of the Incompatibilist Determinist—how can the causal determination of all human action be compatible with a conception of free will necessary for us coherently and fairly to hold people morally and legally responsible for their behavior?—was well recognized far in advance of specific knowledge of those causal antecedents. As we noted earlier, the incompatibilist position regarding free will and determinism is a metaphysical thesis, not an empirical thesis, and so it cannot be proved or even given further support by the genetic knowledge coming from the HGP.

There may be, nevertheless, an important effect of the HGP in undermining many people's belief in human freedom and responsibility and in the justification of responsibility attitudes and practices. Coming to understand the specific genetic influences on specific kinds of behavior, although not strengthening the theoretical case for the incompatibilist position, is likely to increase the general public awareness of the problem and in the eyes of many to undermine our practices of moral and legal responsibility. A debate that heretofore was largely confined to philosophers and philosophy classrooms may reach a much larger public with a focus on our actual social and legal practices. When we can point to the specific genetic influences on the behavior in question, instead of just raising a general theoretical worry that the behavior must have causal determinants, even if we do not know what they are, the potential impact on cultural attitudes and our practices of moral and legal responsibility is much greater. Many participants in the public debate of these issues will urge just what defenders of Compatibilist Determinism in the philosophical debate over free will and determinism have argued—causal determinism is not incompatible with our view of ourselves as responsible agents and with our practices of moral and legal responsibility. But as we have seen with our increasing understanding of the social influences on behavior for which we hold individuals morally and legally responsible, understanding those specific external influences on individuals' behavior leads many persons to question our judgments of individual responsibility and accountability. If it was conditions external to the individual—the broken home or squalid urban ghetto in which he or she was raised—that led to the behavior, then how can the individual fairly be held responsible for it? Tracing proscribed actions to unfortunate social and environmental influences raises doubts about whether the person who did them morally deserves blame or punishment. Yet many features of our social and legal practices of responsibility and accountability are intended to establish that persons who are held responsible and accountable deserve the praise or blame, rewards or punishments, they receive. And the same doubts will be raised about whether praise or blame, rewards or punishments, are deserved for behavior that has important genetic causes.

We have been suggesting how an increasing understanding of the genetic influences on behavior may make more pressing to a wider public the traditional problem of whether, if all behavior is causally determined, we can be justified in holding people morally and legally responsible for that behavior. As we have emphasized, because this is a metaphysical issue, not an empirical issue, no amount of new genetic knowledge and understanding of human behavior from the HGP could possibly settle it. But there is a different, though closely related, argument of why this new genetic knowledge rightly calls into question not just people's *beliefs* in our responsibility attitudes and practices, but the *justification* of those attitudes and practices. This argument was suggested in the previous discussion and has been made most explicitly in a paper by P. S. Greenspan.[9]

Greenspan noted that the linkages that it will likely be possible to make from the fruits of the HGP will not be between genes and a particular action or instance of behavior but rather between genes and particular character or motivational traits, such as shyness or a disposition to violent behavior. These linkages will *not* show, for any particular item of behavior, that given the person's genes the behavior was causally necessary or "compelled" by the genes. A genetic predisposition to shyness, for example, makes it more difficult for a person to speak forcefully in a public context, but we do not believe it typically makes it causally impossible for him or her to do so; sometimes people overcome their shyness on particular occasions. Nor if we find, what some people already believe, that particular individuals have a genetic predisposition to violent behavior, does it follow that any particular instance of violent behavior by such an individual was causally determined or compelled by her genes; here again, such individuals sometimes overcome their predisposition to violent behavior and control their violent impulses.

Character traits like shyness or violence typically make it more difficult for those who have them to avoid acting shyly or violently in a variety of circumstances than it would be for persons without those character traits. If there are genetic bases of such character traits, then those persons' genes make it more difficult for them to avoid or overcome their shyness or disposition to violence. When they fail to do so, knowledge of their genetic predisposition should not lead us to say, "he/she couldn't help it," but rather to recognize that it was harder for them "to help it" than it would have been for most people. But why should not this greater difficulty often mitigate the moral blame we assign to them for their behavior? As we noted previously, it will seem unfair to blame and punish them for behavior that their genes made it much harder for them to resist than it is for most people. This problem for our responsibility attitudes and practices does not appeal to any general incompatibility between the causal determination of behavior and human freedom and responsibility, but to the way general predispositions to particular kinds of behavior make it more difficult for some people to act in particular ways and to control their behavior than it is for the rest of us; the HGP will increase our understanding of the genetic origins of many such behavioral predispositions.

Greenspan pointed out as well that if such character traits or behavioral predispositions are caused by one's genetic inheritance, this threatens a common account of why we can fairly be held responsible, at least in large part, for our character or motivational traits. On an Aristotelian account of character formation, as well as on a common everyday account, we acquire our character traits principally by habituation—that is by habitual behavior that is subject to rational evaluation and control by the person in question. In this way, we have the capacity to reflect on what sorts of persons we want to become and to control, within limits, the development of our character traits, and in turn what kinds of persons we are. If character traits are acquired in this way by habitual action subject to the person's rational reflection and control, then the person can within limits be reasonably or fairly held responsible for them and for the actions to which they dispose him or her. But if these character traits are instead principally the result of or caused by the person's genes, then apart from the possibilities of genetic manipulation to change either the genes or their expression, it would seem the person cannot reasonably or fairly be held responsible for them and the actions to which they dispose him or her.

It still may be necessary for society to protect itself from persons whose genes dispose them to be dangerously violent, but that is to shift from the responsibility framework that we apply with persons to the protection framework that we apply with dangerous but nonresponsible animals or things. And that shift will have a deeply worrisome mixed message when it occurs with human beings, because although it may mitigate the blame and responsibility assigned to persons for behavior to which they are genetically predisposed, it does so by subtly removing the persons from the class of responsible human agents, with the risk of many of the horrible consequences that we know from historical experience have attended viewing some humans as sub- or nonhuman.

It is at least conceivable, however, that many people will come to experience a greater confidence in the reality of human freedom as we come to have a greater mastery over those aspects of our existence that are influenced by our genes. And herein lies the irony of the remark by Watson with which we began this chapter: If we come not only to understand but to be able to modify the functioning of our genes, then our "fate" will increasingly be within our control—which is to say that it will no longer be fate but rather a matter of human choice. Here, too, however, it is important to distinguish between the psychological effect of our increasing knowledge of genes and the truth or falsity of the metaphysical thesis of Incompatibilist Determinism. Even if human beings come to have hitherto unimaginable control of themselves as a result of the application of genetic knowledge, this will do nothing to show either that Determinism is false or that Determinism is compatible with freedom. For even if human beings come to be able to cause significant changes in themselves that were not previously within their control, this will not show either that their choices to use genetic technology to make these changes is not itself determined by a sequence of causal events extending back in time to before they existed and that are therefore beyond their control, or that if their choices are thus determined, they are nonetheless free.[10]

## Notes

1. *See, e.g.*, Peter Van Imwagen, *An Essay on Free Will* (Oxford: Oxford University Press, 1983).
2. *See, e.g.*, Daniel C. Dennett, *Elbow Room: The Varieties of Free Will Worth Wanting* (Cambridge: MIT Press, 1984); Philippa Foot, "Free Will as Involving Determinism." In *Free Will and Determinism,* ed. B. Berofsky (New York: Harper and Row, 1968).
3. *See, e.g.*, Gerald Dworkin, *The Theory and Practice of Autonomy* (Cambridge: Cambridge University Press, 1988).
4. H. L. A. Hart, "Legal Responsibility and Excuses." In *Punishment and Responsibility* (Oxford: Oxford University Press, 1968).
5. *See* Allen E. Buchanan and Dan W. Brock, *Deciding for Others: The Ethics of Surrogate Decision Making* (Cambridge: Cambridge University Press, 1989).
6. *See* John Rawls, *A Theory of Justice* (Cambridge, MA: Harvard University Press, 1971), 485–490.
7. *See, e.g.*, B. F. Skinner, *Beyond Freedom and Dignity* (New York: Knopf, 1971), 3–25.
8. P. F. Strawson, "Freedom and Resentment." In *Proceedings of the British Academy* (1962), 187–210.
9. P. S. Greenspan, "Free Will and the Genome Project," *Philosophy & Public Affairs* 22 (1993): 31–43.
10. This paper draws on an earlier paper, by D. W. Brock, "The Human Genome Project and Human Identity," *Houston Law Review* 29 (1992):7–22, and on later material jointly authored by us, and part of a larger project on which Norman Daniels and Daniel Wikler are collaborators. The project is funded by a grant from the ELSI Program of the National Center for Genome Research, whose support we gratefully acknowledge.

# ELABORATION
## Genetics, Social Responsibility, and Social Practices

Lisa S. Parker

A number of authors have focused on the potentially devastating social and psychological effects that may result from increased knowledge of genetics and both the glamorized reporting and misunderstanding of this information.[1] In the previous chapter, one of Brock and Buchanan's contributions, like that of P. S. Greenspan,[2] is to focus less on the effects of the Human Genome Project (HGP) on the public's beliefs about free will and responsibility and more on the implications of increased genetic knowledge for the justification of these widely held beliefs and the social practices founded on them. I should like to elaborate some of the implications of Brock and Buchanan's argument. In particular, I shall suggest how increased genetic knowledge requires, in some cases, and perhaps surprisingly, an increase in social rather than individual responsibility. It demands a social choice between either reassessing the applicability of some notions of responsibility, or mustering social resources to enable and to justly require individuals to act responsibly according to our notions as they are traditionally applied. Therefore increased genetic knowledge prompts, as Brock and Buchanan suggest, a reexamination of the social practice of holding responsible. This reexamination invites critical reflection on the sharp line that Brock and Buchanan draw between metaphysical and empirical matters. It may be that empirical information—not about practices peculiar to the genetic laboratory or genetic research findings but about our practices of holding people responsible and treating them as agents—constrains the formulation of and attempts to answer metaphysical questions about causation, control, and freedom.

Brock and Buchanan approach the HGP's implications for free will through concerns about (self-)identity and self worth.[3] Information that leads us to question our agency, our ability to be responsible choosing beings, leads us to question our identities and, in turn, our self-worth to the extent that we value our agency and consider it to be a part of what constitutes us as the persons we are. Although the HGP may lead more members of the public to consider questions of free will, whether determination is true and whether it constitutes a threat to free will remains, Brock and Buchanan argue, a metaphysical question that empirical data cannot help to address. But, they argue, because we may learn that our dispositions or character traits have a perhaps substantial genetic component, we may be led to question the freedom of, or our degree of control over, actions prompted or prevented by these dispositions. And thus we may be led to question, with increasing justification, whether our existing social practices of holding responsible are justified, at least to the extent that they are partly predicated on the assumption that we are responsible for our characters.

This argument leaves us with several questions. Brock and Buchanan suggest that we may be prompted to question the applicability of our concepts of legal and moral responsibility and our understanding of excusing conditions, as well as our self-concepts as free agents. Therefore I will pursue the possibility that we decide not to abandon, or find it impossible to abandon, our current practices of holding people responsible for their behaviors and treating them as responsible for many of their traits—for example, gregariousness, even in light of new knowledge of genetic influences on our behavioral dispositions.

Would we treat these genetically influenced dispositions as substantially different from those environmental factors over which people have little control? Either might be thought to relieve individuals of responsibility for their actions because their actions are, to a substantial degree, the result of factors—genetic or environmental—that are outside of their control. Although simply having an unhappy childhood does not relieve a person of responsibility for actions reflecting this background (e.g., difficulty in treating intimates kindly or without violence), we may be more inclined to excuse the mean or violent behavior of someone actually brainwashed or "programmed" as a child to respond with violence to slight provocations. Increased genetic knowledge is likely to reveal similar degrees of influence, the genetic version of the difference between having poor role models and being brainwashed. Just as our legal system has struggled to determine the degree to which learned helplessness, Self-Defeating Personality Disorder, or battered woman syndrome may serve as a defense or mitigating circumstance in cases of violence against abusive domestic patterns, it will likely have to face consideration of the degree of genetic influence on dispositions in determining or excusing liability.

Society will have to determine the degree of effort and resources it will demand from individuals who are found to have genetically influenced behavioral dispositions. Again, this is not a new question. Society has been willing to admit, for example, that it would be asking too much to ask a person faced with a gunman to seek a peaceful, nonviolent resolution to the situation; in this case we accept a defense of self-defense. We have been more willing to require that battered partners seek a nonviolent escape from their abusive situation, unless they are faced with an immediate and serious physical threat. This normative decision is based in large measure on how much effort, resolve, ingenuity, and financial and physical resources it is reasonable to expect an abused partner to expend in effecting an alternative to the abusive social situation.

Similarly, we will need to decide how much effort it is reasonable to demand from someone whose genetic situation disposes him, for example, toward violent behavior. Do we demand that he undergo gene therapy should it be developed?[4] What if that therapy is available only at great financial or personal cost? Many new and emerging medical therapies are too expensive for the average wage earner. Moreover, several drug therapies for mental illness suppress desirable behaviors, feelings, and traits (e.g., sexual appetite, spontaneity, and creativity), as well as undesired symptoms; if gene therapy or other intervention in behavioral dispositions had similar trade-offs, would it be reasonable to require it? Finally, for those who have religious or other personal objections to medical intervention, or to gene therapy in particular, would such intervention be required, especially as a preventive measure? If gene therapy were available to alter the genetic influence on a disposition to violent behavior, for example, and if someone chose not to avail him- or herself of it, would this failure be deemed culpable, even if possession of the initial disposition were not?

If gene therapy were not available, society might still use information about genetically influenced dispositions to require that individuals so disposed avoid environmental triggers or be held liable for any harm that resulted. Workers with genetically based increased risk to develop disease in response to chemicals in the workplace may be required or encouraged to find alternate employment, perhaps by considering their knowledge of their increased risk as the basis for a valid assumption of risk defense. Similarly, individuals with a genetically influenced disposition—for example, to violence—might be required or encouraged to avoid environmental triggers or else be held (strictly?) liable for any violence which they initiate. In the context of discussing implications of possible genetic components of alcoholism, for example, the argument is frequently made that those at increased risk for becoming alcoholics—for example, those with a family history of the disease—have an increased responsibility to avoid situations in which liquor consumption is encouraged. Instead of relieving such at-risk individuals of responsibility for their disease should it develop, their knowledge of their risk places the burden on them to avoid environmental triggers for their disease. In short, they should know better and take steps to avoid alcohol. Although it might be reasonable to require that those disposed toward alcoholism avoid situations involving liquor, it is much less likely that those allegedly disposed toward violence could pursue their life plans in society while avoiding all situations likely to trigger a violent response.

In addition, just as being able to identify and suggest alternate employment for "risky workers" shifts attention and resources from the possibility of eliminating risky workplaces,[5] being able to identify individuals with various behavioral dispositions may shift attention and resources from attempts to eliminate environmental triggers. In the case of a disposition toward violence, attention may be drawn from problems of poverty and unemployment; in the case of a disposition toward shyness, teachers may determine that communication problems lie with the individual so disposed and fail to seek to create a classroom environment in which students of various dispositions may thrive.

Thus society will have to determine how much of its resources to expend to address both the genetic influences on dispositions and the environmental triggers with which they interact. This assessment raises additional concerns of fairness and justice. If society expends resources to develop gene therapy to treat a disposition toward violence, but this therapy is made available only through market mechanisms or standard health insurance schemes, economically disadvantaged persons will be unable to afford this advantage in the social practice of holding and being responsible. Those only somewhat less well off will still have to expend a relatively larger proportion of their economic resources than their richer counterparts (and thus have fewer resources available to pursue their personal life plans). Alternatively, the economically disadvantaged will have to expend more of their noneconomic resources (e.g., self-control) or incur additional costs (e.g., loss of social relationships) to withstand environmental triggers or to avoid those triggers. These alternatives would exacerbate their already economically disadvantaged position.

In short, learning that genetic factors influence behavioral dispositions and character development will raise in sharp relief the questions of how much society may justly demand of its members and how much justice demands that society provide to its disadvantaged members. (Here there are two relevant notions of "disadvantage": first, those disadvantaged by their genetic endowment for successful participation in the social practice of being responsible and being held responsible; second, economic disadvantage.) Although the initial tendency may be to use genetic factors to locate causal responsibility, and thus legal liability—if not moral responsibility—in individuals (as in the case of the risky worker–risky workplace trade-off), critical reflection on the just allocation of legal and moral responsibility suggests an expansion of society's responsibilities. Knowledge of both the genetic influences on behavioral dispositions and the influence of environmental factors on behavior should prompt examination of whether individuals should be expected to overcome or alter their dispositions and to avoid or alter environmental factors, as well as a determination of the social resources to be made available to assist them.[6] The question of how much may be justly demanded of individuals cannot be separated from a determination of how much of our social resources should and will be expended to assist them.

Several important issues remain with respect to genetic influences. First, safe and effective interventions—genetic, psychological, or social—may not be developed to address those who have genetically influenced behavioral dispositions that disadvantage them in the practice of holding and being responsible. Second, if such interventions should become available, many ethical concerns will attend their use. Among them are whether their use should be voluntary or mandatory; whether a person's decision to avail him- or herself of such interventions could be voluntary or will almost always be substantially pressured in light of various social incentive structures like the criminal liability system or industry criteria for hiring and promotion; whether their use should be preventive or "curative" (i.e., used following evidence of the undesired behavior); and who should be entrusted to make these macro- and microdecisions. Another set of risks arises from the availability or use of interventions for some dispositions but not others, or by some people but not others. A form of victim blaming will likely occur; those who do not or cannot avail themselves of available interventions will suffer not only their initial disposition (e.g., shyness) but also blame for failing to "treat" it. Those with behavioral dispositions for which there are few or no interventions may suffer from a phenomenon feared by people with disabilities: As the number of people with disabilities diminishes because of the use of genetic testing and pregnancy termination, social tolerance for the disabled or for diversity of traits and abilities may diminish.

Finally, as Greenspan and Brock and Buchanan allude, knowledge of genetic components of our behavioral dispositions or characters may sap us of one source of self-esteem. For many, their characters are viewed as an accomplishment, a source of pride; for some, sharing causal responsibility for character with genetic "causes" may reduce their sense of personal responsibility for their characters, their sense of achievement or authorship. For others, perhaps those who possess genetically influenced dispositions toward behavior deemed socially undesirable, this new knowledge may provide an opportunity for esteem and a new sense of responsibility and authorship as they take steps to shape their behavior in the face of their contrary disposition. Brock and Buchanan suggest that our increasing genetic knowledge is likely to afford new means to control, within limits, the development of our characters. In addition, it affords new understanding of what it means to shape our characters within normative constraint—in other words, in socially desired ways—in the face of environmental and genetic influences, and in this new understanding may lie a type and degree of responsibility we had not previously contemplated. We may be prompted to reconceptualize responsibility, to consider it as an artifact of our social practices and the norms that constitute them. We may come to think of agency or responsibility as an artifact, rather than as an objective fact about ourselves that is true of (or false about) us quite apart from our social practices surrounding agency, control, and causal determination.

Herein lies the opportunity to challenge Brock and Buchanan's claim that empirical information has no bearing on the metaphysical questions of free will and determinism. Granted, genetic information may have no direct bearing on whether human action is determined and whether determinism excludes freedom. But genetic information and a variety of other information (e.g., about our environment and social conditioning and behaviorism) influence our beliefs and our social practices. For those whose philosophical orientation leads them to adopt a social practice approach to answering metaphysical questions, this empirical information can have a profound, albeit indirect effect on metaphysical questions, because such empirical information affects social practices such as holding people responsible and treating people as free agents.[7] Empirical information about these social practices—for example, a description of the norms implicit in them (the norms according to which the practices proceed and by which they are implicitly constituted and constrained)—may resolve metaphysical questions. On this view, people have free will just in case they are treated as possessing free will in accordance with the norms of the relevant community (the community deciding the question).

Whether people have free will is an objective fact about them; however, the distinction between the objectively factual and the social is itself a social distinction—in other words, a distinction made according to the norms of the community engaged in the social practice of attributing free will. The community treats possession of free will as an objective fact about people but not as a fact or attribute distinct from the fact of how people are treated, the social practice of attributing free will to them as an objective fact about them.

Even for those with different philosophical leanings, as well as for those not philosophically concerned, it would seem that information about our social practices—for example, being (or being treated as) a moral agent and holding responsible—is pertinent at least to the formulation of metaphysical questions like those surrounding agency and determinism. Were it not for those social practices it is inconceivable that the metaphysical questions would arise.

## Notes

1. R. Hubbard, and E. Wald, *Exploding the Gene Myth* (Boston: Beacon Press, 1993); D. Nelkin and L. Tancredi, *Dangerous Diagnostics: The Social Power of Biological Information* (New York: Basic Books, 1989).
2. P. S. Greenspan, "Free Will and the Genome Project," *Philosophy and Public Affairs* 22 (1993):31–43.
3. It is interesting to note that some of the other self-identification characteristics Brock mentions—for example, being Muslim or Iranian—are characteristics over which people frequently have very little choice. Some choose or choose to affirm as authentic these identifying characteristics, but most are born or "fall into" them as a result of social conditions or conditioning. It is significant that these characteristics can nevertheless be deemed important, even essential, to self-identity. Indeed some genetically based traits—for example, membership in a Huntington's disease family or physical characteristics—serve a similar role. This is not to say that information that calls our agency into question will not threaten our senses of identity and self-worth if indeed possessing free choice, or being uncaused, is a part of our self-identities. At most, this observation suggests that many of us may overrate being uncaused in our sense of self-identity, or it may be used to lend credence to Kantian compatibilist solutions suggesting that freedom consists in being constrained by norms (as well as, presumably, by causes).
4. Although I speak of gene therapy, it is equally possible, and in the near future more likely, that genetic screening might be used to identify those at increased risk for particular behaviors (i.e., those with genetically influenced dispositions), but that the intervention be nongenetic. Most of the considerations I raise would still apply. It is interesting to note that people, perhaps especially those involved in medicine, tend to view some types of intervention as inherently "more invasive" (and thus, usually, as more risky) than others. Surgery is usually deemed more invasive than pharmacological interventions; both are deemed more invasive than psychotherapy. On the other hand, some patients, for example those seeking a surgical solution as in the case of aesthetic surgery, find psychotherapy more invasive in the pertinent sense, and thus objectionable. Determinations of invasiveness and of riskiness are value laden; different people will evaluate different options differently. Especially insofar as ethical concerns arise because a particular intervention may contravene a person's values or put at risk plans, traits, and abilities that a person values, it should not be assumed that concerns arise only with encouraging or mandating *genetic* interventions. Mandating pharmacological or psychotherapeutic interventions, for example, could equally threaten personal autonomy or place at risk what individuals value.
5. E. Draper, *Risky Business: Genetic Testing and Exclusionary Practices in the Hazardous Workplace* (Cambridge: Cambridge University Press, 1991).
6. I do not mean to imply that society should undertake to develop genetic technologies to screen for and to intervene in behavioral dispositions. My argument suggests that expanded understanding of environmental factors, like poverty in the case of violence, should prompt reexamination of social responsibility to use resources to address these factors.
7. Robert Brandom ("Freedom and Constraint by Norms," *American Philosophical Quarterly* 16:187–196) employed a social practice approach to elaborate a conception of agency founded on

the notion that positive freedom or agency, as opposed to the mere absence of causal constraint, lies in acting in accordance with norms. Brandom located the source of these norms in empirically discoverable social practices. In his early paper, he used this approach to address debate between naturalists and nonnaturalists concerning the relationship of facts and norms.

# ELABORATION
## Natural-Born Defense Attorneys

Robert F. Schopp

Dan Brock and Allen Buchanan remind us that the relationship between free will and determinism is a metaphysical or conceptual issue rather than an empirical one. They also caution that some of the most significant effects of genetic research might be psychological in that the manner in which we understand this research and the significance we attach to it might alter our attributions of responsibility for behavior and character. Insofar as our understanding of these matters informs our attributions of responsibility, it takes on normative as well as conceptual significance. That is, we often seek a conception of free will that explains and justifies the defensible parameters of the imposition of moral and legal sanctions, including blame, condemnation, and punishment. Roughly, we attempt to understand free will as a necessary condition for the justified imposition of such sanctions. Consider, for example, three defendants charged with sexual assault on children.

Anderson is a socially anxious and inadequate individual who finds adult relationships uncomfortable because he lacks social skills and fears rejection and criticism. Anderson seeks small children in parks or on the street, befriends them, and lures them into isolated situations to engage in sexual activity.[1] Baker demonstrates no serious impairment of psychological capacities, but since adolescence he has engaged in a broad pattern of criminal activity including a variety of crimes against persons. He regularly engages in sexual activity with adults, adolescents, and children, largely on the basis of convenience and often involving abuse of his partner.[2]

Cook is a moderately retarded individual who lives in a group home, works at a sheltered workshop, and engages in sexual activity with children and young adolescents in the neighborhoods around both the home and the workshop. He engaged in similar activity in previous neighborhoods. Sometimes he responds to advances by the adolescents, but on other occasions he approaches younger children in playgrounds and parks to solicit sexual activity.[3]

Assume that research identifies some relatively strong genetic predisposition to pedophilic arousal, the social anxiety and inadequacy manifested by Anderson, the tendency toward aggression and lack of empathy characteristic of Baker, and mental retardation such as that illustrated by Cook. In what way would or *should* this information alter our attribution of moral or legal responsibility to these actors?

I will explore that question. The following discussion examines the intuitive idea reflected in discussions of responsibility and free will, and it sketches a plausible interpretation of that intuitive idea as applied to questions of legal and moral responsibility, including those raised by the three cases described. The next two parts of the discussion suggest that progress in our understanding of genetic predispositions might increase the scope of individual responsibility under certain circumstances.

### Free Will and Responsibility

The timeless debate regarding the metaphysical and ethical significance of free will remains active. In this elaboration, I sketch an interpretation of the conception of free will that we need for moral and legal purposes.[4]

### *The Intuitive Idea*

Parents sometimes report that their children's first words were "mama" or "papa." It often seems, however, that the first words children actually learn are "I couldn't help it!" Kids sound like natural-born defense attorneys because they share the widely held intuition that people are not responsible for conduct and thus not liable to punishment if they could not help performing that conduct. It seems natural to say that a person is not responsible for engaging in particular behavior if she was unable to do otherwise or could not control her conduct. Thus she did not do it of her own free will.

### *Refining the Intuitive Idea*

Unfortunately, it becomes very difficult to explain exactly what we mean by the claim that one was unable to do otherwise or unable to control one's conduct. We can identify some relatively clear cases. Suppose that Davis watches and does nothing as a small child toddles into traffic. We might immediately condemn Davis but revise our judgment when we learn that she suffers quadriplegia or that she was restrained by a defective seatbelt she was unable to release: She couldn't help it. Similarly, we might condemn Ellis for knocking an elderly woman to the ground, breaking her hip. We would likely alter that judgment, however, when we learn that Ellis did so by falling to the ground in her path during a grand mal seizure: He was unable to control himself.

These cases involve external restraints (the malfunctioning seatbelt) or internal disabilities (quadriplegia, seizure disorder) that sever the usual connection between these individuals' decision-making processes and their behavior. These individuals were not free to act on their wills because these restraints or disabilities prevented them from translating their decisions into action. Thus their bodily movement did not represent them as decision makers or as agents. The presence or absence of genetic bases for these conditions seems to have no significance for these attributions of responsibility. The malfunctioning seatbelt certainly does not represent a genetic defect in Davis. The quadriplegia and the grand mal seizure may or may not be the products of disorders with genetic etiologies, but this factor would not ordinarily be central to the actors' responsibility for their conditions or for these specific events.

Certain legal cases extend liability to individuals who were not able to direct their conduct through decision at the time they caused harm. Some cases, for example, hold drivers liable for reckless or negligent homicide when they cause fatal collisions by driving automobiles despite prior warning that they suffer disorders that render them vulnerable to sudden and unpredictable episodes of seizure, dizziness, or impaired consciousness.[5] The courts do not hold the drivers liable for having the seizure or for the manner in which they operate or fail to operate the vehicles during the seizures. Rather, the courts hold them liable for initially choosing to operate the vehicle despite prior episodes as well as medical diagnosis and prognosis that provide notice that they are susceptible to such sudden disability. Thus these cases converge with the intuitive judgments of moral responsibility regarding Davis and Ellis by holding the drivers liable only for decisions made with the relevant information and unimpaired capacities. The cases do not address the presence or absence of genetic bases for these disorders, and there appears to be no reason to attribute moral significance to such information.

### *The Three Pedophilic Cases*

Consider the three pedophilic cases. All three defendants differ from those who are clearly not responsible as a result of seizure or quadriplegia. All three pedophilic defendants are free to act on their wills in the sense that no external restraint or internal disability prevents them from translating their decisions into action. Yet the three seem intuitively to differ from one another in morally relevant ways. Baker engages in a variety of criminal behavior, and there is no obvious reason to think that he is less responsible for his sexual crimes against children than he is for any of his other types of criminal conduct against persons or property. He engages in various types of criminal conduct to fulfill personal desires, and he pursues this conduct in an organized, goal directed manner calculated to maximize satisfaction of his desires while minimizing the risk of capture.

Cook, in contrast, does not engage in other criminal behavior, and his sexual conduct tends to be clumsy, unplanned or poorly planned, and conducted in a manner that renders it likely to come to the attention of authorities. Although he understands that it is wrong, he explains that this means "mama will yell," and he is genuinely perplexed that the police remain angry at him, even after he has said "I'm sorry."

Anderson is a more sympathetic offender than Baker. Unlike Baker, Anderson does not engage in a wide variety of criminal conduct, and he appears to care about the welfare of others, including his victims. He does them no physical harm, and he appears either to be genuinely fond of them or at least to bear them no ill will. Although Baker willingly injures his victims to get what he wants, Anderson avoids causing physical injury, relying instead on coaxing, cajoling, and bribery with candy or toys, and he appears to honestly want to believe that he is causing no psychological injury.

Yet Anderson and Baker are like one another and unlike Cook in ways that seem central to the attribution of responsibility. Anderson and Baker, in contrast to Cook, understand that their conduct violates law and at least conventional morality, and they realize that their conduct is widely considered detrimental to the children they victimize. They also understand that their conduct places them at risk of serious legal punishment, and they pursue their goals in a manner calculated to avoid that punishment. We might hold Anderson less blameworthy than Baker, because he is not vicious and malevolent, but both act with the knowledge and capacities that render them accountable for their actions in a manner that differentiates them from Cook.

These similarities between Anderson and Baker and their dissimilarities to Cook appear to be independent of the etiology of their sexual attraction to children. Whether this attraction is attributable to genetic endowment, early conditioning, or some other experience, they act with the knowledge, understanding, and decision-making capacities of accountable agents, just as the drivers described earlier decided to drive with the information and capacities that rendered them liable for the harm they caused, regardless of the etiology of their vulnerability to seizures or similar episodic disability. Cook, in contrast to all of these accountable actors, suffers impairment of the capacities for comprehension and reasoning that are central to the process of deliberation and decision making that characterize the accountable agent. This impairment undermines his standing as a responsible agent, regardless of the source of this sexual attraction to children.

Some readers might object that a genetic predisposition to be sexually attracted to children must be understood as undermining free will and responsibility because those who are born with such a genetic structure experience desires that they cannot change and that compel them to engage in sexual activity with children. One might say that they could have refrained from such activity if they had chosen to, but that because of the genetically determined desire, they were not able to chose to do otherwise.

Imagine three individuals who experience strong urges to punch their opponents in the nose whenever they lose at tennis. They do not understand why they experience their urges, and they find these desires socially disadvantageous and potentially costly in that they face social condemnation and legal penalties if they act on them. Suppose further that we are able to ascertain that Fisher experiences this desire because, unknown to her, an evil neurosurgeon triggers it through an electrode surreptitiously implanted in her brain; that Gable experiences it because of his genetic structure; and that Hill experiences it because of the manner in which she was raised in a very competitive family.

If all three act on these desires, it may seem intuitive to say that Fisher is not responsible for her action because she could not help it. The neurosurgeon compelled Fisher to act as she did and is responsible for that action. Yet we have no explanation of what we mean by the claim that she was "compelled" to act or that she "could not help it." Shorn of unsupported assertions that she was "compelled," "could not help it," or "could not control herself," Fisher remains a person who experiences a desire to do something that she understands is a violation of law and conventional socital morality.[6] The mere fact that she is unaware of the source of this desire does not differentiate her from Hill or many other people who may experience desires to perform socially unacceptable conduct without understanding the etiology of these desires. Similarly, Gable's genetic structure provides him with a desire to engage in the same sort of unacceptable conduct, and he is also unaware of the etiology. Thus Fisher, Gable, and Hill all experience inconvenient and uncomfortable desires without understanding the etiology. Nothing about the different sources of the desires entails that the individuals are any more or less capable of refraining from acting on those desires.

Although it would be reasonable to conclude both that the evil neurosurgeon was the cause of Fisher's desire and that the neurosurgeon should be held liable for that intrusion, Fisher is responsible for acting on her desire, rather than for experiencing it. Absent some explanation of the artificially induced desire that renders Fisher less capable of directing her conduct through the ordinary process with which we demand that people conform their conduct to social limits, the mere fact that someone else is responsible for her experiencing the desire does not alter Fisher's responsibility for acting on it. Fisher, Gable, and Hill all experience desires that can be traced to some origins for which they are not responsible, but the central question regarding their responsibility for acting on their desires involves the psychological capacities that allow them to understand the social norms that prohibit their acting on these desires and to participate in these rule-based social institutions through a process of practical reasoning.

Anderson, Baker, and Cook, like Fisher, Gable, and Hill, experience desires to engage in socially proscribed conduct. Like them, these pedophilic actors may experience these desires for a variety of reasons that they do not understand, and these reasons may include their genetic endowments. Similarly, however, they will be held responsible for acting on their desires rather than for experiencing them, and the critical questions regarding their responsibility address the capacities they possessed for participating in the rule-based social institutions as competent practical reasoners. Cook differs from Anderson and Baker because he lacks the capacities needed to participate in a rule-based institution as a competent practical reasoner, regardless of the source of his desire to engage in prohibited activity.

## *A Will Free of Impairment*

The intuitive notion that Cook differs from Anderson, Baker, Fisher, Gable, and Hill reflects a conception of free will that we need for moral and legal responsibility. On this interpretation, the decision-making processes that generate decisions to act must be free of impairment that prevents the individual from participating in a legal or moral system in a manner that is unique to moral agents as competent practical reasoners. The individual whose will is free in this sense generates the decision to act (will) through a process of practical reasoning. Although space precludes thorough discussion of this process, suffice it to say that it involves a complex set of psychological operations through which the competent adult comprehends circumstances as well as legal and moral standards, anticipates the likely consequences of various actions, evaluates them by the applicable legal and moral standards, and interprets the significance of these evaluations in the context of his own priorities and his conception of a valuable human life.

The capacity to perform these operations enables the individual to engage in a process of practical reasoning involving self-evaluation, self-instruction, and self-direction that is available only to competent moral agents. Those whose wills are free in this sense have the capacities such as orientation, comprehension, reasoning, and reality relatedness that allow them to direct their behavior in such a manner as to elicit preferred consequences and avoid unwanted ones and to direct their lives according to their conceptions of the kind of persons they want to be.

If determinism is true, genetic and environmental factors determine the conduct of those whose wills are free in this sense in a manner that differentiates them significantly from those whose wills are not free. Members of the latter group, including nonhuman animals and humans who suffer substantial impairment of the relevant psychological capacities, have their behavior shaped primarily through the direct experience of concrete consequences of their conduct. Competent practical reasoners, in contrast, have the capacities to direct their conduct in light of the anticipated consequences and to substantially alter the effects of certain consequences through the cognitive processes that allow them to vest significance in circumstances and events through a process of self-evaluation and symbolic interpretation. Thus they can alter the significance of external events in light of their own standards, including the principles in which they vest value and the type of life they pursue.

Some individuals, for example, would ordinarily expend great effort to avoid the shame they would associate with committing a crime or being convicted of one. Yet they might interpret conviction and punishment for crimes of conscience as affirmation of their conduct. Thus conviction and punishment for trespassing in protest of abortion at a women's health clinic or in protest of immigration policy as part of the sanctuary movement might fortify their resolve to continue engaging in civil disobedience. That is, institutional sanctions designed to deter conduct might either deter or encourage these individuals, according to the manner in which they interpret the significance of their conduct and of the sanctions.

This conception of free will differentiates the unimpaired human adult from the nonhuman animals in a manner that relies on the uniquely human capacities that are directly relevant to one's standing as an accountable moral agent. Thus it explains why humans are the only species of which we are aware that needs morality. That is, this conception of free will supports attributions of responsibility that reflect the special role of ethics as the field of study that attempts to understand and order the pursuit of a uniquely human life. One whose will is free in this sense and who is free to act on his will acts of his own free will. If such a person's conduct is determined by genetic and environmental forces, it is determined in a uniquely human manner that is consistent with the special significance of morality for accountable agents. According to this interpretation, the central issue is not whether a person's acts are caused but the manner in which causal influences affect action.

Some readers might object that this conception of free will is too restrictive in that only Davis, Ellis, and others who are physically unable to translate their decisions into action qualify as unable to act on their wills. These readers might contend that this notion should accommodate individuals for whom it would have been very difficult to have done otherwise because they experienced some strong inclination to perform the forbidden conduct. These actors might include, for example, those who experience rage, addiction-driven craving, or preoccupying urges to engage in various types of ritualistic or fetishistic behavior. According to this interpretation, individuals would be considered unable to act on their wills if they were physically unable to translate their decisions into action or if they experienced very strong urges such that it would have been very difficult for them to do otherwise.

Recall Ellis, who caused bodily injury to an elderly woman when he fell to the ground in her path during a grand mal seizure. Ellis provides a clear case of one who is not morally blameworthy because he was unable to act on his will. He would qualify for legal exculpation because his behavior did not include a voluntary act. Compare Ellis to Ibson, who experiences strong fetishistic urgers to expose himself in subway stations. If we believe that these urges increase in strength such that it becomes extremely anxiety producing for him to resist, we might conclude that we should excuse Ibson from moral blame and criminal punishment because it would have been so difficult for him to refrain that we could not reasonably expect him to do otherwise. One might argue that our conception of free will should exclude Ibson from the category of people who are responsible for their behavior because they act of their own free wills.

Consider slight variations of the stories of Ellis and Ibson. In the first variation of Ellis, no one is injured during the seizure, but Ellis loses his grip on a candy wrapper leading to a charge of littering. In the second, the seizure causes Ellis to fall to the ground in a manner that trips a mother carrying an infant, and both suffer fatal injuries when they hit their heads on the pavement, resulting in Ellis being charged with two counts of homicide. Thus all three variations involve involuntary movement produced by a grand mal seizure, but the offenses involved range from trivial to extremely serious. The legal result would be the same in all three, however, and most readers would probably find Ellis morally blameless in all three variations, assuming that Ellis did not culpably precipitate the seizure. That is, the involuntary movement exculpates regardless of the severity of the harm caused.

Compare the following variation of Ibson. In this variation, Ibson experiences fetishistic urges of identical strength to those in the original story, and he experiences anxiety of identical strength when he resists the urges. In this variation, however, he can relieve the anxiety only by acting on his urge to torture and kill small children. Many readers might find that the intuitive inclination to excuse Ibson for acting on his urge to expose himself because we could not reasonably expect him to do otherwise does not apply in the same manner when he acts on the urge to torture and kill children. Although the severity of harm done seems irrelevant to the exculpatory force of Ellis's seizure, it seems quite relevant to the intuitive judgment that we might not expect Ibson to resist his urge to expose himself, but we surely expect him to resist his urge to torture and kill children. These judgments suggest that the intuitive exculpatory significance of Ibson's anxiety-producing fetishistic urges differs from that of Ellis's seizure.

This difference reflects the complexity of our intuitive judgments of culpability and the corresponding complexity of a satisfactory account of responsibility. A fully satisfactory account must articulate a theory of responsibility; provide a clear description of any impairment, circumstance, or condition thought relevant to attributions of responsibility; and explain the relevance of that description to that theory of responsibility. The account sketched in this elaboration addresses conditions that render individuals unable to act on their wills and impairment that renders individuals unable to form their wills through the ordinary processes of practical reasoning available to competent moral agents.

That a person acts of his own free will does not exhaust the set of factors relevant to attributions of responsibility. Ellis, Cook, Ibson, and many others who experience various types of impairment or encounter a variety of difficult circumstances may represent a complex array of exculpatory or mitigating conditions. We are likely to obscure rather than illuminate these issues, however, if we attempt to accommodate them all under a broad general notion of free will articulated in terms such as "could have done (or been expected to do) otherwise." Rather, we need a precise explanation of the significance of each putative exculpatory or mitigating factor for a defensible theory of responsibility.

This requirement of specificity in our theory of responsibility extends beyond conceptual rigor to serve an important normative function. Vague standards of responsibility increase the risk that preexisting expectations or biases will contaminate determinations of culpability, exacerbating the danger of unequal justice for those who are viewed with suspicion, aversion, or fear. By requiring of ourselves that we articulate our criteria of responsibility with precision, we impose a form of discipline intended to reduce this danger.

## Genetic Research Extending Liability Responsibility

Discussion of the significance of genetic research for responsibility often suggests that increased understanding of the genetic basis for behavior would undermine attribution of responsibility. The analysis in this elaboration suggests, however, at least two ways in which such understanding might expand personal responsibility.

### *Driving Cases*

The drivers discussed previously[6] were found liable for the harm caused by their vehicles after they suffered seizures or similar disability because they had notice from prior episodes and medical advice that they were at risk for such episodes. Thus they engaged in reckless or negligent behavior by driving despite this awareness of risk. These cases raise the possibility that enhanced understanding of genetic endowment could increase the range of risk for which the individual might appropriately be held accountable. Suppose, for example, that an individual has never experienced seizures but that she has been reliably diagnosed as manifesting a genetic predisposition that renders grand mal seizures virtually certain at some point in her life. Recall that the prior episodes and medical advise of risk in the driving cases contributed to the basis for liability only insofar as they provided the notice of risk that supported conviction for recklessness in operating the vehicle despite awareness of an unreasonable and unjustifiable risk. If one assumes that such genetic diagnosis and prognosis could be made with sufficient reliability, they could provide notice prior to the first seizure.

Although these cases involved liability for harm caused, one can be convicted of reckless endangerment for recklessly engaging in conduct that places another at risk of death or serious bodily injury.[8] On the assumption of sufficiently reliable genetic evidence of risk, there seems in principle to be no reason to preclude liability for reckless endangerment for engaging in highly risky conduct prior to the first overtly dangerous episode. Suppose, for example, that a particular genetic endowment raises an extremely high probability that the individual will become violent when drunk. One might argue that an individual who has been informed that he manifests that genetic pattern fulfills the offense elements for reckless endangerment by drinking alcohol.[9]

If one accepts the conception of free will sketched in this elaboration, individuals are responsible not for uncaused conduct but rather for conduct that is properly attributable to them in their status as competent practical reasoners. Thus advances in genetic mapping can increase the range of behavior for which one is responsible by providing additional information that brings additional risks and harms within the range of understanding, deliberation, and decision.

### *Responsibility for Character*

Traditionally, one can be thought of as responsible for one's character to the extent that one shapes one's own character by engaging in a pattern of behavior.[10] Thus insofar as one encourages development of an aggressive character by engaging in an extended series of events in which one responds to frustration or conflict in an aggressive manner, one can be held responsible for nurturing that aggressive character trait. Some might suggest that to the extent that genetic mapping demonstrates that character traits are the product of genetic endowment, it decreases the degree to which one can be held responsible for developing those traits through a pattern of behavior.

This argument seems to assume the causal model of responsibility, however, insofar as it reflects the premise that responsibility decreases as causal explanation (or causal explanation by reference to genetic endowment) increases. If one accepts the notion that one is responsible for the manner in which one uses one's uniquely human decision-making capacities, then the availability of information regarding one's genetic propensities can increase the degree to which one is responsible for one's character.

Suppose that some individuals are provided notice that their genetic endowment renders them particularly likely to respond aggressively to frustration or conflict. They might appropriately be held responsible for taking action that promotes development of that trait in ways that increase risk to others or for failing to take action that would promote countervailing character traits. One such person, for example, might exploit that trait to gain advantage by habitually engaging in threatening or aggressive responses to conflict, whereas another might study Zen or other methods of anger control to moderate this trait. Given the assumption of reliable genetic information, the first might reasonably be held responsible for nurturing the risky trait, and the second might deserve praise for actively pursuing moderation.

## Genetic Research Extending Role Responsibility

The discussion to this point emphasizes responsibility as accountability. That is, we often say a person is responsible for his conduct in the sense that he is appropriately held accountable for it and, particularly in legal contexts, that he is liable to sanctions for it. We often use the term *responsibility* in this sense retrospectively in evaluating the person and the conduct for the purpose of distributing punishment or blame or requiring compensation. When we speak of a person as having certain responsibilities, in contrast, we identify certain duties or expectations as attaching to them by virtue of some role they fulfill.[11] Thus we might say of parents that it is their responsibility to raise their children, of police that it is their responsibility to enforce the law, or of practitioners of various professions that it is their responsibility to practice to the standards of their professions.

Knowledge of genetic tendencies, or the availability of such knowledge, might be seen as broadening the range of responsibilities that attach to individuals as moral agents. Consider, for example, two circumstances in which the availability of genetic information might support special duties to avoid certain circumstances or to take positive action. One who receives notice that he has a high genetic propensity toward pedophilic attraction to children might properly be held to have a duty to avoid circumstances or professions that provide access to children. It seems likely, however, that the subjective experience of such attraction would provide such notice as effectively as would genetic information. Thus high genetic risk for certain sorts of illness might provide a more realistic basis for a positive duty.

Suppose that some persons manifest a genetic endowment that establishes a high risk of certain kinds of physical illness that are likely to render them comatose, incompetent, and dependent on other persons or artificial life support equipment. Perhaps, for example, certain genetic patterns are highly predictive of cerebral insults or of severe senescent decline. Arguably this information would support a positive duty to take certain action in preparation for relatively predictable events. The availability of such information might support a responsibility to prepare for these circumstances by executing a living will or other advanced directives or to secure insurance to provide for one's dependents. Notice that even if there were good reasons to refrain from establishing such responsibilities as legal duties, these reasons would not necessarily preclude attribution of moral responsibility.

## Conclusion

On one interpretation, persons act of their own free wills when they are free to act on their wills and their wills are free of impairment. According to this view, persons' wills are free when they have the unimpaired capacities that allow them to engage in practical reasoning and thus to participate in rule-based social institutions in the manner that is available to competent adults. Increased understanding of the causal role of genetic endowment in human behavior would not decrease individual responsibility for behavior because personal responsibility depends primarily on the capacities one possesses rather than on the presence or absence of causal influences. To the extent that genetic research increased the availability of information relevant to the decision-making process regarding conduct for which we are responsible, it might increase the range of individual liability responsibility and role responsibility.

## Notes

1. Anderson qualifies for the diagnoses of pedophilia and avoidant personality disorder under the most common diagnostic nosology. American Psychiatric Association, *Diagnostic and Statistical Manual of Mental Disorders,* 4th ed. (Washington, DC: Author, 1994), 527–528, 662–665. According to the *DSM-IV,* a diagnosis of pedophilia requires that the individual experiences intense urges or fantasies of sexual activity with children and that the individual either is distressed by those urges and fantasies or acts on them. Anderson, Baker, and Cook have been adapted from Robert F. Schopp and Barbara J. Sturgis, ''Sexual Predators and Legal Mental Illness for Civil Commitment,'' *Behavioral Sciences and The Law* 13:437–458.
2. Baker qualifies for the diagnoses of pedophilia, sexual sadism, and antisocial personality disorder according to the *DSM-IV,* 527–528, 530, 645–650.
3. Cook qualifies for the diagnoses of pedophilia and moderate mental retardation according to the *DSM-IV,* 39–46, 527–528.
4. Robert F. Schoop, *Automatism, Insanity, and the Psychology of Criminal Responsibility* (New York: Cambridge University Press, 1991) §§ 4.3, 7.3.
5. People v. Decina, 138 N.E.2d 799 (N.Y. App. 1956); State v. Gooze, 81 A.2d 811 (N.J. App. Div. 1951); Tift v. State, 88 S.E. 41 (Geo. App. 1916).
6. Clinically, compulsive acts are motivated by the temporary anxiety reduction that follows them. The clinical notion provides no reason to believe the individual is unable to control compulsive conduct. *DSM-IV,* 418.
7. People v. Decina, State v. Gooze, Tift v. State.
8. American Law Institute, MODEL PENAL CODE (official draft and revised comments) (Philadelphia: Author, 1985), § 211.2.
9. David B. Wexler, ''Inducing Therapeutic Compliance Through the Criminal Law.'' In D. B.

Wexler and B. J. Winick, eds., *Essays in Therapeutic Jurisprudence* (Durham: Carolina Academic Press, 1991).

10. *See* chapter 3, this volume; also Patricia S. Greenspan. ''Free Will and the Genome Project,'' *Philosophy & Public Affairs* 22 (1993):31–43.
11. H. L. A. Hart, *Punishment and Responsibility* (Oxford: Oxford University Press, 1968).

# Part II

# The Complex Interface of Clinical Psychiatry and Genetic Research

# INTRODUCTION TO PART II

This part summarizes the state of the art of psychiatric genetics, particularly in relation to the topics important to the philosophy of free will and criminal law that are addressed in the other parts of this book. Basic issues and concepts from the disciplines of psychiatry, psychology, and genetics challenge the reader to cross the boundaries from one discipline to another.

## Overview

Two themes recur throughout this part: complexity and subtlety. Apparent causal relationships, from genetic mutation to psychiatric illness to diminished free will to criminal action, are probabilistic, not deterministic. Genes influencing behavior interact with environment over time, making some possible outcomes more or less likely. Furthermore, genetic influences on behavior, although potentially powerful, may be difficult to measure in the context of any particular act at any particular time.

The relationship between psychiatric diagnosis, criminal behavior, and genetics is a complex and indirect one. Psychiatric disorders such as schizophrenia and mood disorders do not directly result in criminal behavior but do result in impaired judgment and impulse control. Alcohol and drug abuse are more frequently associated with criminal behavior. Only the antisocial personality disorder includes illegal behavior among its diagnostic criteria, and it is possible to meet diagnostic criteria for that disorder without committing a felony. In the first chapter of part II, Samuel Guze describes the categories of psychiatric illness frequently associated with violence and criminal behavior. Guze, a prominent research psychiatrist, also introduces the topic of genetic risk for these disorders, a topic expanded by the other authors of this section.

Mark Leppert, a geneticist with a history of success in finding genes for colon cancer, epilepsy, blindness, and diverse other genetic disorders, highlights the limitations of genetic knowledge related to criminal behavior. He emphasizes that the links between psychiatric diagnosis, genetic causes of psychiatric disorders, and criminal behavior are not currently established. He introduces important genetic concepts and vocabulary that may explain the slow progress in finding genes for common psychiatric disorders. Rare genetic diseases causing childhood epilepsy or heart disease illustrate these concepts. Finally, he raises the question that environmental factors may actually be a more powerful factor than genes in causing criminal behavior.

The perspective of the expert witness in forensic psychiatry, represented by Robert M. Wettstein, brings further concern about the current limits of knowledge. Wettstein raises an awareness of complicating factors, such as the multifactorial nature of behavior, the frequent cooccurrence of multiple psychiatric disorders, the multiple causes of violence, the modest influence of mental illness on the total occurrence of violence in society, the nature of impulsivity and its influence on violence. He emphasizes these factors in the practical task of assessing criminal responsibility. Finally, Wettstein identifies potential harm caused by the misuse of genetic information to predict future violence, resulting in preventive detention.

Steven O. Moldin, an expert in designing and analyzing genetic studies of mental disorders, explains the methods used for such studies, including twin, adoption, and molecular-linkage studies. He summarizes genetic research on schizophrenia, mood disorders, antisocial behavior, and alcoholism. Finally, Moldin describes statistical tests (positive and negative predictive values) and specifies criteria for future admission of genetic evidence for criminal proceedings.

Hilary Coon, a psychologist with expertise in mathematical genetics, expands on the topic of genetic research methods by describing two newer approaches. The first method is that of broadening the assessment of families affected with a specific disease to include some trait that may indicate the presence of genetic susceptibility, even when overt disease is not present. Such a trait is called an *endophenotype,* and she describes two neurophysiologic traits that appear useful for genetic studies of schizophrenia. The second method applies standard psychiatric diagnosis and molecular linkage techniques to unique populations known as genetic isolates. Such populations are descendants of a small number of founding ancestors, isolated from marriage to outsiders by virtue of geography (e.g., an island) or religion (e.g., the Amish). Genetic mutations that are rare in larger populations can thus be common in isolated populations, speeding progress in linkage studies.

## Concepts and Vocabulary

The genetic study of any psychiatric disorder begins with the definition of that disorder with operational definitions that can be reliably applied and supported by various measures of validity. The *Diagnostic and Statistical Manual*[1] lists the accepted categories of psychiatric disorders. If evidence from twin, adoption, and family studies supports a hypothesis that the disorder is genetic, the inference is made that the disorder is the observable manifestation, or *phenotype,* that reflects the underlying genetic defect. The word *phenotype* derives from the Greek roots for *pheno,* to show, and *type,* character. Thus the phenotype shows the underlying genetic character, or genotype. It should be added that a phenotype usually reflects an interaction between genetic and environmental factors. An important concept is that a phenotype caused by one genetic defect may appear indistinguishable from the phenotype resulting from another cause. Thus a mental disorder caused by a particular mutation may also result from some psychosocial stressor, head injury, toxin, or other genetic mutation. Such apparently identical disorders resulting from other causes are referred to as *phenocopies.* Subtle clinical differences between cases observed by astute clinical researchers may discriminate between phenocopies. For example, it is possible that a future study would show that schizophrenia, resulting from a mutation in a gene on chromosome 1, is associated with onset at an average age of 20 years. Another form of schizophrenia, resulting from a mutation of a different gene on chromosome 2, onsets at age 25, but only in individuals who abuse alcohol. The two forms of schizophrenia, each resulting from a different mutation and clinically distinguishable only in the subtle differences in age of onset and exposure to risk from alcohol abuse, would be considered phenocopies of each other.

Note that in this example, individuals with the chromosome 2 mutation may not develop schizophrenia if they do not abuse alcohol. Such an individual, carrying the disease mutation but unaffected by the disease, illustrates the concept of reduced penetrance. *Penetrance* is defined as that proportion of a population of individuals possessing a disease-causing genotype who express the disease phenotype. When the proportion is less than 100%, the disease genotype is said to have reduced penetrance. Reduced penetrance is a characteristic of diseases with complex phenotypes, such as those involving mental illness. Penetrance of a mutation may be influenced by environmental risks, such as exposure to toxins such as alcohol, and by a number of other factors, including age, gender, diet, and other genes.

This hypothetical schizophrenia example also illustrates another genetic concept important for understanding part II: genetic heterogeneity. When different genetic mutations

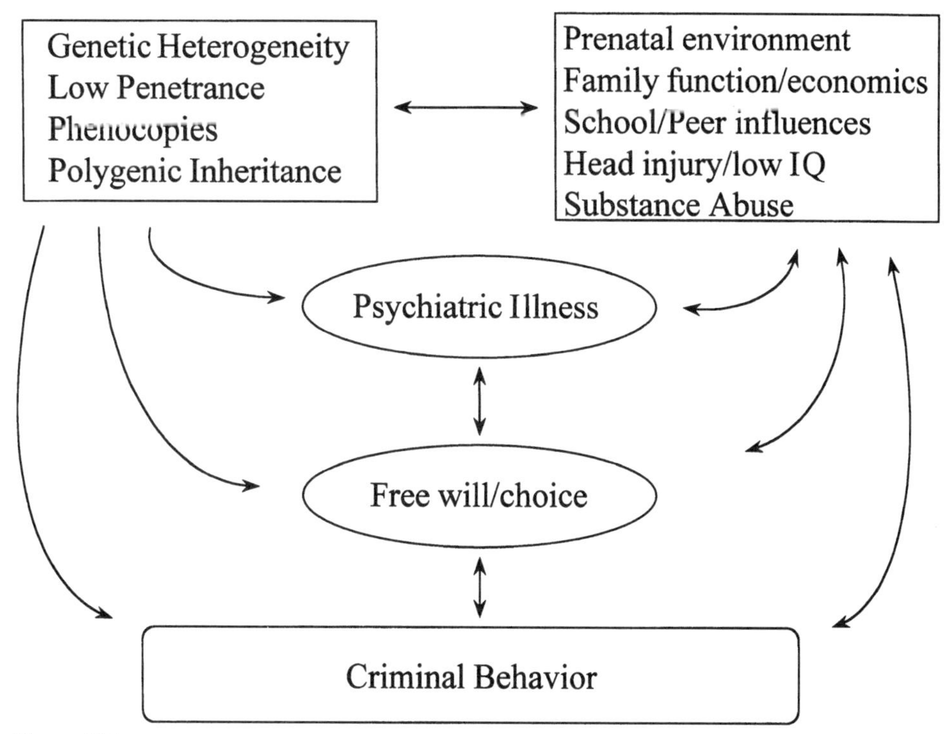


**Figure II-1**
Flow chart illustrating the interaction of variables related to criminal behavior.

result in the same phenotype, *genetic heterogeneity* is said to exist. There are two types of genetic heterogeneity. *Locus heterogeneity,* illustrated by the schizophrenia example discussed earlier, refers to the occurrence of causal mutations on different chromosomal locations in different families. *Allelic heterogeneity* refers to different mutations within the same gene that may cause the disorder in different families. For example, the hypothetical risk for the schizophrenia gene on chromosome 1 may be found to have mutations at 5 different sites, each mutation occurring in different families with schizophrenia.

## Complex Interactions Leading to Criminal Behavior

The complex interaction between genetic factors, environmental risks, psychiatric illness, free will, and criminal behavior is illustrated in Figure II-1. The probability of criminal behavior may increase or decrease in response to increased or decreased risk factors that accumulate over the developmental course of an individual. Genetic factors are present at conception. However, the expression of genes depends on environmental factors. In other words, environmental factors, such as substance abuse, may make penetrance of a mutation more likely, whereas abstinence from alcohol or drugs may decrease the probability that a mutation is expressed. Figure II-1 lists some environmental factors, but this is not an exhaustive list and could include such variables as nutrition or maltreatment ranging from emotional neglect to physical or sexual abuse. It should also be pointed out that genetic factors might also increase or decrease the risk for environmental risks. For example, a child born with a gene that results in impulsivity may be at greater risk for accidents and head injury. Once the child experiences a head injury, her judgment may further deteriorate, and the risk for psychiatric illness, impaired free will, and criminal behavior will increase.

Vicious cycles can occur at all levels, so that an increased risk for psychiatric illness results in schizophrenia, which then increases risk for substance abuse or other environmental problems. Legal difficulties and limited job or social opportunities may then further compromise free will, which is eroded by paranoid delusions and chronic intoxication. Adverse events accelerate each other.

On the other hand, genetic, environmental, and psychiatric factors need not be necessary for an individual to choose to commit a criminal act. Because human fallibility appears as universal as mortality, everyone may succumb to temptation or suffer a lapse of judgment. Furthermore, probability and certainty are different concepts. The prediction of human behavior is an inexact science.

In conclusion, the following chapters outline what is known about the interaction of genetics and psychiatric disorders that are relevant to considering criminal behavior. Much is unknown, but an impressive body of knowledge has accumulated.

## Note

1. *DSM-IV.* (Washington, DC: American Psychiatric Association, 1994).

# Chapter 4
# PHENOMENOLOGY OF PSYCHIATRIC ILLNESSES WITH SPECIAL REFERENCE TO RISK OF VIOLENCE AND OTHER CRIMINAL BEHAVIOR

Samuel B. Guze

Psychiatric illness has been a subject of interest to philosophers, jurists, and physicians from the beginning of recorded history. Although no molecular mechanism has been established for any single psychiatric illness, significant progress has been made in understanding the biologic basis of these conditions. The types of psychiatric disorders, their symptoms, courses, and effective treatments are known. Methods of diagnosis have been standardized. Genetic and nongenetic risk factors have been identified. These aspects of major psychiatric illnesses will be discussed in this chapter, as well as the relationship of psychiatric illness with criminal behavior.

## Schizophrenia

In many ways, schizophrenia may be the paradigmatic psychiatric disorder because it characteristically manifests some of the most disturbing and extreme psychological disturbances: delusions, hallucinations, and other disorders of thinking. It is a very serious condition, affecting approximately 1% of the population, though the prevalence may vary somewhat depending on the specific diagnostic criteria used.

The illness usually begins early in life, frequently in adolescence or early adulthood. Only infrequently does it begin after the age of 40, though there are cases that appear to be indistinguishable from early-onset ones that do not become manifest until old age. Whether these are the same basic disorders remains to be clarified. Typically, the illness begins gradually and insidiously. Social withdrawal, social isolation, difficulty in making and keeping friends, and unusual shyness are very often early manifestations in patients who develop the full spectrum of symptoms. The hallmark of the illness involves disordered thinking: difficulty in expressing ideas in a logical and coherent fashion; a tendency to misinterpret what others are saying or mean; and a sense that it is very hard to communicate with the afflicted individual. Sooner or later, these clues are followed by open delusions and hallucinations. Auditory hallucinations are the most common—for example, hearing voices speaking to the patient, commenting on the patient's behavior, suggesting or commanding certain actions, and so on.

The patient typically becomes more withdrawn, irritable, suspicious, unable to function at school or at work as well as previously, and often begins to manifest deviant or even bizarre behavior. This manifests itself in poor personal hygiene, uncouth or strange patterns of eating, unreasonable reactions to criticism or suggestions from parents and friends, and often a striking incongruity between the things that the patient will talk about and the patient's apparent emotional state.

In the great majority of cases, the untreated illness proceeds without prolonged remission, though perhaps as many as 5 to 10% of patients who warrant the diagnosis of schizophrenia will ultimately show what appears to be a complete or nearly complete recovery. The typical pattern, on the other hand, is one of continued impairment or even further deterioration in the ability of the individual to lead a normal life, including a normal range of thinking, emotion, and behavior. Sooner or later, the typical patient is unable to continue in school or hold a job or fulfill family responsibilities.

In the past, before the introduction of modern psychopharmacologic agents, half of the patients in chronic mental hospitals were suffering from schizophrenia, and a substantial proportion of these people spent most of their lives in such institutions. Today only a small percentage of patients with schizophrenia have such a bleak course and outcome, though the illness continues to be very serious, and continues to be associated with severe personal and social disability.

In recent years, as a result of the high percentage of schizophrenic patients spending long periods outside of mental hospitals, all forms of substance abuse have become common among these patients, and often seriously complicate the underlying illness. Estimates suggest that as many as half of all schizophrenic patients seen today by psychiatrists are also suffering from alcoholism or other substance abuse. These complications, of course, contribute importantly to the disability associated with the underlying illness.

The average schizophrenic patient presents very severe problems for the family. The illness interferes seriously in all efforts to integrate the patient into family life, and many patients are extremely resistant to psychiatric care and, especially, to taking psychiatric medication as prescribed. In many cases, the illness makes it impossible for the patient to live in the family home, and yet the patient does not always require psychiatric hospitalization. Various living arrangements have been developed to house and care for such patients, including various forms of group homes. Some of these have been quite successful.

The treatment of schizophrenia involves two main features: the use of various psychotropic medications and working with the patients and their families to achieve greater understanding and greater success in coping with the disorder. Occasionally, in cases in which the patient becomes quite depressed, as many do, antidepressant medication or electroconvulsive therapy can tide the patient over a particularly difficult interval.

That schizophrenia runs in families has been known for several generations. Studies of kindreds, twins, and adoptees, in various combinations and in many countries all over the world, have resulted in highly consistent findings. There is little or no room for doubt as to an important hereditary factor, though thus far modern efforts to identify any specific gene or genes have not yet been successful. At the same time, a variety of modern neuropathological studies of the brains of deceased patients reveals a range of gross and microscopic deviations from normal that are not yet explained and that vary somewhat from one study to another. The explanation for such findings is still problematic. At the same time, modern neuroimaging techniques, including positron emission tomography (PET) and magnetic resonance imaging (MRI), reveal an increased frequency of gross changes in the relative volumes of brain tissue in different areas as well as enlarged ventricular spaces. The full significance of these changes is also still not clear.

Schizophrenia appears to be a worldwide disorder. Wherever it has been studied by modern methods, the illness has been found in approximately similar frequencies. Some studies have revealed a difference in the distribution of the time of the year schizophrenics were born compared to controls, though again, no satisfactory explanation has been found.

Turning to the matter of violence and crime, although the frequency of violent or criminal behavior is probably increased in patients suffering from schizophrenia, the disorder is encountered infrequently in studies of consecutive series of serious criminals. In part, this may be a result of the fact that patients suffering from schizophrenia are diverted from the criminal justice system early, but, in at least some studies, this did not prove to be a significant explanation. Nevertheless, because of the nature of the illness with its frequent difficulty in establishing normal human relationships and its frequently associated suspiciousness and even frank delusions, patients with schizophrenia are at an increased risk for violence of one form or another or for other criminal behavior.

## Manic Depressive Disease

This category encompasses a range of disorders primarily manifested by striking alterations in mood, affect, energy, concentration, appetite, sexuality, and sleep. The mood disturbance frequently includes alterations in thinking manifested chiefly by pessimism, discouragement, hopelessness, and guilt, or the obverse: enthusiasm, overconfidence, and grandiosity. When the thinking alterations are severe enough, they lead to frank delusions and hallucinations, characteristically involving the emotional state. So, for example, patients may insist that they have committed unforgivable sins or that they have special powers and can solve all kinds of difficulties. Currently, the official approach to subclassifying these disorders is to divide them into two main categories: one characterized by the presence of mania (with or without periods of depression) and the other by the presence of depression (without periods of mania). The first category is frequently referred to as *bipolar disorder.*

These conditions collectively probably represent the largest group of patients seen by practicing psychiatrists. The lifetime prevalence of these disorders, at least in developed countries, may be as high as 5 or 6%. These disorders can begin at anytime in life, though bipolar disorder is much more likely to begin earlier. About half of the depression-only group does not show clear-cut manifestations until they reach at least the mid- to late thirties or early forties. A minority of patients may experience only one episode of affective illness during a lifetime. Most patients will experience recurrent episodes, and a significant minority may suffer from a more or less chronic condition—nearly all of these, however, are without periods of mania. About half of the patients with affective disorders present initially with physical complaints, including insomnia, loss of appetite, weight loss, headaches, constipation, and reduced energy. They may not volunteer information about their mood, though they nearly always will discuss such things if the physician is interested and patient when it comes to eliciting a full history. The typical patient will report, in addition to the physical symptoms just enumerated, low mood and loss of interest in work, friends, and sex. The disturbed thinking noted previously will fluctuate, but, when present, is very important diagnostically. A typical episode of manic depressive disease will last for months, though, unfortunately, sometimes even years. Manic episodes are nearly always briefer than depressive episodes. As with schizophrenia, a significant percentage of these patients will abuse alcohol or other drugs, though this is not as clearly a recent phenomenon as is true with schizophrenia. Bipolar illness is especially associated with substance abuse. The substance abuse nearly always intensifies and aggravates manic symptoms.

Treatment of affective disorders involves the use of many new antidepressant and mood stabilizing drugs, a variety of psychotherapeutic techniques designed primarily to help the patient cope more effectively with the illness, and, not rarely, the use of electroconvulsive treatments. In general, the manifestations of affective disorders are more satisfactorily moderated by available treatments than is true in schizophrenia. Most patients with affective illness can lead reasonably normal lives, if they understand the nature of their condition, have competent medical help, and are cooperative and compliant with the recommendations of their physicians. Marital discord, difficulties on the job or in school, and difficulties in interpersonal relationships with family members are very common during periods of illness.

Sometimes, unfortunately, behavior during an episode of illness will lead to lasting disruptions in important relationships and continue to plague the patient for years.

As is true for schizophrenia, evidence from many studies around the world strongly indicate a genetic contribution to the development of an affective illness. Again, the use of kindreds, twins, and adoptees have provided convincing data concerning this. But again, specific genes have not yet been convincingly identified. What was noted earlier concerning schizophrenia with regard to neuroimaging findings and disturbances in neurophysiology must be said for this group of disorders as well. There have been many reports of deviations in the apparent anatomy and physiology of the brains in such patients, but not all the reports are consistent about the findings, leaving us with uncertainty about the significance of these reports.

Turning to the question of violence and criminal behavior, the evidence is again fairly clear: Manic depressive illness is not frequent if one studies consecutive series of convicted criminals, but the risk of such behavior is somewhat increased in affective disorder, especially in manic patients. Again, it is important to keep in mind that patients who are recognized early as suffering from an affective illness will frequently be diverted from the criminal justice system, thus affecting the findings.

The most frequent kind of violent or criminal behavior will be seen in individuals during manic episodes, when their high energy level, overconfidence, severely disturbed judgment, and striking tendency to irritability and anger can precipitate fights, nonviolent illegal acts, and, in some cases, homicide. Among patients whose illness is manifested only by depression, there may be a slight increase in fights and in a variety of nonviolent crimes, but the principal risk for homicide is seen in those cases in which the homicide is followed by suicide, and both are manifestations of the altered thinking in manic depressive illness. More will be said about this later.

There have been many epidemiologic studies with somewhat differing findings, but there continues to be a recurrent theme in many of the findings that bipolar illness may be associated with increased creativity and related abilities. Thus there are a number of reports suggesting that bipolar illness is more common in writers, artists, and perhaps scientists.

## Substance Abuse Disorders

The category of substance abuse disorders involves a wide range of agents whose use can result in serious medical consequences, including, very importantly, a wide range of psychiatric manifestations. Among the most frequently encountered substances of abuse are alcohol, opiate derivatives, amphetamine derivatives, barbiturates, a wide range of other drugs used to calm or sedate patients, cocaine, and marijuana. The chemical and chronological effects of these various substances are not identical and, therefore, the range of physiological and other clinical manifestations associated with abuse will vary considerably. For the lay person, the important point to emphasize is that all of these agents can be used to a limited and controlled extent by most people, but when the use begins to dominate the individual's life and leads to recurrent medical, social, or legal complications, a diagnosis of substance abuse disorder is warranted.

Some substances lead to unmistakable and very uncomfortable or even life-threatening withdrawal symptoms. Other substances, on the other hand, may result in much more subtle disturbances on withdrawal, which are often difficult to define precisely and separate from other existing psychiatric disturbances. Of all substance abuse, alcoholism has been consistently recognized for the longest period and has been most thoroughly studied. In fact, the voluminous literature dealing with studies of alcohol and alcoholism makes it difficult for any single individual to be fully knowledgeable about the subject. A good deal is known about the difficulties associated with legally prescribed sedative and antianxiety agents, but much less is known about the illegal drugs, because of the difficulty in carrying out appropriate experiments and even nonexperimental studies. Experienced psychiatrists are well aware that nearly all other psychiatric disorders can lead to a complicating substance abuse and nearly all forms of substance abuse can lead to clinical pictures that closely resemble the whole range of other psychiatric disorders. Substance abuse of all kinds, whether recognized and diagnosed or not, is frequent in nearly all medical practice. One or another form of substance abuse can lead to serious complications involving nearly all organ systems, including the nervous system, the cardiovascular system, the gastrointestinal tract, the endocrine system, and the skin. A significant percentage of individuals with recurrent or chronic substance abuse disorder suffer premature disability and death from one or another specific medical condition. Substance abuse is greatly involved in all forms of accidents, from automobile accidents to falls, as well as from a wide range of injuries associated with the use of various tools and other equipment.

Violence leading to homicide and other criminal behavior, as well as suicide, are common complications of nearly all forms of substance abuse. More will be said about this later.

Substance abuse is worldwide. It is nearly always found to be more common in men than in women. It typically begins in adolescence or early adulthood, and it often lasts for years or even a lifetime. It is associated with reduced longevity. Substance abuse interferes with all sorts of important human activities from school, marriage, career, family life, and friendships.

Treatment of substance abuse involves a number of important elements. The first is the development of insight and concurrence in the diagnosis by the individual afflicted. Without such insight and concurrence, other forms of treatment rarely prove helpful. Most forms of treatment involve the use of a wide range of rehabilitation techniques and programs, all designed to increase the individual's insight about her substance abuse, the context in which it has arisen, the impact it has had on the individual's life and on important people in the individual's life, and the cultivation of coping strategies that will strengthen the individual's self-discipline, so as to avoid relapse. Often, several attempts of treatment are required over a period of months to years before success is achieved. For too many individuals, satisfactory success is never achieved. The cost to society of all substance abuse is very great and probably can be measured in many billions of dollars each year in the United States alone.

All forms of substance abuse carry strong risks that the individual will run afoul of the criminal justice system, either because the use of the particular substance is illegal or because the abuse of the substance leads to all forms of violence or to behavior that is otherwise illegal or likely to attract attention from the police.

## Antisocial Personality Disorder

The diagnosis of antisocial personality disorder is used for individuals who present a recurrent history, beginning in childhood or adolescence, of striking delinquent, antisocial, and criminal behavior. Such individuals often show childhood inattention, hyperactivity, and restlessness, though only a minority of children with such features go on to develop antisocial personality disorder. Typically, these children are difficult to discipline, they disrupt school, they fight with other children, and they often fail to meet other normal behavior expectations of parents, teachers, and neighbors. As they get older, such difficulties intensify and are complicated by early and often irresponsible sexuality, experimentation with alcohol and other drugs, more serious fighting, and overt criminal behavior, some of it violent. Many such individuals seem to have a craving for excitement and risk taking, and many act as though they have little in the way of conscience and little appreciation for the consequences of their behavior.

School truancy, academic failure, school dropout, and failure to graduate from high school are common. Special private schools, military academies, and other such institutions are frequently used by parents of increased means.

Antisocial personality disorder appears to be worldwide. It is more common in males than in females, though in recent years, at least in industrialized societies, the prevalence in females may be rising. The prevalence of antisocial personality disorder appears to be greater in poorer, especially less educated populations. It is very clear, however, that it is encountered as well in well-to-do and even highly educated families.

One of the most serious complications of antisocial personality is multisubstance abuse. Individuals with antisocial personality disorder are much more likely to experiment with illicit drugs, begin using illicit and other drugs, including alcohol, at an earlier age than comparison groups, and, at the same time, begin to show concomitant behavior that ends up in their being repeatedly involved with the law enforcement and criminal justice system.

A difficult diagnostic problem obviously exists when antisocial personality and substance abuse occur in the same individual, as frequently happens. Because substance abuse can itself lead to a wide range of antisocial, delinquent, and criminal behavior, the question arises as to which is more basic: the antisocial personality pattern or the substance abuse? In general, experienced psychiatrists make the diagnosis based on an extensive and careful history and review of records. If the individual displayed the early and typical features of antisocial personality disorder well before the substance abuse had reached a significant level, antisocial personality disorder will be considered the primary diagnosis and the substance abuse will be considered secondary. If such temporal distinction is not possible, one is left with a good deal of uncertainty. The matter is compounded by the fact that antisocial personality disorder and substance abuse run in the same families to a considerable extent.

The evidence for a genetic effect is strong, but not yet as conclusive as for the other psychiatric conditions described previously. At the same time, a growing range of studies indicates that all sorts of personality features, including those characteristic of antisocial personality, show significant genetic predispositions. It should be noted, of course, that there are many individuals who abuse various substances without ever really demonstrating the pattern of recurrent behavior difficulties characteristic of antisocial personality disorder.

Because of these diagnostic difficulties, it is not clear just how high a prevalence of antisocial personality disorder is present in many populations, but figures as high as 10% of males and 3 to 4% of females are not out of line. For many decades, psychiatrists and others have emphasized that there is a strong tendency for the antisocial personality disorder pattern to subside and become attenuated when the individual reaches middle age (over age 40). The subsequent life course is typically determined by just how many "bridges have been burned" before the reduction in the characteristic manifestations.

Treatment of antisocial personality disorder is generally of limited help, though occasional patients do show dramatic improvement, especially if all substance abuse can be stopped. Much attention is now being focused on beginning therapeutic intervention early in the hope that change can be effected before all sorts of unacceptable behavior leads to extreme rejection by families and communities.

## Homicide Followed by Suicide

Although not a frequent occurrence, the combination of homicide followed by suicide is dramatic and, of course, disturbing. Past studies have suggested that as much as 10% of consecutive suicides in Great Britain are associated with homicide, though the percentage in the United States has always been considerably less. The typical clinical situation is that of an individual with severe depression, often of psychotic proportions, with overwhelming pessimism, guilt, and self-condemnation. A particularly disturbing circumstance for this combination of behaviors is to be seen in women with postpartum affective illness. The individual suffering from the psychiatric illness nearly always has had a clear history of depression, typically beginning in the first few weeks after giving birth, but sometimes occurring in individuals who have had depressions under other circumstances as well. The history indicates that these individuals become progressively more disturbed and come to assess the world and its future in a most pessimistic way. The rationalization leading to homicide, frequently of close relatives, such as newborn infants, spouses, or parents, leads to the conclusion that there is nothing to live for, that everyone will suffer terribly in the future, and that death is the best solution for all concerned. Clinical awareness of such situations is fundamental to preventing these outcomes. In the great majority of cases, appropriate treatment, applied early, can result in a marked reduction in the frequency of such painful outcomes. We bring it up because homicide followed by suicide is nearly always associated with defined and recognizable psychiatric disorders.

## Conclusion

The following conclusions appear consistent with the literature. Some psychiatric illnesses are associated with increased risks for violence and other criminal behavior. For illnesses such as schizophrenia, manic depressive disease, anxiety disorders, and obsessional disorders, the risk for such behavior is relatively small, perhaps only a few percentages. It is difficult to be more accurate because of the possibility that people with such psychiatric disorders involved in criminal behavior will be recognized as ill and diverted from the criminal justice system.

Antisocial personality disorders, alcoholism, and other substance abuse—jointly and separately—whether or not they are complications of other preexisting psychiatric disorders, are the psychiatric conditions clearly and definitely associated with violence and other criminal behavior. Probably a substantial majority of all serious personal crimes are committed by individuals suffering from antisocial personality, alcoholism, or other substance abuse. If it were not for these three conditions, the crime problem in the United States would be vastly reduced. Unfortunately, however, these conditions are not easily treated and efforts to prevent or forestall their development are only being pursued in early and rudimentary studies.

Finally, evidence pointing to some significant hereditary predisposition is available for antisocial personality and for alcoholism. Probably the same conclusion is true for other substance abuse, but the results have not been as extensive and have not been based on as many different kinds of studies.

# ELABORATION
## Genetic Research and the Clinical Subtleties of Mental Illness

Mark Leppert

This book attempts to answer fundamental questions concerning causality of human behavior and then relate this knowledge to our legal system. Some key questions from a geneticist's viewpoint are as follows. How does the phenomenology of psychiatric illness relate to the causality of criminal behavior? Specifically, have major genes been identified as genetic components of specific psychiatric illnesses, and what are the immediate prospects of success in identifying such genes? Should our legal system take into account the genetic and clinical manifestations of criminal defendants for purposes of determination of guilt or innocence or for mitigating sentences? Has science progressed far enough in the understanding of genetic risk factors to allow or promote genetic testing as robust evidence in today's courtroom? Chapter 4 by Guze contributes to our understanding by describing in some detail those conditions most commonly diagnosed by psychiatry, the overlap of these disorders with criminal and antisocial violence, and the evidence for underlying genetic causes.

In this chapter, several important points are emphasized. The two illnesses for which genetic factors are most established—schizophrenia and manic depression—are encountered only infrequently in studies of consecutive series of serious criminals. Although these diseases account for a small fraction of violent crimes, individuals with these illnesses nevertheless are apparently at increased risk for exhibiting violent or criminal behavior. Another aspect not explored in this chapter is the likelihood that individuals afflicted with schizophrenia and manic depression are also at increased risk for becoming a *victim*, rather than a perpetrator of violence and crime. Is this risk even greater than the risk for causing violence? Do such individuals choose an environment more likely to include violence? These questions are of equal importance to the legal profession and to society.

Unlike schizophrenia and bipolar disorder, a third psychiatric class of disorders highlighted in the chapter, substance abuse disorders, commonly includes violence and criminal activity as part of its phenomenology. Unfortunately, even less is known about the genetics of substance abuse disorders. This current lack of scientific understanding will prove to be a difficult hurdle to overcome for legal experts interested in submitting into tomorrow's judicial proceedings evidence for genetic determinism.

How difficult will it be to increase our genetic knowledge of these disorders? Despite significant efforts, very little progress has been accomplished, although large-scale genetic projects with many collaborators are ongoing and hold considerable promise. However, it is relatively simple to identify the involvement of genetics in the causality of a particular disorder via twin or adoption studies compared to the identification of specific disease-causing mutations. Some of the difficulties in disease gene studies are commonly known and include *locus heterogeneity*—the ability of more than one gene to produce an identical or nearly identical disorder—*multigenic traits*—the final phenotype or clinical description of a disorder resulting from the addition and combination of effects from multiple genes, and *phenocopies*—the presence of affected individuals within the study population or family caused by nongenetic or environmental factors. These principles are well-known from other human genetic traits, and there is no reason to believe that they will not be operative in psychiatric disorders as well. Indeed, the lengthy descriptions of disorders given by Guze emphasize the complexity of diagnostic symptoms and their overlap with other related disorders, especially in the areas of substance abuse, antisocial personality, and suicides following homicides. A few clinically homogeneous genetic diseases outside of behavior genetics, such as the eye disorder retinitis pigmentosa (RP), the cardiovascular disorder Long QT Syndrome (LQT), and the childhood epilepsies are enough to remind us that even with relatively rare human diseases (frequencies of about 1/10,000) there are many paths to the same clinical endpoint. The genetic catalogue of RP lists at least 20 types and includes autosomal dominant, autosomal recessive, and X-linked forms. LQT syndrome can be caused by five different genes. Rare forms of benign familial childhood epilepsies involve a minimum of three genetic loci. It makes one wonder how many genetic pathways could be involved in common behavioral disorders (frequencies of 1 to 5% in the population) with violence and criminality as part of the phenomenology. And to make matters even more complex, nearly every simple single gene disorder in humans does not lead to absolute genetic determinism with regard to the observation of the phenotype—that is, the penetrance of the gene (the probability of observing the disorder given the occurrence of a particular mutation) is *not* 100%. For most behavior traits and psychiatric disorders in humans the accepted range of penetrances for postulated major genes is approximately 30 to 70%. Finally, "phenocopies" (conditions that look the same clinically but may have very different causes), such as a head injury leading to aggressiveness, can cause misclassification of an individual within a given study, adding to the problems of analysis.

How can a detailed study of the clinical phenomena of psychiatric illnesses associated with violence best answer the questions posed by this book? To this writer, the answer must lie in the contribution of clear thinking and sagacious clinician–researchers to the identification of exceptional families for inclusion into disease-gene searches and to the identification of subtypes of schizophrenia, antisocial personality disorders, alcoholism, and so on for inclusion into large-scale genetic association or affected sib–pair studies. The role of understanding the subtleties of clinical diagnosis cannot be underestimated. Progress in understanding the genetic basis of these conditions is required before any general use of genetic testing in the judicial arena. One celebrated case illustrates this point. In 1993 researchers identified a large Dutch family with disturbed regulation of impulsive aggression and borderline mental retardation.[1] The inheritance pattern, affecting males, suggested that the gene would lie on the X-chromosome. Biochemical information from the urine of family members indicated a disturbance in the metabolism of brain chemicals, focusing research efforts that subsequently demonstrated the behavioral syndrome in males was caused by a mutation in the structural gene for monoamine oxidase type A (MAOA). Without the careful identification of this family by skilled clinicians the gene obviously would not have been found and our understanding of the genetics of human behavior would have been less.

This example also illustrates the immense work ahead for researchers in this area. Despite a large effort no one has identified a defect in the MAOA gene in any additional family or in any individual selected from a high-risk population of violence-prone individuals. In other words, this mutation may be unique to this family and *not* transfer to the general population. If this is a harbinger of the world of psychiatric genetics then we are quite a way off from offering scientific evidence to the courts, except in very isolated circumstances.

It should also be noted that the type of mutation identified in the Dutch family represents a "null mutation," a mutation that produces an incomplete protein product and one that predicts a rather drastic effect. Everything we know about human disease-causing mutations tells us that the majority are not null, or knock-out, mutations but rather mutations with more subtle effects. If this holds then the complex behavioral patterns must reflect a spectrum of mutations, each with a slightly different effect and each interacting somewhat differently with other modifying genes and with changing environments.

Finally, those carefully studying the phenomenology of psychiatric disorders might want to delineate the environmental factors predicting violent-prone or criminal behavior, for is it not quite possible in some cases that environmental determinism is stronger than, or necessary for, genetic determinism?

## Note

1. H. G. Brunner, M. Nelen, X. O. Breakfield, H. H. Ropers, and B. A. van Oost, "Abnormal Behavior Associated With a Point Mutation in the Structural Gene for Monoamine Oxidase A." *Science* 262 (1993):578–580.

# ELABORATION
# Violence and Mental Illness: Additional Complexities

Robert M. Wettstein

A fundamental issue in the study of violence and mental illness is that human behavior is multifactorial in origin. This is true whether or not a psychiatric disorder is present. It is therefore proper to refer to "individuals with schizophrenia" rather than "schizophrenics." There is often a tendency, by mental health clinicians and nonclinicians alike, to reduce the individual to her symptoms and mental disorder. It is well-known that behavior has biological (genetic, neurologic, neurochemical), cultural, environmental–familial, as well as psychopathological origins, but this is often overlooked. There is no bright line that demarcates these various contributing factors in the development of a particular trait, symptom, or disorder. It is further likely that these influences are not static but are dynamic, changing over time, and interacting in a complex manner. One factor may attenuate or exaggerate another, depending on the presence or timing of other factors.

Medicine generally employs a biopsychosocial etiological model, with an understanding of multiple contributing causes or influences. The prevailing diagnostic approach in psychiatry, using the *Diagnostic and Statistical Manual,*[1] uses five axes to incorporate intraindividual, interpersonal, social, and familial influences on the development and course of a mental disorder.

Reductionism and hard determinism, whether psychoanalytic, genetic, neurologic, or neurochemical are perilous. The history of psychiatry in general and psychotherapy in particular has been marked by tension and conflict among the various schools of thought or practice, each claiming a unique understanding of the development and treatment of a problematic behavior. This is as present today as ever before. This concept has important implications for clinical intervention, risk assessment, and criminal responsibility, as no one school of thought in psychiatry or psychology can as yet rightfully claim to be the only "correct" one.

## Comorbidity

The issue of comorbidity of psychiatric disorders is critical in many respects. It is quite common to diagnose an individual seeking treatment, or an individual referred by the civil or criminal courts, as having more than a single psychiatric disorder. It might only be a slight exaggeration to state that comorbidity is the rule rather than the exception. Many varieties of comorbid psychiatric illness occur: depressive disorders with anxiety disorders, depressive disorders with eating disorders, and depressive disorders with personality disorder are common examples. Of greatest prevalence and concern, perhaps, is the comorbidity with substance abuse disorders. Substance use (intoxication and withdrawal), abuse, and dependence complicate the presentation, assessment, and treatment of all psychiatric illness. Patients, especially those with bioplar disorder or schizophrenia, commonly are noncompliant with their prescribed psychotropic regimen and then self-medicate their symptomatic distress with alcohol and street drugs, usually unsuccessfully. It is known that schizophrenia increases the risk of substance abuse, because schizophrenics are more likely to abuse substances than those individuals without schizophrenia.

Comorbidity of psychiatric illness presents problems to clinicians, researchers, expert witnesses, policy makers, and theoreticians alike. One comorbid disorder (e.g., personality disorder) can affect the presentation, diagnosis, treatment, and outcome of a coexisting disorder. Comorbid illnesses can each wax and wane, or even come and go, over time, independently or in parallel. It is often difficult to determine whether comorbid illnesses are related, dependent, or independent of one another. Determining which disorder is "primary" and which is "secondary," if any such relationship between them can be established, is problematic. Equally problematic is the process of conducting civil or criminal responsibility evaluations for court, because the expert witness may need to identify which disorder is causally related to the outcome behavior in question (e.g., criminal conduct), which occurred at a particular point in time, usually some time ago. Cognitive or volitional impairments of a defendant as a result of substance intoxication, as opposed to a major mental disorder, could be irrelevant to a finding of criminal nonresponsibility, though both conditions may have been simultaneously present at the time of the alleged crime.

## Violence in General

Just as behavior in general is multifactorial, violence in particular has multiple contributing influences. This truism need not be detailed, because it is so well-accepted, at least by most researchers and observers.[2] Thus apart from medical- or illness-related contributions to violence, we must recognize social, environmental, and cultural factors, whether in an individual case or collectively in a community or society. Domestic violence, one of the most prevalent forms of violence in the United States, is an excellent example of this.[3]

Although this is not the place for a critique of research method, it is important to note that we do not have a sound nosology for violent behavior and that many definitions are used. We too often use the terms *crime, aggression, violence,* and *criminality,* as if they were interchangeable. Further, there surely are many types of violence (sexual, nonsexual, homicidal, instrumental, defensive, predatory, territorial, fear-induced, loss-related, between strangers or acquaintances, impulsive, or premeditated), involving differences that are no doubt significant at many levels. Violence is not a homogeneous form of behavior. Research tools designed to distinguish among these are still evolving.

## Mental Illness and Violence

Guze notes that those with serious mental illnesses are at higher risk for interpersonal violence than those without such illness, having controlled for the influence of other variables. This issue has been a matter of intense debate for years, and there are many methodological difficulties in the studies in this area. Two decades ago we proclaimed that the seriously mentally ill were not any more likely to be violent than anyone else, but today the evidence points in the opposite direction.

Epidemiological data of serious mental illness among jail and prison inmates reveals serious illness in many such inmates, though there are many reasons for such a prevalence.[4] In community samples, outside of a corrections setting, data suggest that the risk of violence increases when an individual has more than one psychiatric diagnosis, with the largest risk among individuals who had both schizophrenia and substance abuse disorders.[5] Samples of mentally ill persons reveal at least equal if not greater arrest and conviction rates for violence compared to community samples of persons who are not mentally ill.

However, the relationship between mental disorder and violence is a modest rather than potent one.[6] Risk of violence may be statistically better explained by demographic variables such as age, gender, race, and socioeconomic status rather than mental disorder, but the evidence on this point remains inconclusive. Violence is not unique to those with serious mental illness, and Guze properly emphasized the role of antisocial personality disorder and substance abuse in criminal behavior, in contrast to that of serious mental illness. Though mentally ill persons may be at somewhat higher risk of violence, most of the violence in the United States has nothing to do with mental disorder. In countries with lower homicide rates (e.g., Finland[7]), violence and mental disorder are more clearly linked. Mental disorder is neither a necessary nor, perhaps, even a sufficient condition for interpersonal violence. Many individuals can be mentally ill for years before, and after, a single violent act. Further, there is no single uniformly successful treatment or intervention for violence. It is unknown to what degree violence would be prevented were mental illness maximally prevented and treated in our society, though some have argued that a substantial result would accrue.[8]

This is not to indicate that at least some cases of serious violence directly appear to be symptoms of a mental disorder. Those individuals who kill a loved one in response to a command hallucination from God, or in response to a paranoid delusion that the victim will destroy the world unless killed first, or in the midst of the severe despair and guilt of psychotic depression, ostensibly act on their mental disorders. Torrey has asserted that, although most seriously ill individuals are not violent, there exists a violent mentally ill subgroup who have a history of medication noncompliance, substance abuse, and a violent history.[9] Mulvey has viewed the relationship between mental disorder and violence as a "dynamic process in which people are at increased risk at different times, possibly as the result of the emergence of particular types of symptoms or beliefs."[10]

Still, the coexistence of mental disorder and violence at a particular point in time does not necessarily indicate a direct causal relationship between the two. Perhaps they just coexist, without the mental disorder directly leading to the violent act, or there may be an indirect link between the two phenomena.

## Impulsivity and Neuroscience of Violence

An important but poorly understood clinical phenomenon is impulsivity. Sometimes a symptom of individuals with an antisocial personality disorder, impulsivity is associated with a variety of clinical syndromes involving food intake, substance abuse, self-destructive or mutilative behavior, hair-pulling, sexual behavior, or shopping. The impaired impulse control, or its consequences, can readily place the individual in legal jeopardy (e.g., kleptomania, pyromania, pathological gambling). The controversial diagnosis of intermittent explosive disorder is given when several discrete episodes of violent or destructive behavior occur that are "grossly out of proportion to any precipitating psychosocial stressors" (*DSM-IV*), and are not attributable to another mental disorder. This is the only example of violence as constituting a mental disorder by itself. Impulsivity may also be a personality trait or element of temperament.

Recent research has identified neurochemical (i.e., serotonin and glucose metabolism), and neuropsychological test (i.e., frontal lobe dysfunction) results thought to correlate with impulsivity.[11]

But what sense can we make out of the growing neuroscience of violence and impulsivity?[12] Structural brain damage (e.g., cysts, tumors), as seen through abnormalities in CT scans, MRI scans, or cerebral angiography does not necessarily indicate deficits in brain function. Further, it may be unclear what constitutes an abnormality of brain function (e.g., a finding on computerized EEG), given the absence of information about the prevalence of such findings in the general population. Structural brain damage and impairment of brain function, like serious mental illness, are present in many individuals, most of whom are not violent. Do such test findings explain or account for a person's violence? It is well-known that correlations are not causes and causes are not excuses, at least legally.[13] By comparison, it is well-known that structural anatomical abnormalities in the lumbar spine (e.g., herniated discs) do not correlate well with the presence and extent of the patient's pain or disability. Similar brain injuries in different people can result in, or express themselves in, different ways.

Most contemporary observers view violence as originating from a constellation and interraction of risk factors including cultural, environmental (e.g., witness to, or victim of, severe child abuse), serious mental illness or personality disorder, neurologic deficit, genetic, substance abuse, external life stressors, and availability of weapons.[14]

## Criminal Responsibility

The credibility of mental health professionals is never more in question than in court, where their credibility is all they have to market.[15] Credibility is especially a problem in criminal responsibility evaluations in which, as Resnick has noted, experts are called on to ''excuse sin.'' But prosecution experts in criminal cases, as well as experts in civil litigation, are equally vulnerable to undue influences on their striving for objectivity and nonpartisanship, which are the guiding ethical obligations for forensic psychiatrists and psychologists. Expert witnesses are widely accused of bringing their personal political, social, or moral agendas into court in the guise of their scientific or professional expertise. The dangers, with regard to the introduction of genetic evidence in court, are no less significant.

Any criminal responsibility evaluation is beset by the problem of retrospectively assessing a defendant's thought process, perhaps months or even years following an alleged criminal act. If multiple crimes are alleged that occurred over a period of time (e.g., multiple homicides or multiple acts of embezzlement), the examiner must evaluate each act individually with regard to the defendant's state of mind. The examiner must identify and specify any disturbances in cognitive capacity and any impairment in the regulation of affect, anxiety, and impulse control. It is necessary to distinguish the impact of comorbid illnesses, especially intoxication, emotional states, psychosocial stressors (e.g., provocation, compulsion) from any major mental disorders.

The difficulty in assessing volitional incapacity, usually in the course of criminal responsibility evaluations, is well-known. This difficulty was one of the reasons that the American Psychiatric Association abandoned the criminal responsibility standard based on volitional incapacity in favor of a cognitive–affective standard. Despite the controversy in this decision, it truly is difficult to determine whether a defendant was unable to resist an impulse to commit a criminal act or simply failed to resist it. Such difficulty exists regardless of whether the basis for the incapacity is neurochemical, genetic, neurologic, psychiatric, or psychodynamic.

In the end, it may be as interesting and important to understand why individuals *are able* to resist criminal or other impulses as why are they not able to do so, particularly as we gain more understanding of underlying volitional impairments, whether genetic or neurochemical.

## Preventive Detention

Apart from whether a mental disorder (whatever the ''cause'') is used at trial to excuse the act or diminish the defendant's capacity to act, mental disorders are sometimes invoked at the time of sentencing to mitigate the sentence severity. But it is not so clear that testimony that the defendant has an antisocial personality disorder, a serious mental illness, or any genetic predisposition to act violently will serve to mitigate the defendant's sentence (e.g., life without parole rather than death) in the jury's mind.[16] The jury could rightly conclude that, retributive considerations aside, such a defendant is *more* likely than others not so afflicted to constitute a danger to society, especially in cases in which future danger is an aggravating criteria for a capital sentence, and no clinical intervention is available to negate that additional risk factor.

The larger concern is that data regarding the genetic predisposition to violent behavior will be used to look forward rather than backward in time, possibly against the interests of the identified individual. At present, violence risk assessments are made routinely in psychiatric hospitals (civil and forensic) and emergency rooms, as well as by probation and parole agencies. Our ability to predict whether an individual will be violent in the future is limited, and genetic data might be helpful in this regard. Of course, we would need to know the strength of such a risk factor for prediction, especially when compared to other known risk factors (e.g., past history of violence, age–sex, substance abuse, mental disorder).[17] But the pitfall is that the state could conceivably use such data to preventively detain individuals at "high genetic risk" of violence, whether in hospital, prison, or a special facility for genetically defective individuals. Lest this sound too draconian, the states have rapidly moved in the past 10 years to preventively and indefinitely detain sexual offenders through labeling them as "sexual predators," or at least requiring registration and community notification of their whereabouts.[18] Mandatory DNA testing of all convicted sexual offenders, at least for purposes of identifying and prosecuting them for additional crimes, is already underway in the United States. One can readily imagine the political call for preventive detention, or worse, of individuals thought to have a genetic predisposition to violence.

## Notes

1. *DSM-IV* (Washington, DC: American Psychiatric Association, 1994).
2. A. J. Reiss and J. A. Roth (eds.), *Understanding and Preventing Violence.* Vol. 1. (Washington, DC: National Academy Press, 1993).
3. ———. *Understanding and Preventing Violence: Social Influences.* Vol. 3. (Washington, DC: National Academy Press, 1994).
4. L. A. Teplin, "The Prevalence of Severe Mental Disorder Among Male Urban Jail Detainees: Comparison With the Epidemiologic Catchment Area Program," *American Journal of Public Health* 80 (1990):663–669; ———. "Psychiatric and Substance Abuse Disorders Among Male Urban Jail Detainees," *American Journal of Public Health* 84 (1994):290–293.
5. J. F. Swanson, C. E. Holzer, V. K. Ganju and R. T. Jono, "Violence and Psychiatric Disorder in the Community: Evidence from the Epidemiologic Catchment Area Surveys," *Hospital and Community Psychiatry* 41 (1990):761–770.
6. J. Monahan and H. J. Steadman, *Violence and Mental Disorder* (Chicago: University of Chicago Press, 1994).
7. M. Eronen, J. Tiihonen and P. Hakola, "Schizophrenia and Homicidal Behavior," *Schizophrenia Bulletin* 22 (1996):83–89.
8. E. F. Torrey, "Violent Behavior by Individuals With Serious Mental Illness," *Hospital and Community Psychiatry* 45 (1994):653–662.
9. *Id.*
10. E. P. Mulvey, "Assessing the Evidence of a Link Between Mental Illness and Violence," *Hospital and Community Psychiatry* 45 (1994):663–668.

11. D. J. Stein, E. Hollander, and M. R. Liebowitz, ''Neurobiology of Impulsivity and the Impulse Control Disorders,'' *Journal of Neuropsychiatry* 5 (1993):9–17.
12. L. R. Tancredi and N. Volkow, ''Neural Substrates of Violent Behavior: Implications for Law and Public Policy,'' *International Journal of Law and Psychiatry* 11 (1988):13–49. P. Y. Blake, J. H. Pincus, and C. Buckner, ''Neurologic Abnormalities in Murderers,'' *Neurology* 45 (1995):1641–1647. C. M. Epstein, ''Computerized EEG in the Courtroom,'' *Neurology* 44 (1994):1566–1569. A. J. Reiss and J. A. Roth (Eds.), *Understanding and Preventing Violence: Biobehavioral Influences.* Vol. 2. (Washington, DC: National Academy Press, 1994).
13. S. J. Morse, ''Brain and Blame,'' *Georgetown Law Review* 84 (1996):527–549.
14. Reiss, *Understanding and Preventing Violence*, Reiss and Roth, ''The Prevalence of Severe Mental Disorder Among Urban Jail Detainees''; Reiss and Roth, *Understanding and Preventing Violence*; R. J. Cadoret, W. R. Yates, E. Troughton, G. Woodworth, and M. A. Stewart. ''Genetic-Environmental Interaction in the Genesis of Aggressivity and Conduct Disorders,'' *Archives of General Psychiatry* 52 (1995):916–924.
15. P. J. Resnick, ''Perceptions of Psychiatric Testimony: A Historical Perspective on the Hysterial Invective,'' *Bulletin of the American Academy of Psychiatry and the Law* 14 (1986):203–219.
16. S. H. Dinwiddie, ''Genetics, Antisocial Personality, and Criminal Responsibility,'' *Bulletin of the American Academy of Psychiatry and the Law* 24 (1996):95–108.
17. *Ciba Foundation Symposium: Genetics of Criminal and Antisocial Behaviour* (Chicester, MA: John Wiley, 1996).
18. ''Predators and Politics: A Symposium on Washington's Sexually Violent Predators Statute,'' *University of Puget Sound Law Review,* 15 (1992):507–911.

# Chapter 5
# GENETIC RESEARCH ON MENTAL DISORDERS

Steven O. Moldin

Evidence has accumulated in recent years to indicate that some individuals who suffer from mental disorders are at increased risk to engage in both violent and nonviolent criminalality. In the largest unselected birth cohort studied to date, persons who had psychiatric hospitalizations in Denmark were 3 to 11 times more likely to have criminal convictions than persons who had never been hospitalized.[1] In addition, a study of more than 600 convicted murderers in Finland showed that the odds ratios of homicidal violence were substantially increased for both men and women with the diagnoses of schizophrenia, antisocial personality disorder, and alcoholism as compared with the general population.[2]

A considerable body of empiric evidence now exists to support the contributory role of genes in the etiology of major mental disorders. Given the relationship between mental disorders and criminality, it is highly likely that genetic information will play an increasingly more important role in cases in which the presence of mental illness is a consideration. This chapter presents a review of methodologies used in the genetic analysis of mental disorders, as well as an evaluation of the available empiric evidence. I discuss the impact of genetic research and findings on mental disorders for the criminal justice system, as well as guidelines for the interpretation of genetic information in criminal cases.

## Genetic Explanations of Disease and Other Phenotypes

Human nuclear DNA is organized into 22 pairs of autosomal chromosomes and two sex-specific chromosomes (called X and Y). A particular gene exists at a specific position, or *locus,* on a chromosome. There are three broad categories of genetic explanations of human diseases: chromosomal alterations, single major gene effects, and polygenic/oligogenic effects (Table 5-1). A complex interaction between genetic susceptibility and nongenetic (e.g., pre- and postnatal environment) factors is typically responsible for phenotypic expression or modulation.

### *Chromosomal Alterations*

*Chromosomal alterations* are abnormalities in the morphology or number of autosomes or sex chromosomes, and are detected through karotyping. They likely account for only a small proportion of the familial cases of mental disorders.

### *Single Major Gene Effects*

More than 4000 distinct single major gene diseases exist,[3] but each is rare. The specific gene for several disorders has been identified and a genetic test developed—for example, Huntington disease.[4] Single gene effects may contribute to the etiology of a small number of cases of mental disorders, and will be seen in relatively few families.

Table 5-1—*Genetic Explanations of Disease*

| Etiology | Characteristics | Examples |
|---|---|---|
| Chromosomal aberration | found in > .5% of newborns<br>important cause of congenital malformations detected through karotyping | Down syndrome<br>cri-du-chat syndrome<br>Turner syndrome<br>Fragile X syndrome |
| Single major gene | familial transmission follows simple patterns identical or similar to those described by Mendel for discrete physical characteristics in plants<br>genetic variation is a result of one locus<br>classification is based on where the mutant gene resides and how many copies confer disease (autosomal or sex-linked, dominant or recessive) | polycystic kidney disease<br>Huntington's disease<br>cystic fibrosis<br>retinoblastoma<br>neurofibromatosis<br>Duchenne muscular dystrophy |
| Polygenic/oligogenic | familial transmission follows irregular patterns<br>multiple genes interact to produce disease<br>liability (susceptibility) to illness is transmitted | cleft lip + cleft palate<br>pyloric stenosis<br>hypertension<br>type 1 diabetes<br>late-onset Alzheimer's disease<br>multiple sclerosis<br>schizophrenia<br>bipolar disorder<br>alcoholism |

## *Polygenic/Oligogenic Effects*

For most common human disorders and complex traits, transmission in families follows irregular, unpredictable patterns. Genetic influences on these phenotypes are a result of the effects of multiple genes in interaction. Such multiple genes of relative equal effect that cannot be individually differentiated because there are so many are called *polygenes.* An *oligogenic* phenotype is influenced by a limited number of genes of relative equal effect called *quantitative trait loci* (*QTLs*). The "trait" influenced by QTLs can be actually quantitative (like height or intelligence) but also can be a discrete event (the presence or absence of a disease).

Genetic *liability* or susceptibility to a polygenic/oligogenic illness is transmitted, rather than illness per se; thus the same genetic constitution (*genotype*) can result in different observable outcomes (*phenotypes*), and different genotypes can produce the same phenotype. This lack of simple correspondence between genotype and phenotype for polygenic/

oligogenic disorders reflects the effects of chance (environmental effects not shared among family members), environmental effects shared among family members, and interactions among several genes (*epistasis*).

## Research Designs for Genetic Inquiry

A variety of research designs may be used to identify genetic contributions to the familial aggregation of mental disorders (Table 5-2). These are discussed in greater detail elsewhere.[5]

### *Population (Epidemiologic) Studies*

Population studies permit establishment of prevalence and incidence of a disorder in the general population or in isolated subpopulations. Estimation of such rates is essential for genetic analyses, and also can document temporal or other changes that have far-reaching scientific and health policy implications.

### *Family Studies*

Familial coaggregation of a disorder is a necessary but not sufficient criterion for implicating a genetic mechanism. Large-scale family studies permit determination of morbid risks to illness for different classes of relatives; such estimates not only provide evidence of familial transmission but also provide important information for genetic counseling. Quantitative genetic methods for the analysis of family study data—for example, segregation analysis—are used to discriminate among genetic models and identify the mode of inheritance of a disorder.

### *Twin Studies*

Twin studies permit the implication or exclusion of genetic factors in the familial transmission of a disorder. Given that monozygotic (MZ) twins have identical genotypes, dissimilarities between pair members must be a result of the influence of the pre- or postnatal environment. As a consequence, less than 100% concordance for a disease among MZ pairs living through the period of risk excludes genetic factors as exclusive determinants of that disorder. Dizygotic (DZ) twins share on average 50% of their genes, as do nontwin siblings.

When the MZ concordance rate is higher than the DZ concordance rate, a genetic basis is the most likely explanation if MZ twins are not more predisposed to developing the disorder and if MZ and DZ twins are equally correlated for their exposure to environmental influences of etiologic importance (the equal-environment assumption). If this assumption is violated, excess resemblance among MZ twins truly attributable to shared environmental exposure would be falsely attributed to genetic factors. The equal-environment assumption appears valid, with differential treatment of MZ and DZ twins by their parents unlikely to represent a significant bias, in twin studies of several major mental disorders.[6]

A measure of the degree of potential genetic control over a phenotype that can be calculated from twin studies is *heritability,* the ratio of genetic variance to the total variance in the population.

Table 5-2—*Major Methodologies for Genetic Research on Human Disease*

| Methodology | Unit of analysis | Goal |
|---|---|---|
| Population (epidemiologic) studies | Individuals in the general population | Establish lifetime incidence |
| Family studies | Pedigrees | Establish familiality and risks to relatives; estimate the mode of transmission |
| Twin studies | Monozygotic and dizygotic twin pairs | Distinguish genetic from environmental effects |
| Adoption studies | Adoptees; adoptive and biologic relatives of adoptees | Distinguish genetic from environmental effects |
| Linkage studies | Nuclear and extended pedigrees | Establish chromosomal location for a disease susceptibility locus |

## *Adoption Studies*

The effects of genes and environment may be teased apart by studying individuals who have been raised by biologically unrelated individuals and by studying their adoptive and biological relatives. Adoption study designs permit comparisons of the effects of different types of familial environments on groups with similar genetic susceptibilities to illness. The only controlled environmental factor in adoption study designs is the pre- and postnatal presence or absence of a parent with the disorder. Separation of genetic and environmental factors that contribute to susceptibility to a disorder are successfully achieved in adoption studies to the degree that: (a) separation of a child from biological parents happens early enough that exposure to a putative environmental risk factor is avoided or minimized; (b) the trauma and stress of the adoption process is not a specific risk factor for the disorder; and (c) the child is not selectively placed in an adoptive rearing environment that resembles the rearing environment of the biological parents with regard to putative environmental risk factors.

## *Linkage Studies*

The highest level of statistical proof for establishing the existence of a genetic influence in the familial aggregation of a disorder involves *linkage analysis,*[7] a statistical procedure by which family data is examined to determine whether the disorder cosegregates with a genetic marker of known chromosomal location. The purpose of linkage analysis is to infer that two loci (the marker locus and a putative disease-susceptibility locus) are located close enough together on the same chromosome such that they tend to be transmitted together from parent to child more frequently than would occur by random assortment. Demonstration of linkage between a putative susceptibility locus and a genetic marker thereby allows one to determine on which chromosome that putative disease susceptibility lies. A statistical measure of the strength of linkage is the *lod score,* the logarithm of the odds that two loci are linked closely enough to segregate together in families.

In linkage analysis of simple single-locus disorders, a lod score of 3 has traditionally been accepted as convincing evidence for linkage.[8] However, the genetic analysis of complex diseases such as diabetes or mental disorders is complicated by unknown modes of transmission and the typical unavailability of candidate loci that can target chromosomal regions of interest.

Current strategies for analyzing genetically complex diseases now include scanning the entire genome with a dense collection of genetic markers. Lander and Kruglyak[9] developed guidelines for the interpretation of linkage results derived from such whole-genome scans to minimize false-positive claims. Although there is no simple cutoff for determining the true significance of linkage results in the analysis of genetically complex diseases, and the process of scientific inference is never completely objective,[10] these guidelines provide a useful frame of reference for enhancing communication among scientists. Lander and Kruglyak consider a lod score of 1.9 in parametric analyses of genetic markers across the whole genome as suggestive of linkage, and a value of 3.3 as significant evidence. Corresponding values using nonparametric linkage methods are 2.2 and 3.6, respectively. Furthermore, these authors argue that linkage confirmation requires a two-step process: (a) significant linkage evidence is found in at least one study, and (b) evidence of linkage to the same region is obtained by another investigator in a new, independent sample.

Linkage analysis is a necessary first step in the process of isolating a disease gene on the basis of its chromosomal position. This process of *positional cloning*[11] permits identification of genetic abnormalities that contribute to a disorder before pathophysiology is fully understood. The availability of high resolution maps of the human genome[12] have made possible the positional cloning of genes for many simple single-locus diseases that include Duchenne muscular dystrophy, Huntington's disease, cystic fibrosis, and neurofibromatosis.

## Genetic Dissection of Complex Traits

Quantitative genetic research has built a strong case for the importance of genetic factors in many complex behavioral disorders. This review will focus on the genetic evidence for schizophrenia, mood disorders, alcoholism, antisocial personality disorder, and adult criminality.

### *Schizophrenia*

Data from more than 40 family and twin studies spanning seven decades of research consistently show that risk to relatives of affected individuals is greater than the worldwide population risk of approximately 1%;[13] hence schizophrenia clearly aggregates in families. Risk varies as a function of the degree of genetic relatedness to an affected individual (Figure 5-1). The highest risk (48%) is to the MZ cotwin of an affected individual—these two individuals share 100% of their genes in common. Lower morbid risks to parents likely reflect reduced fertility.[14]

An MZ : DZ concordance ratio of more than 3 : 1 is obtained by pooling the results of the six twin studies published in the past 25 years, and this strongly implicates genetic factors in the etiology of schizophrenia. MZ concordance is less than 100% and implicates the involvement of nongenetic factors. The heritability of schizophrenia has consistently been estimated as high (0.8), with minimal contribution of shared environmental effects to liability.[15] Thus the nongenetic component to liability for schizophrenia most likely reflects unshared (random) environmental effects.

In 1963 a compilation of adoptees was begun in Denmark by several American investigators to study the prevalence of schizophrenia in the biologic and adoptive nuclear families of individuals who eventually had developed schizophrenia.[16] A combined analysis

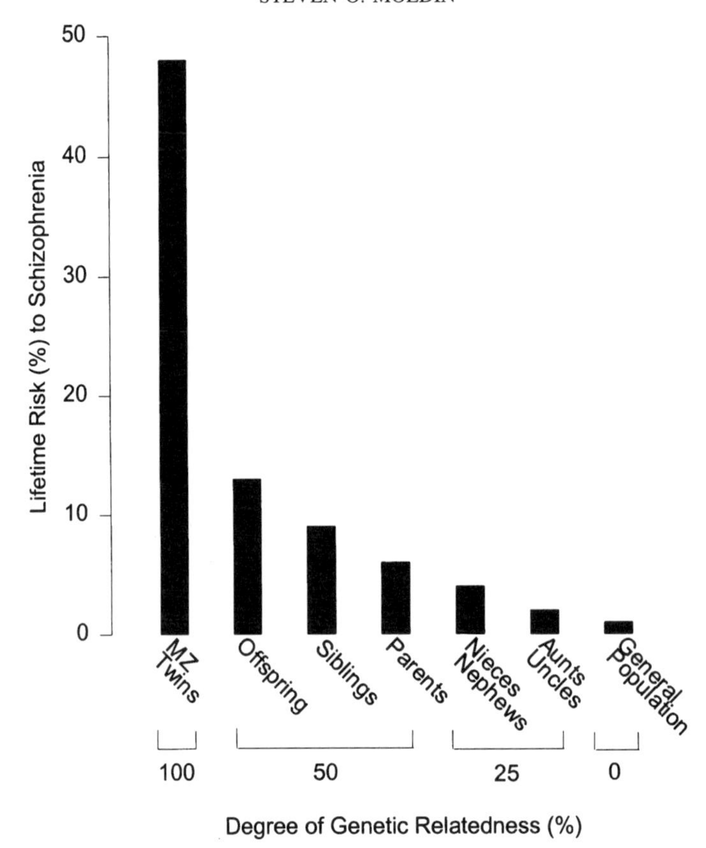


of data collected in the city and county of Copenhagen, as well the remainder of Denmark, showed that the prevalence of *DSM-II*-defined chronic schizophrenia and latent schizophrenia was significantly greater in biological relatives of adoptees with chronic schizophrenia compared to the biologic relatives of control adoptees.[17] Comparable results were obtained when *DSM-III* diagnostic criteria were used.[18]

A multilocus (polygenic) model has been consistently supported—and a single major-locus model consistently excluded—in the quantitative analysis of actual and simulated schizophrenia family data.[19] Risk ratios for classes of relatives of schizophrenic probands in pooled Western European twin and family studies were consistent with the influence of two or three major loci in epistatic interaction.[20] This analysis and other work[21] suggest that the genetic dissection of individual loci of small relative effect on liability to a complex trait like schizophrenia is feasible.

Despite the inability of the single-locus model to explain the familial aggregation of schizophrenia, investigators used available informative DNA polymorphisms and conducted traditional linkage studies under single-locus model assumptions. A summary of the recent history of linkage studies in schizophrenia is shown in Table 5-3, which only lists those showing positive evidence for linkage to defined chromosomal regions. Linkage studies of schizophrenia have also been reviewed elsewhere.[22]

Based on Lander and Kruglyak's[23] guidelines, two findings meet the criterion for significant schizophrenia linkage evidence: (a) the 6p linkage first reported in the Irish data set.[24] However, augmentation of this sample by more than 400 individuals and 79 new

Table 5-3—*Reported Chromosomal Locations for Schizophrenia Susceptibility Loci*

| Location | Year | Reference |
|---|---|---|
| 5q13-11 | 1988 | R. Sherrington, J. Brynjolfsson, H. Petursson, M. Potter, K. Dudleston, B. Barraclough, J. Wasmuth, M. Dobbs, and H. M. D. Gurling, "Localization of a Susceptibility Locus for Schizophrenia on Chromosome 5," *Nature* 336 (1988):164–167. |
| 11q21-22 | 1990 | D. St. Clair, D. Blackwood, W. Muir, A. Carothers, M. Walker, G. Spowart, C. Gosden, and H. J. Evans, "Association Within a Family of a Balanced Autosomal Translocation With Major Mental Illness," *Lancet* 336 (1990):13–16. |
| Pseudoautosomal | 1991 | J. Collinge, L. E. DeLisi, A. Boccio, E. C. Johnstone, A. Lane, C. Larkin, M. Leach, R. Lofthouse, F. Owen, and M. Poulter, "Evidence for a Pseudo-Autosomal Locus for Schizophrenia Using the Method of Affected Sibling Pairs," *British Journal of Psychiatry* 158 (1991):624–629. |
| Pseudoautosomal | 1992 | T. d'Amato, G. Waksman, M. Martinez, C. Laurent, P. Gorwood, D. Campion, M. Jay, C. Petit, C. Savoye, and C. Bastard, "Pseudoautosomal Region in Schizophrenia: Linkage Analysis of Seven Loci by Sib-Pair and Lod-Score Methods," *Psychiatry Research* 52 (1994):135–147. |
| 22q13 | 1994 | A. E. Pulver, M. Karayiorgou, P. S. Wolyniec, V. K. Lasseter, L. Kasch, G. Nestadt, S. Antonarakis, D. Housman, H. H. Kazazian, and D. Meyers, "Sequential Strategy to Identify a Susceptibility Gene for Schizophrenia: Report of Potential Linkage on Chromosome 22q12-q13.1: Part 1," *American Journal of Medical Genetics* 54 (1994):36–43; V. K. Lasseter, A. E. Pulver, P. S. Wolyniec, G. Nestadt, D. Meyers, M. Karayiorgou, D. Housman, S. Antonarakis, H. Kazazian, and L. Kasch, "Follow-Up Report of Potential Linkage for Schizophrenia on Chromosome 22q: Part 3," *American Journal of Medical Genetics* 60 (1995):172–173. |
| 22q13 | 1994 | H. Coon, J. Holik, M. Hoff, F. Reimherr, P. Wender, M. Myles-Worsley, M. Waldo, R. Freedman, and W. Byerly, "Analysis of Chromosome 22 Markers in Nine Schizophrenia Pedigrees," *American Journal of Medical Genetics* 54 (1994):72–79. |
| 22q13-12 | 1994 | M. H. Polymeropoulos, H. Coon, W. Byerley, E. S. Gershon, L. Goldin, T. J. Crow, J. Rubenstein, M. Hoff, J. Holik, and A. M. Smith, "Search for a Schizophrenia Susceptibility Locus on Human Chromosome 22," *American Journal of Medical Genetics* 54 (1994):93–99. |
| 6p24-22 | 1995 | S. Wang, C. E. Sun, C. A. Walczak, J. S. Ziogle, B. R. Kipps, L. R. Goldin, and S. R. Diehl, "Evidence for a Susceptibility Locus for Schizophrenia on Chromosome 6pter-p22," *Nature Genetics* 10 (1995):41–46; R. E. Straub, C. J. MacLean, F. A. O'Neill, J. Burke, B. Murphy, F. Duke, R. Shinkwin, B. T. Webb, J. Zhang, D. Walsh, and K. S. Kendler, "A Potential Vulnerability Locus for Schizophrenia on Chromosome 6p24-22: Evidence for Genetic Heterogeneity," *Nature Genetics* 11 (1995):287–293. |

Table 5-3—*Continued*

| Location | Year | Reference |
|---|---|---|
| 6p23 | 1995 | H. W. Moises, L. Yang, H. Kristbjarnarson, C. Wiese, W. Byerley, F. Macciardi, V. Arolt, D. Blackwood, X. Liu, B. Sjogren, H. N. Aschauer, H.-G. Hwu, K. Jang, W. J. Livesley, J. L. Kennedy, T. Zoega, O. Ivarsson, M.-T. Bui, M.-H. Yu, B. Havsteen, D. Commenges, J. Weissenbach, E. Schuringer, I. I. Sottesman, A. J. Pakstis, L. Wetterberger, K. K. Kidd, and T. Helgason, "An International Two-Stage Genome-Wide Search for Schizophrenia Susceptibility Genes," *Nature Genetics* 11 (1995):321–324. |
| 6p24-23 | 1995 | S. E. Antonarakis, J.-L. Blouin, A. E. Pulver, P. Wolyniec, V. K. Lasseter, G. Nestadt, L. Kasch, R. Babb, H. H. Kazazian, and B. Dombroski, "Schizophrenia," *American Journal of Medical Genetics* 81 (1995):81–91. |
| 6p24-22 | 1995 | S. G. Schwab, M. Albus, J. Hallmayer, S. Honig, M. Borrmann, D. Lichtermann, R. P. Ebstein, M. Ackenheil, B. Lerer, and N. Risch, "Evaluation of a Susceptibility Gene for Schizophrenia on Chromosome 6p by Multipoint Affected Sib-Pair Linkage Analysis," *Nature Genetics* 11 (1995):325–327; B. J. Mowry, D. J. Nancarrow, D. P. Lennon, L. A. Sandkuijl, R. R. Crowe, J. M. Silverman, R. C. Mohs, L. J. Siever, J. Endicott, and L. Sharpe, "Schizophrenia Susceptibility and Chromosome 6p24-22," *Nature Genetics* 11 (1995):233–234. |
| 22q12 | 1995 | H. P. Vallada, M. Gill, P. Sham, L. C. Lim, S. Nanko, P. Asherson, R. M. Murray, P. McGuffin, M. Owen, and D. Collier, "Linkage Studies on Chromosome 22 in Familial Schizophrenia," *American Journal of Medical Genetics* 60 (1995):139–146. |
| 22q12 | 1995 | H. W. Moises, L. Yang, T. Li, B. Havsteen, R. Fimmers, M. P. Baur, X. Liu, and I. I. Gottesman, "Potential Linkage Disequilibrium Between Schizophrenia and Locus D22S278 on the Long Arm of Chromosome 22," *American Journal of Medical Genetics* 60 (1995):465–467. |
| 22q13 | 1995 | S. G. Schwab, B. Lerer, M. Albus, W. Maier, J. Hallmayer, R. Fimmers, D. Lichtermann, J. Minges, B. Bondy, and M. Ackenheil, "Potential Linkage for Schizophrenia on Chromosome 22q12-q13: A Replication Study," *American Journal of Medical Genetics* 60 (1995):436–443. |
| 9p, 20p | 1995 | Moises et al., "An International Two-Stage Genome-Wide Search for Schizophrenia Susceptibility Genes." |
| 3p26-24 | 1995 | A. E. Pulver, V. K. Lasseter, L. Kasch, P. Wolyniec, G. Nestadt, J.-L. Blouin, M. Kimberland, R. Babb, S. Vourlis, and H. Chen, "Schizophrenia: A Genome Scan Targets Chromosomes 3p and 8p as Potential Sites of Susceptibility Genes," *American Journal of Medical Genetics* 60 (1995):252–260. |
| 8p | 1995 | *Id.* |

Table 5-3

| Location | Year | Reference |
| --- | --- | --- |
| 22q12 | 1996 | M. Gill, H. Vallada, D. Collier, P. Sham, P. Holmans, R. Murray, P. McGuffin, S. Nanko, M. Owen, S. Antonarakis, D. Housman, H. Kazazian, G. Nestadt, A. E. Pulver, R. E. Straub, C. J. MacLean, D. Walsh, K. S. Kendler, L. DeLisi, M. Polymeropoulos, H. Coon, W. Byerley, R. Lofthouse, E. Gershon, L. Goldin, T. Crow, R. Freedman, C. Laurent, S. Boodeau-Pean, T. d'Amato, M. Jay, D. Campion, J. Mallet, D. B. Wildenauer, B. Lerer, M. Albus, M. Ackenheil, R. P. Ebstein, J. Hallmayer, W. Maier, H. Gurling, D. Curtis, G. Kalsi, J. Brynjolfsson, T. Sigmundson, H. Petursson, D. Blackwood, W. Muir, D. St. Clair, L. He, S. Maguire, H. W. Moises, H.-G. Hwu, L. Yang, C. Wiese, L. Tao, X. Liu, H. Kristbjarnason, D. F. Levinson, B. J. Mowry, H. Donis-Keller, N. K. Hayward, R. R. Crowe, J. M. Silverman, D. J. Nancarrow, and C. M. Read, ''A Combined Analysis of D22S278 Marker Alleles in Affected Sib-Pairs: Support for a Susceptibility Locus for Schizophrenia at Chromosome 22q12,'' *American Journal of Medical Genetics* 67 (1996):40–45. |
| 13q | 1996 | J. Williams, G. Spurlock, P. McGuffin, J. Mallet, M. M. Nothen, M. Gill, H. Aschauer, P.-O. Nylander, F. Macciardi, M. J. Owen, and the European Multicentre Association Study of Schizophrenia, ''Association Between Schizophrenia and T102C Polymorphism of the 5-Hydroxytryptamine Type 2a-Receptor Gene,'' *Lancet* 347 (1996):1294–1296; J. Williams, P. McGuffin, M. Nothen, and M. J. Owen, ''Meta-Analysis of Association Between the 5-$HT_{2a}$ Receptor T102C Polymorphism and Schizophrenia,'' *Lancet* 349 (1997):1221. |
| 5p | 1996 | J. M. Silverman, D. A. Greenberg, L. D. Altstiel, L. J. Siever, R. C. Mohs, C. J. Smith, G. Zhou, T. E. Hollander, X.-P. Yang, M. Kedache, G. Li, M. L. Zaccario, and K. L. Davis, ''Evidence of a Locus for Schizophrenia and Related Disorders on the Short Arm of Chromosome 5 in a Large Pedigree,'' *American Journal of Medical Genetics* 67 (1996):162–171. |
| 6p23 | 1996 | S. Wang, S. Detera-Wadleigh, H. Coon, C.-E. Sun, L. R. Goldin, D. L. Duffy, W. F. Byerley, E. S. Gershon, and S. R. Diehl, ''Evidence of Linkage Disequilibrium Between Schizophrenia and the SCA1 CAG Repeat on Chromosome 6p23,'' *American Journal of Human Genetics* 59 (1996):731–736. |
| 6p, 8p | 1996 | Schizophrenia Linkage Collaborative Group for Chromosomes 3 6 and 8, ''Additional Support for Schizophrenia Linkage on Chromosomes 6 and 8: A Multicenter Study,'' *American Journal of Medical Genetics* 67 (1996):580–594. |
| 8p | 1996 | K. S. Kendler, C. J. MacLean, A. O'Neill, J. Burke, B. Murphy, F. Duke, R. Shinkwin, S. M. Easter, B. T. Webb, J. Zhang, D. Walsh, and R. E. Straub, ''Evidence for a Schizophrenia Vulnerability Locus on Chromosome 8p in the Irish Study of High-Density Schizophrenia Families,'' *American Journal of Psychiatry* 153 (1996):1534–1540. |

Table 5-3—*Continued*

| Location | Year | Reference |
|---|---|---|
| 5q | 1997 | R. E. Straub, C. J. MacLean, F. A. O'Neill, D. Walsh, and K. S. Kendler, "Support for a Possible Schizophrenia Vulnerability Locus in Region 5q22-31 in Irish Families," *Molecular Psychiatry* 2 (1997):148–155. |
| 5q | 1997 | S. G. Schwab, G. N. Eckstein, J. Hallmayer, B. Lerer, M. Albus, M. Borrmann, D. Lichtermann, M. A. Ertl, W. Maier, and D. B. Wildenauer, "Evidence Suggestive of a Locus on Chromosome 5q31 Contributing to Susceptibility for Schizophrenia in German and Israeli Families by Multipoint Affected Sib-Pair Linkage Analysis," *Molecular Psychiatry* 2 (1997):156–160. |
| 6q | 1997 | Q. Cao, M. Martinez, J. Zhang, A. R. Sanders, J. A. Badner, A. Cravchik, C. Markey, E. Beshah, J. J. Guroff, M. E. Maxwell, D. Kazuba, R. Whiten, L. R. Goldin, E. S. Gershon, and P. V. Gejman, "Suggestive Evidence for a Schizophrenia Susceptibility Locus on Chromosome 6q and a Confirmation in an Independent Series of Pedigrees," *Genomics* 43 (1997):1–8. |
| Xp | 1997 | J. Dann, L. E. DeLisi, M. Devoto, S. Laval, D. J. Nancarrow, G. Shields, A. Smith, J. Loftus, P. Peterson, A. Vita, M. Comazzi, G. Invernizzi, D. F. Levinson, D. Wildenauer, B. J. Mowry, D. Collier, J. Powell, R. R. Crowe, N. C. Andreasen, J. M. Silverman, R. C. Mohs, R. M. Murray, M. K. Walters, D. P. Lennon, N. K. Hayward, M. Albus, B. Lerer, W. Maier, and T. J. Crow, "A Linkage Study of Schizophrenia to Markers Within Xp11 Near the *MAOB* Gene," *Psychiatry Research* 70 (1997):131–143. |
| 13q14-32 | 1997 | M. W. Lin, P. Sham, H.-G. Hwu, D. Collier, R. Murray, and J. F. Powell, "Suggestive Evidence for Linkage of Schizophrenia to Markers on Chromosome 13 in Caucasian but not Oriental Populations," *Human Heredity* 99 (1997):417–420. |
| 6p24-22 | 1997 | M. Maziade, L. Bissonnette, E. Rouillard, M. Martinez, M. Turgeon, L. Charron, V. Pouliot, P. Boutin, D. Cliche, C. Dion, J. P. Fournier, Y. Garneau, J. C. Lavallee, N. Montgrain, L. Nicole, A. Pires, A. M. Ponton, A. Potvin, H. Wallot, M. A. Roy, le groupe IREP, and C. Merrette, "6p24-22 Region and Major Psychoses in the Eastern Quebec Population," *American Journal of Medical Genetics* 74 (1997):311–318. |
| 6q24,8pter-8q12, 9q32-q34, 15p13-12 | 1998 | C. A. Kaufmann, B. Suarez, D. Malaspina, J. Pepple, D. Svrakic, P. D. Markel, J. Meyer, C. T. Zambuto, K. Schmitt, T. C. Matise, J. M. Harkavy Friedman, C. Hampe, H. Lee, D. Shore, D. Wynne, S. V. Faraone, M. T. Tsuang, and C. R. Cloninger, "NIMH Genetics Initiative Milliennium Schizophrenia Consortium: Linkage Analysis of African-American Pedigrees," *American Journal of Medical Genetics* 81 (1998):282–289. |

Table 5-3

| Location | Year | Reference |
|---|---|---|
| 10p14 p13 | 1998 | S. V. Faraone, T. C. Matise, D. Svrakic, J. Pepple, D. Malaspina, B. Sudrey, C. Hampe, C. T. Zambuto, K. Schmitt, J. Meyer, J. Merkel, H. Lee, J. Harkavy Friedman, C. A. Kaufmann, C. R. Cloninger, and M. T. Tsuang, "Genome Scan of European-American Schizophrenia Pedigrees: Results of the NIMH Genetics Initiative and Millennium Consortium," *American Journal of Medical Genetics* 81 (1998):290–295. |
| 10p15-p11 | 1998 | R. E. Straub, C. J. MacLean, R. B. Martin, Y. Ma, M. V. Myakishev, C. Harris-Kerr, B. T. Webb, F. A. O'Neill, D. Walsh, and K. S. Kendler, "A Schizophrenia Locus May Be Located in Region 10p15-p11," *American Journal of Medical Genetics* 81 (1998):296–301. |
| 10p14-p11 | 1998 | S. G. Schwab, J. Hallmayer, M. Albus, B. Lerer, C. Hanses, K. Kanyas, R. Segman, M. Borrman, B. Dreikorn, D. Lichtermann, M. Rietschel, M. Trixler, W. Maier, and D. B. Wildnauer, "Further Evidence for a Susceptibility Locus on Chromosome 10p14-p11 in 72 Families With Schizophrenia by Nonparametric Linkage Analysis," *American Journal of Medical Genetics* 81 (1998):302–307. |
| 15q14 | 1998 | S. Leonard, J. Gault, T. Moore, J. Hopkins, M. Robinson, A. Oliney, L. E. Adler, C. R. Cloninger, C. A. Kaufmann, M. T. Tsuang, S. V. Faraone, D. Malaspina, D. M. Svrakic, and R. Freedman, "Further Investigation of a Chromosome 15 Locus in Schizophrenia: Analysis of Affected Sibpairs From the NIMH Genetics Initiative," *American Journal of Medical Genetics* 81 (1998):308–312. |
| 14q13, 7q11, 22q11, 8p21, 13q32 | 1998 | J.-L. Blouin, B. A. Dombroski, S. K. Nath, V. K. Lasseter, P. S. Wolyniec, G. Nestadt, M. Thornquist, G. Ullrich, J. McGrath, L. Kasch, M. Lamacz, M. G. Thomas, C. Gehrig, U. Radhakrishna, S. E. Snyder, K. G. Balk, K. Neufeld, K. L. Swartz, N. DeMarchi, G. N. Papadimitriou, D. G. Dikeos, C. N. Stefanis, A. Chakravarti, B. Childs, D. E. Housman, H. H. Kazazian, S. E. Antonarakis, and A. E. Pulver, "Schizophrenia Susceptibility Loci on Chromosomes 13q32 and 8p21," *Nature Genetics* 20 (1998):70–73. |

pedigrees resulted in a reduction in the evidence for linkage in subsequent analyses.[25] Significant evidence of linkage to 6p of a quantitative trait of positive psychotic symptoms, but not schizophrenia per se, has been reported in Canadian pedigrees.[26] Suggestive evidence of 6p linkage has been reported in another data set,[27] but evidence is lacking in at least one other study;[28] and (b) the 13q32 linkage reported in 54 multiplex pedigrees by Blouin et al.[29] Less than suggestive linkage evidence, according to Lander and Kruglyak's guidelines, was found by Blouin et al. in another sample but evidence is lacking in at least one other study.[30] Suggestive evidence has been found for linkages to 5q,[31] 22q,[32], 8p,[33]

6q,[34] and 10p.[35] Additional analyses conducted in other data sets are needed to confirm which of these findings are real and to reduce in size the implicated targeted regions.

The National Institute of Mental Health (NIMH) launched a human genetics initiative in 1989 to collect family data for the linkage analysis of Alzheimer disease, schizophrenia, and bipolar disorder. The goal was to create a national resource of demographic, clinical, and diagnostic data, as well as DNA extracted from immortalized cell lines, that would be available for the scientific community. These data are available to qualified investigators conducting genetic research on schizophrenia. As of February 1999, clinical–diagnostic data and DNA samples are available on 866 affected and unaffected individuals; these constitute 128 families that contain 132 affected sib-pairs ("affected" is defined as *DSM-III-R* schizophrenia or schizoaffective disorder–depressive type). Another project has recently been funded to collect additional pedigrees to augment existing resources distributed by NIMH for genetic analyses by the scientific community. Such data may permit independent replication and extension of the findings discussed previously. Information about access to data and biomaterials, as well as descriptive information on the sample, is available on the World Wide Web at http://www-grb.nimh.nih.gov/gi.html.

### *Mood Disorders*

In the modern era of genetic studies of mood disorders, it is customary to follow Leonhard's suggestion and subdivide mood disorders into bipolar (BP) disorder—episodes of both mania and depression occur—and unipolar depressive (UP) disorder—only episodes of depression occur.[36] Perris[37] and Angst[38] applied Leonhard's distinction in the first family studies. Perris found a striking degree of homotypia, whereas Angst found an excess of both BP and UP disorders in the relatives of BP probands.

Subsequent work has consistently found that mood disorders aggregate in families; risks to relatives of UP and BP probands are significantly greater than risks to relatives of normal controls.[39] The risks to both BP and UP disorders are increased in the relatives of BP probands, whereas the first-degree relatives of UP probands have a higher rate of UP disorder only.[40] As many as 12% of patients with UP in the general population and as many as 74% of UP cases among relatives of BP probands may have a genetic predisposition to BP disorder.[41]

The risks to mood disorders in the first-degree relatives of UP and BP probands as compiled from 19 large-scale studies are as follows: Risk to BP in relatives of BP probands, 7.8%; Risk to BP in relatives of UP probands, 0.6%; risk to UP in relatives of BP probands, 11.4%; risk to UP in relatives of UP probands, 9.1%. Rates for UP disorder vary depending on the specific diagnostic criteria employed, as well as histories of multiple episodes, somatic treatment, or functional impairment.[42] A recent report of population-based epidemiologic studies using *DSM-III* criteria and similar methods from 10 countries[43] has found striking similarities across countries in patterns of mood disorders. The lifetime rate for bipolar disorder ranged from 0.3% to 1.5%, with equal risks to males and females. The lifetime rate of major depression varied widely by site, with a range from 1.5% to 19%; a higher risk to females, with a female : male ratio on the order of 2–3 : 1, was found in every country. Heterogeneity within the *DSM-III* major depression diagnostic category may partially account for this; a more genetically homogeneous subgroup may be delineated by requiring the criteria of early age of onset and recurrent episodes to define major depression.[44]

A MZ : DZ concordance ratio of more than 3 : 1 is obtained by pooling the results of nine twin studies (Table 5-4), and this strongly implicates genetic factors in the etiology of mood

Table 5-4—*Twin Studies in Mood Disorders*

| Location and publication year(s) | Monozygotic pairs: Concordance Pairs | Monozygotic pairs: Concordance Rate (%) | Dizygotic pairs: Condordance Pairs | Dizygotic pairs: Condordance Rate (%) |
|---|---|---|---|---|
| Combined figures[a] from six studies conducted in Germany, North America, United Kingdom, and Denmark (1930–1974) | 91 | 69 | 226 | 13 |
| Denmark (1977) | 69 | 67 | 54 | 20 |
| Norway (1986) | 37 | 51 | 65 | 20 |
| United Kingdom (1991) | 62 | 53 | 79 | 28 |
| Weighted mean concordance rate | — | 62 | — | 18 |
| Median concordance rate | — | 60 | — | 20 |

[a]E. S. Gershon, W. E. Bunney, J. F. Leckman, M. Van Eerdewegh, and B. A. DeBauche, ''The Inheritance of Affective Disorders: A Review of Data and of Hypotheses,'' *Behavior Genetics* 6 (1976):227–261.

disorders. As in schizophrenia, MZ concordance is less than 100% and implicates the involvement of nongenetic factors. The heritability of mood disorders is high (close to 0.8), as is the estimation of the contribution of shared environment effects to liability.[45]

Three adoption studies provide support for the involvement of genetic factors,[46] whereas two others are not supportive.[47] These inconsistencies are difficult to reconcile because of methodologic differences among studies; however, the two most methodologically sound provide strong support for genetic factors.[48]

Early pedigree analyses yielded evidence for vertical familial transmission, but results in general were not consistent with inheritance under a single major gene.[49] Although recent studies suggest that familial transmission of BP disorder is consistent with autosomal-dominant single major-locus inheritance,[50] there are inconsistencies across families studied. Likewise, segregation analyses have not produced clear results on the familial transmission of UP.[51] In conclusion, the mode of transmission of mood disorders is complex and remains unresolved.

Few linkage studies of UP have been conducted, and a locus that exclusively influences risk to UP disorder has not been identified.[52]

Several reports[53] produced sizeable lod scores linking biopolar disorder to the X chromosome with color blindness and G6PD deficiency.[54] However, methodologic criticisms have been raised about many of the earlier studies, and multiple failures to replicate have been reported.[55] A lod score of 7.5 for X chromosome linkage was obtained in a methodologically rigorous study of five Israeli families,[56] however, additional subsequent analysis in these families led to a diminution of evidence.[57] Linkage to 11p was reported in an analysis of Amish family data,[58] but the lod score (4.9) was diminished to nonsignificance when pedigrees were extended and members reevaluated.[59]

Suggestive evidence was found for linkage to 18p.[60] Another report found suggestive evidence for 18p linkage and also found significant linkage evidence to another region on 18q; both findings were in 11 ''paternally transmitting'' pedigrees only (i.e., probands' fathers or uncles were affected).[61] Analysis of the full data set resulted in less than suggestive evidence for 18q linkage and suggestive evidence for 18p linkage. The interpretation of these results is difficult, given that evidence has been presented for a maternal effect in the transmission of bipolar disorder in these[62] and other[63] families, and given that linkage evidence in this sample[64] is highly dependent on which age correction is employed.[65] Suggestive evidence for a locus on 18q was obtained in association analyses;[66] however, the implicated region was over a 5 Megabase (Mb) region and other markers in between provided evidence against linkage. This region implicated was at least 15 Mb away from the 18q region for which significant linkage evidence was reported previously.[67] Thus an 80 Mb region encompassing most of both arms of chromosome 18 has been implicated. Several nonreplications of chromosome 18 linkage have been reported.[68]

Significant evidence of linkage to 21q was found in 1 of 47 bipolar families, but with analysis of the entire sample resulting in only suggestive evidence.[69] Suggestive[70] and less than suggestive[71] evidence has been reported in other samples; unfortunately, the strongest evidence of linkage in Detera-Wadleigh et al.'s study (1996) was to a region more than 15 Mb away from that implicated earlier.[72] Three nonreplications have been published.[73]

Significant evidence for linkage to 4p was reported in a single pedigree,[74] but a failure to replicate occurred in other pedigrees from the same population. Suggestive evidence has been reported for linkages to other chromosomal regions: 4q,[75] 5p,[76] 6p,[77] 10q,[78] 11p,[79] 12q,[80] 16p,[81] and 22q.[82]

In summary, the strongest linkage evidence to date is consistent with susceptibility loci for biopolar disorder on chromosomal regions 18p, 18q, and 21q; however, the methodologic issues discussed previously and the existence of nonreplication demonstrate that these are clearly not confirmed findings. The inability to obtain more compelling evidence may have resulted because: (a) genes on 18 and 21 confer susceptibility to biopolar disorder, but they have such a small relative effect on risk that a very large sample is required for detection; (b) genes on 18 and 21 confer susceptibility in a small number of families (failures to replicate reflect the confounding effects of genetic heterogeneity); or (c) the reported positive results are a result of chance. Evidence for linkages to other autosomes (4, 5, 6, 10, 11, 12, 16, 22) and the X chromosome is less compelling. Additional analyses conducted in other data sets are needed to confirm which of these findings are real and to reduce in size the implicated targeted regions.

The clinical–diagnostic data and DNA samples from BP families collected in the NIMH Genetics Initiative are now available to qualified investigators conducting genetic research on mood disorders. As of February 1999, clinical–diagnostic data and DNA samples are available on 931 affected and unaffected individuals; these constitute 153 families that contain 148 affected sib-pairs (''affected'' is defined as *DSM-III-R* bipolar disorder or schizoaffective disorder–bipolar type). Two projects have recently been funded to collect additional pedigrees to augment existing resources distributed by NIMH for genetic analyses by the scientific community. Such data may permit independent replication and extension of the findings discussed previously. Information about access to data and biomaterials, as well as descriptive information on the sample, is available on the World Wide Web at http://www-grb.nimh.nih.gov/gi.html.

## *Alcoholism*

Alcoholism is a strongly familial trait, and risks to first-degree relatives of alcoholics are approximately four to five times that of the general population.[83] When rigorous experimental methods and standardized instruments are used, comparable lifetime cumulative incidence rates of alcoholism are observed across populations. Helzer et al.[84] compared the epidemiology of *DSM-III*-defined alcohol dependence cross nationally in five populations; respective weighted mean rates for men and women are 12% and 2%. The rate of alcoholism is as much as seven times higher in relatives of alcoholics than in controls in some studies,[85] with risks to male relatives greater than risk to female relatives. For example, Reich and Cloninger[86] reported morbid risks of 15% to the sisters and 57% to the brothers of male alcoholic probands; comparable lifetime cumulative general population incidences for alcohol abuse in their study are more than 20% for males and 5% for females. Given that concordances in half-sibs compared to full sibs are comparable[87]—in other words, there is as much resemblance in pairs of relatives sharing one quarter of their genes as in pairs who share one half—family data suggest that environmental effects play an etiologic role.

Exclusion of a twin study that shows no genetic influence for males and females[88] and a study that found 0% heritability in females[89] results in respective MZ and DZ weighted mean concordances of 45% and 25% for men and 35% and 24% for women in seven other twin studies of alcohol abuse or dependence. Despite diagnostic differences, variations in concordance rates and one small study showing no evidence for a genetic effect in males or females,[90] these twins studies generally support the involvement of genetic factors in the etiology of broadly defined alcohol abuse or dependence; however, the MZ : DZ ratio of less than 2 : 1 suggests that genetic factors play a lesser role in the etiology of alcoholism than in the etiologies of schizophrenia or mood disorders. A MZ:DZ concordance ratio of more than 3 : 1 was obtained in one study[91] by using a narrow definition of alcoholism.

Twin studies from Scandinavia, the United Kingdom, the United States, and Australia provide evidence consistent with a substantial genetic contribution to variation in levels of alcohol intake: As much as 67 to 80% of the longitudinally stable proportion of the variation in normal drinking patterns may be accounted for by genetic factors.[92]

Heritability to alcoholism is likely highly specific.[93] Heritability estimates are moderate in men[94] and much lower in women.[95] In recent studies of female same-sex twins from a population-based twin study, Kendler and colleagues[96] have estimated heritability for alcoholism in women to be higher (0.51 to 0.59) and have concluded that liability to alcoholism in women is a result largely of genetic factors. Likewise, data analyses from a large population-based twin registry[97] showed that genetic factors accounted for 58% of the variation in alcohol consumption in females and 45% of the variation in males, with low to moderate heritability estimates for problem drinking under different models for both women (.08–0.44) and men (0.1–0.5).

Although the environmental consequences of parental alcoholism do not appear to influence the risk for alcoholism,[98] other environmental effects may play a role. Childhood parental loss has been shown to be a direct and significant environmental risk factor for alcoholism in women.[99]

Genetic factors likely play a greater role in the transmission of alcohol dependence than in the transmission of alcohol abuse; for example, respective heritabilities for alcohol dependence and alcohol abuse in one study are 0.6 and 0.4 for males and 0.4 and 0.0 for females.[100]

Adoption studies have shown that male and female adoptees with an alcoholic biologic parent are at increased risk for alcoholism.[101] However, one study[102] found that alcoholism was only increased in the daughters of female alcoholics.

Segregation analysis has provided weak evidence for a single major-gene model, but also evidence for nongenetic transmission.[103] A molecular genetic study of alcoholism yielded weak evidence for linkage to the MNS blood group on chromosome 4,[104] but this finding has not been replicated. More recent interest has focused on a *Taq*-I polymorphism at the dopamine $D_2$ receptor (DRD2). An allelic association between the *Taq*-I "A" system A1 allele and either alcoholism or severe alcoholism has been proposed.[105] An analysis of accumulated evidence from the original report and all subsequent studies through 1993 showed (a) no significant difference in DRD2 A1 allele frequency between alcoholics and controls; (b) significant heterogeneity among reported alcoholics and controls; and (c) no significant difference in DRD2 A1 allele frequency between severe and nonsevere alcoholics.[106] Sampling error and ethnic variation has been posited as the most likely causes of the reported positive finding.

The National Institute on Alcohol Abuse and Alcoholism (NIAAA) has funded a large six-center collaborative project to ascertain families with multiple members affected with alcoholism for genetic studies. Recent results of a genomic survey conducted in 987 research participants from 105 pedigrees ascertained for this project yielded suggestive evidence for linkage of alcoholism to chromosomal regions 1p, 4p, and 7p.[107] Suggestive evidence of linkages to 4p and 11p has been reported in a sample from a Southwest American Indian tribe.[108] An analysis in the six-center collaborative project of the P3 event-related brain potential, a risk factor for alcoholism, yielded significant evidence for linkage of this quantitative trait to chromosomes 2q and 6q.[109] Additional analyses conducted in other data sets are needed to confirm which of these findings are real and to reduce in size the implicated targeted regions.

### *Antisocial Personality Disorder and Adult Criminality*

Antisocial personality disorder (ASPD) is a relatively common disorder. The best estimate of lifetime cumulative incidence in the U.S. population was obtained by including extra questions about illegal activities for the St. Louis sample in the Epidemiologic Catchment Area Study;[110] the estimate was 7.3% in males and 1.0% in females. The diagnosis of ASPD is not simply a medicalization of criminality, given that only 47% of those with ASPD have a history of multiple nontraffic arrests and given that only 37% of those with a history of multiple nontraffic arrests have ASPD.[111]

Results of several twin studies of adult criminality have been pooled to give mean pair-wise concordance rates for male adult criminality of 51% in MZ twins and 23% in DZ twins.[112] Heritabilities for adult criminality in men and women are moderate (0.5), as is the estimation of the effect of shared environment.[113] These data are consistent with the significant involvement of genetic factors in adult criminality. This is contrasted with the pooled results for juvenile delinquency,[114] where pair-wise concordance rates of 91% for MZ twins and 73% for DZ twins suggest that genetic factors are of little importance. Comparable results have been obtained from a recent large study of antisocial traits in male twins.[115] More than 3000 pairs of male twins from the Vietnam Era Twin registry were studied, and the results showed that twin resemblance for juvenile antisocial traits (retrospectively reported in adulthood) was largely a result of the shared familial environment, with genetic factors making only a slight contribution. The picture was reversed when adult antisocial traits were examined: Twin resemblance was almost entirely genetic, with the familial environment having only a very modest impact. This pattern (a strong impact of familial environment during childhood and adolescence that weakens in adulthood) is common for many behavioral characteristics.[116]

Despite methodologic differences, adoption studies have been very consistent in supporting the influence of genetic factors on adult antisocial behavior and criminality.[117] Adoption study findings also suggest that criminality is a heterogeneous group of behaviors and that genetic factors differentially influence the type of crime committed: Recividist property offenses such as petty theft are particularly heritable, whereas violent crimes may be secondarily associated or interactive with a genetic predisposition to alcoholism.[118] It is likely that both antisocial behavior and alcohol abuse—to the extent that genetic factors are contributory—share a common diathesis in some cases and in other cases have independent diatheses. Furthermore, both are likely to be genetically heterogenous.

A highly publicized genetic–metabolic finding related to antisocial behaviors and criminality comes from the study of a large Dutch kindred in which several males were affected with a syndrome of borderline mental retardation and abnormal behaviors that included impulsive aggressive outbursts, arson, attempted rape, and exhibitionism.[119] The syndrome was related to a complete and selective deficiency of enzymatic activity of monoamine oxidase A (MAOA). So far, this is the only reported family in which this relationship has been identified.

Overall, it is likely that there exists partially genetically influenced predispositions for basic behavioral tendencies—such as impulsivity—that in certain environments increase the probability of committing certain crimes and engaging in other antisocial behaviors.[120] Single gene effects, such as the MAOA mutation observed in one Dutch family,[121] will likely be very rare, and the role of shared environmental effects and multiple genetic susceptibility loci are likely to be more prototypic.

## Complexities in Unravelling the Genetic Basis of Mental Disorders

The evidence reviewed in this chapter documents the role of genetic factors in the etiology of several mental disorders. While the influence of genes is greater on some conditions (schizophrenia, bipolar disorder) than on others (alcoholism, adult antisocial behaviors), and a single major gene does not contribute to disease susceptibility for most cases. Rather, multiple loci—each of small relative effect—are the most likely genetic susceptibility factors for each mental disorder. Reduced penetrance, etiologic heterogeneity, diagnostic imprecision/error, the involvement of multiple susceptibility loci of small relative effect, the impact of shared and unshared environment, and the existence of nongenetic cases are complications in genetic analysis that likely contribute to further obscure the relationship between genotype and phenotype.

Linkage analyses have so far failed to convincingly identify a well-demarcated chromosomal region as containing a susceptibility locus for any mental disorder. Given that there are likely multiple susceptibility loci of small relative effect, this is not surprising. For example, Suarez et al.[122] found that although detection of an individual QTL for a disease influenced $N$ QTLs is feasible (where $4 \leq N \leq 10$), replication will likely require a sample $N - 1$ times larger than that required for the first detection. Clearly, sufficiently large replication studies have not been employed in genetic analyses of mental disorders.

Chromosomal localization is only the first step in the positional cloning process. An identified 10 to 20 Mb region is large enough to contain hundreds of genes. Although application of molecular genetic methods to study simple monogenic diseases has led to resounding successes as genes have been detected, isolated, cloned, and characterized, the challenge is considerable. For example, it is noteworthy that it took 10 years from the time of establishment of linkage to identification of the disease gene for Huntington's disease.[123]

Typically, successful cloning efforts have depended on the fortuitous presence of chromosomal aberrations, trinucleotide repeat expansions, or previously known candidate genes. Positional cloning is likely to be even harder for complex traits like mental disorders, where cosegregation is far from perfect and the genes involved may be subtle mutations and not gross deletions.[124]

The Human Genome Project promises to make a tremendous contribution to positional cloning efforts for mental disorders by eventually providing a complete catalog of all genes in a relevant region. Continued advances in molecular and statistical genetics will further facilitate identification of susceptibility loci for mental disorders. For example, high-throughput genotyping and full genomic scans have led to identification of putative susceptibility loci for other genetically complex diseases—for example, type 1 insulin-dependent diabetes mellitus[125] and multiple sclerosis.[126]

The next paradigmatic advance in genetic analysis is currently occurring and involves high-density silicon DNA arrays.[127] Potentially hundreds of thousands of oligonucleotides can be synthesized on a single chip. This technology will revolutionize genotyping, DNA sequencing, and mutation analysis as applied to the genetic analysis of mental disorders, and will permit a full understanding of subtleties in the expression, function, and regulation of all human genes under various conditions. The utility of DNA microchip arrays in screening individuals for mutations and in studying expression patterns in breast cancer has been demonstrated.[128]

*Single nucleotide polymorphisms,* or SNPs, are strips of DNA that vary at a single base pair. SNPs are the most common DNA sequence variations found in the human genome,[129] and will form the basis of a new map of genetic markers currently being developed for extensive use in future genetic analyses.[130] A dense panel of SNPs from such a map may be used in future studies to identify associations and narrow chromosomal regions across the genome that may contain a locus influencing susceptibility to a mental disorder. The full potential of SNPs for forensic applications has yet to be realized.

The analysis of biologic traits correlated with liability to illness offers another method for increasing the power to detect susceptibility loci. Simulation studies have shown that simultaneous analysis of correlated trait data and affection status, as well as use of such data in sampling schemes, enhanced the ability to detect QTLs for a genetically complex disease.[131]

Although it is now premature to talk of identifiable susceptibility genes for mental disorders, their ultimate identification will permit realization of several benefits for clinical practice: (a) diagnosis and risk prediction in genetic counseling scenarios will be enhanced; (b) gene products can be identified, and biochemical and pathophysiological bases of diseases elucidated; (c) new medications and other therapeutic agents can be developed; and (d) gene-replacement therapy may be implemented as a preventive intervention.

Realization of these benefits requires cloning of the disease gene and development of a test to screen for alternations or mutations in that gene. Use of such a laboratory test for susceptibility gene detection in mental disorders requires careful consideration of a variety of scientific and ethical issues, given that (a) an identified susceptibility locus may be one of many and only confer a small increase in risk; (b) the influence of a susceptibility locus may be seen in only a small number of families; (c) shared and unshared environmental effects will continue to play an important role in influencing eventual phenotypic outcomes.

The ambiguity of results will undoubtedly limit the utility of such a test. Comparable examples exist for simple monogenetic diseases. Polymerase chain reaction (PCR)-based laboratory methods have been developed to detect pathological cytosine-adenine-guanine (CAG) expansions near the end of the IT-15 gene in Huntington's disease; however, use of PCR primers that amplify both the CAG repeat and a polymorphic cytosine-cytosine-guanine (CCG) repeat sequence six trinucleotides from the CAG repeat can result in errors in Huntington's disease diagnostic testing.[132] Although there is a strong association between the apolipoprotein E (APOE) ● 4 allele and Alzheimer's disease (AD), AD develops in the absence of APOE ● 4 and many with APOE ● 4 escape disease; lack of sufficient specificity or sensitivity has led to the recommendation that APOE genotyping not be used for predictive genetic testing.[133] Likewise, the presence of modifying factors and the fact that the known mutations in the BRCA1 and BRCA2 genes do not account for all cases of disease make susceptibility testing for breast cancer an uncertain endeavor.[134] The availability of commericially available tests for the BRCA1 185delAG mutation—despite the ambiguity of both positive and negative test results—demonstrates the commercial realities that will likely occur in genetic testing for mental disorders and that will likely precede a full understanding of the meaning of test results.

## Personal Responsibility and Genetic Determinism

The potential application and development of genomic technologies and knowledge to the diagnosis, prevention, and treatment of mental disorders raises a variety of issues. One of the most important is the impact on our notions of individual autonomy and legal liability.

It is crucial to acknowledge that genes influence predispositions for complex behaviors that, under certain environmental conditions, make the probability of behavioral outcomes more or less likely. There is no one gene that in itself "causes" a specific complex behavior; the path from gene to behavior is an indirect and complex one via proteins and physiological systems. In addition, genes do not determine destiny, and mental disorders with a genetic component may be bypassed or remediated by environmental interventions. These points have been made regarding the misinterpretation of the state of knowledge and the misuse of genetics to inform social policy in a controversial book on intelligence.[135] Prediction of risk to mental illness based on a single genetic risk factor will be further complicated by the fact that relevant shared or nonshared environmental risk factors for mental disorders elude precise detection and quantification.

Criminal law presumes that behavior is a consequence of free will. Although genetic factors do not directly determine complex behavioral outcomes, they do set a range within which the outcome will occur. The subtleties of genetic influences are reflected in the limits of possible phenotypic outcomes that are determined by a particular genotype. Although the limits of this "reaction range"[136] are set by genes and this reflects an element of determinism (one cannot choose one's genes), the actual outcome within the range is determined by developmental and environmental modifiers. Free will and personal responsibility are operative to the point that one can choose to avoid or not to avoid a set of environmental circumstances leading to a particular behavioral outcome that falls within the reaction range set by one's genetic endowment.

An important question related to genetic determinism is a variant of the traditional question: Is the genetic predisposition so strong that an impulse to commit a particular behavior could not be resisted?[137] The relevant issues are (a) the relationship between the predisposition and a specific behavioral outcome, and (b) the irresistibility of the predisposition. These questions can be indirectly addressed by determining the conditional probability of an individual actually displaying a specific behavioral outcome *given* the presence of a particular finding on a genetic test (which establishes that such a predisposition is present). This is the positive predictive power (PPP)[138] of that genetic defect; negative predictive power (NPP)[139] is the corresponding probability of not having an outcome given the *absence* of the genetic defect. (PPP and NPP are very different from *specificity* and *sensitivity,* which are the respective conditional probabilities of having [or not having] a genetic defect given the presence [or absence] of the behavioral outcome.)

Is genetic predisposition deterministic and thus a consideration in the establishment of a defense or of mitigating factors? Consider the example of a capital crime case, in which a defendant with schizophrenia is accused of homicide. Imagine that a linkage finding in schizophrenia has been confirmed and a cloned susceptibility gene in this region forms the basis of a genetic test that identifies a moderately high number of affected cases. The relevance of a positive result in this criminal case will depend on empirical data provided by the scientific community: (a) the PPP of the genetic test result for predicting schizophrenia—the probability of having schizophrenia *given* that you have a positive test result; and (b) the PPP of schizophrenia for predicting homicide—the probability of committing homicide *given* that you have schizophrenia. If these PPPs are moderately high, one can argue that the genetic predisposition to schizophrenia may confer an impulse to homicide that is so strong that it is virtually irresistible. In this regard, genetic predisposition for a mental disorder could be used to establish a defense or mitigating factor. (The PPP of a positive genetic test result for directly predicting a behavioral outcome is not considered, because it is highly improbable that a genetic defect exists that can *directly* influence a complex behavior of relevance to the criminal justice system.)

However, if the PPP specified in the second question is low, and only a very small number of individuals with schizophrenia actually commit murder, it is clear that for most affected individuals this genetic predisposition does not confer an irresistible impulse to commit homicide. The PPP of two indicators may be then considered—for example, the probability of committing homicide given the presence of a positive test result *and* the ingestion of phencyclidine (PCP) at the time of the crime. High doses of this drug frequently induce behavioral disinhibition and aggressive outbursts, which in combination with schizophrenia may increase the chances of engaging in homicidal behavior.

The individual's knowledge that PCP ingestion is an environmental risk factor that sufficiently "aggravates" his enhanced genetic liability to develop schizophrenia is the crucial consideration. With prior knowledge, one may argue that the individual's willful decision to engage in behavior that would result in the triggering of illness and the subsequent state of behavioral inhibition precludes defense's argument for genetic predisposition as a mitigating circumstance. However, the *absence* of prior knowledge that willful PCP ingestion would interact with genetic susceptibility to produce illness and behavioral disinhibition is consistent with the argument that genetic predisposition is a mitigating circumstance.

Denno[140] recently discussed the case of *Mobley v. State.*[141] The attorney of accused murderer Stephen Mobley requested that he be allowed to present evidence that the crime could be attributable to the defendant's genetic makeup. The goal was not to use this to establish a defense but to establish the existence of mitigating factors during the penalty phase of the trial. Brunner et al.'s study[142] was introduced into evidence, and the defense requested expert assistance and finances to perform the necessary analyses and determine if Mobley had a point mutation in the structural gene that regulated production of MAOA. Despite the tendency of courts' increasingly liberal acceptance of biological and psychological evidence to justify defenses,[143] the State in *Mobley* denied the request on two grounds: (a) Mobley did not have borderline mental retardation, which was a characteristic of the aggressive syndrome identified in a Dutch family by Brunner et al.[144]; and (b) the direct effect of the MAOA defect on aggressive behavior is unclear, as stated by Brunner et al. "It is presently unclear whether all of the biochemical alterations caused by the MAOA deficiency state are required to cause the apparent increase in liability to impulsive aggressive behavior. The inhibition of MAO has not been reported to cause aggressive behavior in adult humans but deficiencies throughout life might have different consequences."[145]

Table 5-5—*Reported Chromosomal Locations for Bipolar Disorder Susceptibility Loci (1994–1996)*

| Location | Year | Reference |
|---|---|---|
| 21q22 | 1994 | R. E. Straub, T. Lehner, Y. Luo, J. E. Loth, W. Shao, L. Sharpe, J. R. Alexander, K. Pas, R. Simon, and R. R. Fieve, ''A Possible Vulnerability Locus for Bipolar Affective Disorder on Chromosome 21q22.3,'' *Nature Genetics* 8 (1994):291–296. |
| 12q23 | 1994 | N. Craddock, M. Owen, S. Burge, B. Kurian, P. Thomas, and P. McGuffin, ''Familial Cosegregation of Major Affective Disorder and Darier's Disease (Keratosis Follicularis),'' *British Journal of Psychiatry* 164 (1994):355–358. |
| 18p11 | 1994 | W. H. Berrettini, T. N. Ferraro, L. R. Goldin, D. E. Weeks, S. Detera-Wadleigh, J. I. Nurnberger, Jr., and E. S. Gershon, ''Chromosome 18 DNA Markers and Manic-Depressive Illness: Evidence for a Susceptibility Gene,'' *Proceedings of the National Academy of Sciences of the USA* 91 (1994):5918–5921; W. H. Berrettini, T. N. Ferraro, L. R. Goldin, S. D. Detera-Wadleigh, H. Choi, D. Muniec, J. J. Guroff, D. M. Kazuba, J. I. Nurnberger, W.-T. Hsieh, M. R. Hoehe, and E. S. Gershon, ''A Linkage Study of Bipolar Illness,'' *Archives of General Psychiatry* 54 (1997): 27–35; D. Blacker, and M. T. Tsuang, ''Unipolar Relatives in Bipolar Pedigrees: Are They Bipolar?'' *Psychiatric Genetics* 3 (1993):5–16. |
| Xq24-26 | 1995 | P. Pekkarinen, J. Terwilliger, P.-E. Bredbacka, J. Lonnqvist, and L. Peltonen, ''Evidence of a Predisposing Locus to Bipolar Disorder on Xq24-q27.1 in an Extended Finnish Pedigree,'' *Genome Research* 5 (1995):105–115. |
| 18p12-18q11 | 1995 | O. C. Stine, J. Xu, R. Koskela, F. J. McMahon, M. Gschwend, C. Friddle, C. D. Clark, M. G. McInnis, S. G. Simpson, and T. S. Breschel, ''Evidence for Linkage of Bipolar Disorder to Chromosome 18 With a Parent-of-Origin Effect,'' *American Journal of Human Genetics* 57 (1995):1384–1394. |
| 16p13 | 1995 | H. Ewald, O. Mors, T. Flint, K. Koed, H. Eiberg, and T. A. Kruse, ''A Possible Locus for Manic Depressive Illness on Chromosome 16p13,'' *Psychiatric Genetics* 5 (1995):71–81. |
| 21q22 | 1995 | H. Gurling, C. Smyth, G. Kalsi, E. Moloney, L. Rifkin, J. O'Neill, P. Murphy, D. Curtis, H. Petursson, and J. Brynjolfsson, ''Linkage Findings for Bipolar Disorder,'' *Nature Genetics* 10 (1995):8–9; C. Smyth, G. Kalsi, D. Curtis, J. Brynjolfsson, J. O'Neill, L. Rifkin, E. Moloney, P. Murphy, H. Petursson, and H. Gurling, ''Two-Locus Admixture Linkage Analysis of Bipolar Affective and Unipolar Affective Disorder Supports the Presence of Susceptibility Loci on Chromosomes 11p15 and 21q22,'' *Genomics* 39 (1997):271–278; A. Malafosse, M. Leboyer, T. d'Amato, S. Amadeo, M. Abbar, D. Campion, O. Canseil, D. Castelnau, F. Gheysen, B. Branger, B. Henrikson, M.-F. Poirier, O. Sabate, D. Samolyk, J. Feingold, and J. Mallet, ''Manic Depressive Illness and Tyrosine Hydroxylase Gene: Linkage Heterogeneity and Association,'' *Neurobiology of Disease* 4 (1997):337–349. |

Table 5-5—*Continued*

| Location | Year | Reference |
|---|---|---|
| 18q23-22 | 1996 | N. B. Freimer, V. I. Reus, M. A. Escamilla, A. McInnes, M. Spesny, P. Leon, S. K. Service, L. B. Smith, S. Silva, E. Rojas, A. Gallegos, L. Meza, E. Fournier, S. Baharloo, K. Blankenship, D. J. Tyler, S. Batki, S. Vinogradov, J. Weissenbach, S. H. Barondes, and L. A. Sandkuijl, "Genetic Mapping Using Haplotype, Association and Linkage Methods Suggests a Locus for Severe Bipolar Disorder (BPI) at 18q22-q23," *Nature Genetics* 12 (1996):436–441. |
| 4p16 | 1996 | D. H. R. Blackwood, L. He, S. W. Morris, A. McLean, C. Whitton, M. Thomson, M. T. Walker, K. Woodburn, C. M. Sharp, A. F. Wright, Y. Shibasaki, D. M. St. Clair, D. J. Porteous, and W. J. Muir, "A Locus for Bipolar Affective Disorder on Chromosome 4p," *Nature Genetics* 12 (1996):427–430. |
| 6p24, 13q13, 15q11 | 1996 | E. I. Ginns, J. Ott, J. A. Egeland, C. R. Allen, C. S. J. Fann, D. L. Pauls, J. Weissenbach, J. P. Carulli, K. M. Falls, T. P. Keith, and S. M. Paul, "A Genome-Wide Search for Chromosomal Loci Linked to Bipolar Affective Disorder in the Older Order Amish," *Nature Genetics* 12 (1996):431–435. |
| 5p | 1996 | J. R. Kelsoe, A. D. Sadovnick, H. Kristbjarnarson, P. Bergesch, Z. Mroczkowski-Parker, M. Drennan, M. H. Rapaport, P. Flodman, M. A. Spence, and R. A. Remick, "Possible Locus for Bipolar Disorder Near the Dopamine Transporter on Chromosome 5," *American Journal of Medical Genetics* 67 (1996):533–540. |
| 21q | 1996 | S. Detera-Wadleigh, J. A. Badner, L. R. Goldin, W. Berrettini, A. R. Sanders, D. Y. Rollins, G. Turner, T. Moses, H. Haerian, D. Muniec, J. I. Nurnberger, and E. S. Gershon, "Affected-Sib-Pair Analyses Reveal Support of Prior Evidence for a Susceptibility Locus for Bipolar Disorder, on 21q," *American Journal of Human Genetics* 58 (1996):1279–1285. |
| 11p15 | 1997 | C. Smyth, G. Kalsi, D. Curtis, J. Brynjolfsson, J. O'Neill, L. Rifkin, E. Moloney, P. Murphy, H. Petursson, and H. Gurling, "Two-Locus Admixture Linkage Analysis of Bipolar Affective and Unipolar Affective Disorder Supports the Presence of Susceptibility Loci on Chromosomes 11p15 and 21q22," *Genomics* 39 (1997):271–278; A. Malafosse, M. Leboyer, T. d'Amato, S. Amadeo, M. Abbar, D. Campion, O. Canseil, D. Castelnau, F. Gheysen, B. Granger, B. Henrikson, M.-F. Poirier, O. Sabate, D. Samolyk, J. Feingold, and J. Mallet, "Manic Depressive Illness and Tyrosine Hydroxylase Gene: Linkage Heterogeneity and Association," *Neurobiology of Disease* 4 (1997):337–349. |
| 11p15 | 1997 | A. Malafosse, M. Leboyer, T. d'Amato, S. Amadeo, M. Abbar, D. Campion, O. Canseil, D. Castelnau, F. Gheysen, B. Granger, B. Henrikson, M.-F. Poirier, O. Sabate, D. Samolyk, J. Feingold, and J. Mallet, "Manic Depressive Illness and Tyrosine Hydroxylase Gene: Linkage Heterogeneity and Association," *Neurobiology of Disease* 4 (1997):337–349. |

Table 5-5

| Location | Year | Reference |
|---|---|---|
| 10q | 1997 | J. P. Rice, A. Goate, J. T. Williams, L. Bierut, D. Door, W. Wu, S. Shears, G. Gopalakrishnan, H. J. Edenberg, T. Foroud, J. T. Nurnberger, E. S. Gershon, S. D. Detera-Wadleigh, L. R. Goldin, J. J. Guroff, F. J. McMahon, S. Simpson, D. MacKinnon, M. McInnis, O. C. Stine, J. R. DePaulo, M. C. Blehar, and T. Reich, "Initial Genome Scan of the NIMH Genetics Initiative Bipolar I Pedigrees: Chromosomes 1, 6, 8, 10, and 12," *American Journal of Medical Genetics* 74 (1997):247–253; M. A. Spence, P. L. Flodman, A. D. Sadovnick, J. E. Bailey-Wilson, H. Ameli, R. A. Remick, "Bipolar Disorder: Evidence for a Major Locus," *American Journal of Medical Genetics* 60 (1995):370–376; D. L. Pauls, J. N. Bailey, A. S. Carter, C. R. Allen, and J. A. Egeland, "Complex Segregation Analyses of Old Order Amish Families Ascertained Through Bipolar I Individuals," *American Journal of Medical Genetics* 60 (1995):290–297. |
| 22q | 1997 | H. M. Lachman, J. R. Kelsoe, R. A. Remick, A. D. Sadovnick, M. H. Rapaport, M. Lin, B. A. Pazur, A. M. A. Roe, T. Saito, and D. F. Papolos, "Linkage Studies Suggest a Possible Locus on Bipolar Disorder Near the Velo-Cardio-Facial Syndrome Region on Chromosome 22," *American Journal of Medical Genetics* 74 (1997):121–128. |
| 6p24-22 | 1997 | M. Maziade, L. Bissonnette, E. Rouillard, M. Martinez, M. Turgeon, L. Charron, V. Pouliot, P. Boutin, D. Cliche, C. Dion, J. P. Fournier, Y. Garneau, J. C. Lavallee, N. Montgrain, L. Nicole, A. Pires, A. M. Ponton, A. Potvin, H. Wallot, M. A. Roy, le groupe IREP, and C. Merrette, "6p24-22 Region and Major Psychoses in the Eastern Quebec Population," *American Journal of Medical Genetics* 74 (1997):311–318. |
| 4q35 | 1998 | L. J. Adams, P. B. Mitchell, S. L. Fielder, A. Rosso, J. A. Donald, and P. R. Schofield, "A Susceptibility Locus for Bipolar Affective Disorder on Chromosome 4q35," *American Journal of Human Genetics* 62 (1998):1084–1091. |
| 11q23, 9p24 | 1998 | B. E. Baysal, S. G. Potkin, J. E. Farr, M. J. Higgins, J. Korcz, S. M. Gollin, M. R. James, G. A. Evans, and C. W. Richard, "Bipolar Affective Disorder Partially Cosegregates With a Balanced t(9;11)(p24;q23.1) Chromosome Translocation in a Small Pedigree," *American Journal of Medical Genetics* 81 (1998):81–91. |

The state court's ruling appears not so much a repudiation of genetic evidence as a mitigating factor in a capital crime case, but rather a refutation of the genetic finding as an adequate predictor of the eventual behavioral outcome. This reflects the low PPP of MAOA deficiency for predicting aggressive behavior.

In summary, establishment of a defense based on a genetic finding may be difficult, even when predictive testing is available in the future. This in part reflects the lack of simple correspondence between genotype and phenotype typical of mental disorders. Even in cases in which genetic susceptibility is very high, and genetic tests are developed for multiple

Table 5-6—*Proposed Guidelines for the Use of Genetic Information About Mental Disorders in the Courtroom*

- Prior to the identification of a specific genetic factor, if the genetic information to be introduced is presence of a family history then the recurrence risk for the subject in a given pedigree is high.
- Acceptable scientific techniques are used to test for the presence of genetic factors conferring susceptibility to mental disorders (once they are identified) and to properly interpret results.
- Scientific data exists to interpret how the risk conferred by the genetic factor varies by ethnicity or racial origin.
- Consideration is given to how the genetic factor interacts with other genetic or nongenetic factors.
- The positive and negative predictive power of the genetic factor for predicting a given mental disorder is moderately high to high in several independent scientific studies.
- The positive and negative predictive power of that mental disorder for predicting a particular behavioral outcome is moderately high to high in several independent scientific studies.

susceptibility loci, the role of shared and nonshared environmental factors and etiologic heterogeneity will likely ensure that a specific genetic risk factor or factors will not predict with very high probability a given disorder. Given that no genetic abnormality will likely have high PPP to directly predict a particular behavior of interest to the criminal justice system, the relationship between genetic predisposition and eventual behavioral outcome is further weakened. However, given that the presence of specific mental disorders may increase the likelihood of engaging in particular behaviors of relevance to the criminal justice system, establishment of genetic susceptibility to that disorder may be used to establish mitigating circumstances.

## Conclusion

The question is not whether genetic evidence of predisposition to a mental disorder will ever be admitted into the courtroom, but when and under what conditions. Table 5-6 presents a set of proposed guidelines for the use of genetic evidence in the courtroom. Determination by the legal and scientific community of what constitutes "moderately high" PPP or "strong or convincing" evidence will of course fluctuate as legal standards change and as genetic research on mental disorders continues and replicated findings emerge.

Replication of a finding in independent data sets is an important consideration. Consistency between the genetic finding and available evidence from molecular and nonmolecular studies is also very important. A primary consideration is the PPP of the genetic finding for a mental disorder and the PPP of that mental disorder for a particular behavioral outcome. As discussed earlier, the PPP of a positive genetic test result for directly predicting a complex behavior is not considered, because it is highly unlikely that such a genetic factor exists. This was a problem in *Mobley,* in which a direct relationship between a biochemical abnormality (MAOA) and a complex behavioral outcome (homicide) was posited.

The breathtaking speed at which dense genetic maps are becoming available and increasingly sophisticated quantitative methods of genetic analysis are being applied to ever-larger data sets suggest that scientifically acceptable findings on the genetic basis of mental disorders eventually will be generated. Given that genetic and other comparable biologic evidence will have a major impact on courts as this information becomes rapidly disseminated within the legal and scientific communities, it is essential that proactive discussions of pertinent ethical, scientific, social, and legal issues take place forthwith. A program announcement[146] cosponsored by NIMH specifically solicits projects to address these issues as they arise from the use of genetic technologies and knowledge related to mental disorders.

## Notes


1. S. Hodgins, S. A. Mednick, P. A. Brennan, F. Schulsinger, and M. Engberg, "Mental Disorder and Crime: Evidence From a Danish Birth Cohort," *Archives of General Psychiatry* 53 (1996):489–496.
2. M. Eronen, P. Hakola, and J. Tiihonen, "Mental Disorders and Homicidal Behavior in Finland," *Archives of General Psychiatry* 53 (1996):497–501.
3. V. A. McKusick, *Mendelian Inheritance in Man: A Catalog of Human Genes and Genetic Disorders* (Baltimore: Johns Hopkins University Press, 1994).
4. Huntington's Disease Collaborative Research Group, "A Novel Gene Containing a Trinucleotide Repeat That Is Expanded and Unstable on Huntington's Disease Chromosomes," *Cell* 72 (1993):971–983.
5. S. O. Moldin and I. I. Gottesman, "Population Genetic Methods in Psychiatry." In H. I. Kaplan and B. J. Sadock, eds., *Comprehensive Textbook of Psychiatry,* 7th ed. (Baltimore: Williams and Wilkins, in press); S. O. Moldin, "Research Methods in Behavioral Genetics." In P. C. Kendell, J. N. Butcher, and G. N. Holmbeck, eds., *Handbook of Research Methods in Clinical Psychology,* 2nd ed. (Washington, DC: American Psychological Association, 1999).
6. K. S. Kendler, M. C. Neale, R. C. Kessler, A. C. Heath, and L. J. Eaves, "Parental Treatment and the Equal Environment Assumption in Twin Studies of Psychiatric Illness," *Psychological Medicine* 24 (1994):579–590.
7. J. Ott, *Analysis of Human Genetic Linkage, Revised Edition* (Baltimore: Johns Hopkins University Press, 1991).
8. N. E. Morton, "Sequential Tests for the Detection of Linkage," *American Journal of Human Genetics* 7 (1955):277–318.
9. E. S. Lander, and L. Kruglyak, "Genetic Dissection of Complex Traits: Guidelines for Interpreting and Reporting Linkage Results," *Nature Genetics* 11 (1995):241–247.
10. J. S. Witte, R. C. Elston, and N. J. Shork, "Genetic Dissection of Complex Traits," *Nature Genetics* 12 (1996):355–356; D. Curtis, "Genetic Dissection of Complex Traits," *Nature Genetics* 12 (1996):356–357; Lander and Kruglyak, "Genetic Dissection of Complex Traits."
11. F. S. Collins, "Positional Cloning: Let's Not Call It Reverse Anymore," *Nature Genetics* 1 (1992):3–6.
12. F. S. Collins, A. Patrinos, E. Jordan, A. Chakravarti, R. Gesteland, L. Walters, and members of the DOE and NIH planning groups, "New Goals for the US Human Genome Project: 1998–2003," *Science* 282 (1998):682–689.
13. I. I. Gottesman, *Schizophrenia Genesis: The Origins of Madness* (New York: W. H. Freeman, 1991).
14. N. J. Risch, "Estimating Morbidity Risks in Relatives: The Effect of Reduced Fertility," *Behavior Genetics* 13 (1983):441–451.
15. D. C. Rao, N. E. Morton, I. I. Gottesman, and R. Lew, "Path Analysis of Qualitative Data on Pairs of Relatives: Application to Schizophrenia," *Human Heredity* 31 (1981):325–333; D. H. O'Rourke, I. I. Gottesman, B. K. Suarez, J. P. Rice, and T. Reich, "Refutation of the General Single Locus Model for the Etiology of Schizophrenia," *American Journal of Human Genetics* 34 (1982):630–649; M. McGue, and I. I. Gottesman, "Genetic Linkage in Schizophrenia: Perspectives From Genetic Epidemiology," *Schizophrenia Bulletin* 15 (1989):453–464.
16. S. S. Kety, D. Rosenthal, P. H. Wender, and F. Schulsinger, "The Types of Prevalence of Mental Illness in the Biological and Adoptive Families of Adopted Schizophrenics." In D. Rosenthal

and S. S. Kety, eds., *The Transmission of Schizophrenia* (Oxford, England: Pergamon Press, 1968).

17. S. S. Kety, P. H. Wender, B. Jacobsen, L. J. Ingraham, L. Jansson, B. Faber, and D. K. Kinney, ''Mental Illness in the Biological and Adoptive Relatives of Schizophrenic Adoptees. Replication of the Copenhagen Study in the Rest of Denmark,'' *Archives of General Psychiatry* 51 (1994):442–445.
18. K. S. Kendler, A. M. Gruenberg, and D. K. Kinney, ''Independent Diagnoses of Adoptees and Relatives as Defined by *DSM-III* in the Provincial and National Samples of the Danish Adoption Study of Schizophrenia,'' *Archives of General Psychiatry* 51 (1994):456–468.
19. I. I. Gottesman and J. Shields, ''A Polygenic Theory of Schizophrenia,'' *Proceedings of the National Academy of Sciences of the USA* 58 (1967):199–205; Rao et al., ''Path Analysis of Qualitative Data on Pairs of Relatives''; O'Rourke et al., ''Refutation of the General Single Locus Model for the Etiology of Schizophrenia''; McGue and Gottesman, ''Genetic Linkage in Schizophrenia''; S. O. Moldin, ''Indicators of Liability to Schizophrenia: Perspective From Genetic Epidemiology,'' *Schizophrenia Bulletin* 20 (1994):169–184.
20. N. J. Risch, ''Linkage Strategies for Genetically Complex Traits: I. Multilocus Models,'' *American Journal of Human Genetics* 46 (1990):222–228.
21. S. O. Moldin and P. Van Eerdewegh, ''Multivariate Genetic Analysis of Oligogenic Disease,'' *Genetic Epidemiology* 12 (1995):801–806.
22. K. S. Kendler, R. E. Straub, C. J. MacLean, and D. Walsh, ''Reflections on the Evidence for a Vulnerability Locus for Schizophrenia on Chromosome 6p24-22,'' *American Journal of Medical Genetics* 67 (1996):124–126; M. Baron, ''Linkage Results in Schizophrenia,'' *American Journal of Medical Genetics* 67 (1996):121–123; ———, ''Further Reflections on Linkage Results in Schizophrenia,'' *American Journal of Medical Genetics* 67 (1996):430–432; S. O. Moldin, ''The Maddening Hunt for Madness Genes,'' *Nature Genetics* 17 (1997):127–129; S. O. Moldin and I. I. Gottesman, ''Genes, Experience, and Chance in Schizophrenia: Positioning for the 21st Century,'' *Schizophrenia Bulletin* 23 (1997):547–561.
23. Lander and Kruglyak, ''Genetic Dissection of Complex Traits.''
24. S. Wang, C. E. Sun, C. A. Walczak, J. S. Ziegle, B. R. Kipps, L. R. Goldin and S. R. Diehl, ''Evidence for a Susceptibility Locus for Schizophrenia on Chromosome 6pter-p22,'' *Nature Genetics* 10 (1995):41–46.
25. R. E. Straub, C. J. MacLean, F. A. O'Neill, J. Burke, B. Murphy, F. Duke, R. Shinkwin, B. T. Webb, J. Zhang, D. Walsh, and K. S. Kendler, ''A Potential Vulnerability Locus for Schizophrenia on Chromosome 6p24-22: Evidence for Genetic Heterogeneity,'' *Nature Genetics* 11 (1995):287–293.
26. L. M. Brzustowicz, W. G. Honer, E. W. C. Chow, J. Hogan, K. Hodgkinson, and A. S. Bassett, ''Use of a Quantitative Trait to Map a Locus Associated With Severity of Positive Symptoms in Familial Schizophrenia to Chromosome 6p,'' *American Journal of Human Genetics* 61 (1997):1388–1396.
27. S. G. Schwab, M. Albus, J. Hallmayer, S. Honig, M. Borrmann, D. Lichtermann, R. P. Ebstein, M. Ackenheil, B. Lerer, and N. Risch, ''Evaluation of a Susceptibility Gene for Schizophrenia on Chromosome 6p by Multipoint Affected Sib-Pair Linkage Analysis,'' *Nature Genetics* 11 (1995):325–327.
28. H. W. Moises, L. Yang, H. Kristbjarnarson, C. Wiese, W. Byerley, F. Macciardi, V. Arolt, D. Blackwood, X. Liu, B. Sjogren, H. N. Aschauer, H.-G. Hwu, K. Jang, W. J. Livesley, J. L. Kennedy, T. Zoega, O. Ivarsson, M.-T. Bui, M.-H. Yu, B. Havsteen, D. Commenges, J. Weissenbach, E. Schwinger, I. I. Gottesman, A. J. Pakstis, L. Wetterberg, K. K. Kidd, and T. Helgason, ''An International Two-Stage Genome-Wide Search for Schizophrenia Susceptibility Genes,'' *Nature Genetics* 11 (1995):321–324.
29. J.-L. Blouin, B. A. Dombroski, S. K. Nath, V. K. Lasseter, P. S. Wolyniec, G. Nestadt, M. Thornquist, G. Ullrich, J. McGrath, L. Kasch, M. Lamacz, M. G. Thomas, C. Gehrig, U. Radhakrishna, S. E. Snyder, K. G. Balk, K. Neufeld, K. L. Swartz, N. DeMarchi, G. N. Papadimitriou, D. G. Dikeos, C. N. Stefanis, A. Chakravarti, B. Childs, D. E. Housman, H. H.

Kazazian, S. E. Antonarakis, and A. E. Pulver, ''Schizophrenia Susceptibility Loci on Chromosomes 13q32 and 8p21,'' *Nature Genetics* 20 (1998):70–73.

30. Moises et al., ''An International Two-Stage Genome-Wide Search for Schizophrenia Susceptibility Genes.''
31. R. E. Straub, C. J. MacLean, F. A. O'Neill, D. Walsh, and K. S. Kendler, ''Support for a Possible Schizophrenia Vulnerability Locus in Region 5q22-31 in Irish Families,'' *Molecular Psychiatry* 2 (1997):148–155.
32. H. Coon, J. Holik, M. Hoff, F. Reimherr, P. Wender, M. Myles-Worsley, M. Waldo, R. Freedman, and W. Byerley, ''Analysis of Chromosome 22 Markers in Nine Schizophrenia Pedigrees,'' *American Journal of Medical Genetics* 54 (1994):72–79.
33. A. E. Pulver, V. K. Lasseter, L. Kasch, P. Wolyniec, G. Nestadt, J. L. Blouin, M. Kimberland, R. Babb, S. Vourlis, and H. Chen, ''Schizophrenia: A Genome Scan Targets Chromosomes 3p and 8p as Potential Sites of Susceptibility Genes,'' *American Journal of Medical Genetics* 60 (1995):252–260; Kendler et al., ''Reflections on the Evidence for a Vulnerability Locus for Schizophrenia on Chromosome 6p24-22''; Blouin et al., ''Schizophrenia Susceptibility Loci on Chromosomes 13q32 and 8p21.''
34. Q. Cao, M. Martinez, J. Zhang, A. R. Sanders, J. A. Badner, A. Cravchik, C. Markey, E. Beshah, J. J. Guroff, M. E. Maxwell, D. Kazuba, R. Whiten, L. R. Goldin, E. S. Gershon and P. V. Gejman, ''Suggestive Evidence for a Schizophrenia Susceptibility Locus or Chromosome 6q and a Confirmation in an Independent Series of Pedigrees,'' *Genomics* 43 (1997):1–8.
35. S. V. Faraone, T. C. Matise, D. Svrakic, J. Pepple, D. Malaspina, B. Suarez, C. Hampe, C. T. Zambuto, K. Schmitt, J. Meyer, J. Merkel, H. Lee, J. Harkavy Friedman, C. A. Kaufmann, C. R. Cloninger, and M. T. Tsuang, ''Genome Scan of European-American Schizophrenia Pedigrees: Results of the NIMH Genetics Initiative and Millennium Consortium,'' *American Journal of Medical Genetics* 81 (1998):290–295; S. G. Schwab, J. Hallmayer, M. Albus, B. Lerer, C. Hanses, K. Kanyas, R. Segman, M. Borrman, B. Dreikorn, D. Lichtermann, M. Rietschel, M. Trixler, W. Maier, and D. B. Wildenauer, ''Further Evidence for a Susceptibility Locus on Chromosome 10p14-p11 in 72 Families With Schizophrenia by Nonparametric Linkage Analysis,'' *American Journal of Medical Genetics* 81 (1998):302–307.
36. K. Leonhard, *Aufteilung der Endopen Psychosen* (Berlin: Akademic Verlag).
37. C. Perris, ''A Study of Biopolar (Manic-Depressive) and Unipolar Recurrent Depressive Psychosis,'' *Acta Psychiatrica Scandinavica* 194(Suppl., 1966):15–44.
38. J. Angst, ''Zur Aetiologie und Nosologie Endogoner Depressiver Psychosen,'' *Monogaphien aus dem Gesamtgebiete der Psychiatrie* 112 (1966):1–118.
39. E. S. Gershon, J. H. Hamovit, J. J. Guroff, E. Dibble, J. F. Leckman, W. Sceery, S. D. Targum, J. I. Nurnberger, L. R. Goldin, and W. E. Bunney, ''A Family Study of Schizoaffective, Bipolar I, Bipolar II, Unipolar, and Normal Control Probands,'' *Archives of General Psychiatry* 39 (1982):1157–1167; G. Winokur, M. T. Tsuang, and R. R. Crowe, ''The Iowa 500: Affective Disorder in Relatives of Manic and Depressive Patients,'' *American Journal of Psychiatry* 139 (1982):209–212; M. M. Weissman, E. S. Gershon, K. K. Kidd, B. A. Prusoff, J. F. Leckman, E. Dibble, J. Hamovit, D. Thompson, D. L. Pauls, and J. L. Guroff, ''Psychiatric Disorders in the Relatives of Probands With Affective Disorders,'' *Archives of General Psychiatry* 41 (1984):13–21.
40. *See* previous reviews: T. Reich, C. R. Cloninger, B. K. Suarez, and J. P. Rice, ''Genetics of the Affective Psychoses. In L. Wing and J. K. Wing, eds., *Handbook of Psychiatry, Volume 3, Psychoses of Uncertain Aetiology* (Cambridge: Cambridge University Press, 1982); P. McGuffin and R. Katz, ''Nature, Nurture, and Affective Disorder.'' In J. F. W. Deakin, ed., *The Biology of Affective Disorders* (London: Gaskell Press, Royal College of Psychiatrists, 1986); ———, ''The Genetics of Depression and Manic-Depressive Disorder,'' *British Journal of Psychiatry* 155 (1989):294–304; M. T. Tsuang and S. V. Faraone, *The Genetics of Mood Disorders* (Baltimore: Johns Hopkins University Press, 1990); S. O. Moldin, T. Reich, and J. P. Rice, ''Current Perspectives on the Genetics of Unipolar Depression,'' *Behavior Genetics* 21

(1991):211–242; S. O. Moldin and T. Reich, ''The Genetic Analysis of Depression: Future Directions,'' *Clinical Neuroscience* 1 (1993):139–145.

41. D. Blacker and M. T. Tsuang, ''Unipolar Relatives in Bipolar Pedigrees: Are They Biopolar?,'' *Psychiatric Genetics* 3 (1993):5–16.
42. P. McGuffin, M. J. Owen, M. C. O'Donovan, A. Thapar, and I. I. Gottesman, *Seminars in Psychiatric Genetics* (London, England: Royal College of Psychiatrists, 1994); Moldin et al., ''Current Perspectives on the Genetics of Unipolar Depression''; Moldin and Reich, ''The Genetic Analysis of Depression.''
43. M. M. Weissman, R. C. Bland, G. J. Canino, C. Faravelli, S. Greenwald, H.-G. Hwu, P. R. Joyce, E. G. Karam, C.-K. Lee, J. Lellouch, J.-P. Lepine, S. C. Newman, M. Rubio-Stipec, J. E. Wells, P. J. Wickramaratne, H.-U. Wittchen, and E.-K. Yeh, ''Cross-National Epidemiology of Major Depression and Bipolar Disorder,'' *Journal of the American Medical Association* 276 (1996):293–299.
44. Moldin et al., ''Current Perspectives on the Genetics of Unipolar Depression''; Moldin and Reich, ''The Genetic Analysis of Depression.''
45. Tsuang and Faraone, *The Genetics of Mood Disorders;* P. McGuffin, and R. Murray, *The New Genetics of Mental Illness* (Oxford, England: Butterworth-Heinemann, 1991).
46. R. Cadoret, ''Evidence for Genetic Inheritance of Primary Affective Disorder in Adoptees,'' *American Journal of Psychiatry* 133 (1978):463–466; J. Mendlewicz, and J. D. Rainer, ''Adoption Study Supporting Genetic Transmission in Manic-Depressive Illness,'' *Nature* 268 (1977):326–329; P. H. Wender, S. S. Kety, D. Rosenthal, F. Schulsinger, J. Ortmann, and I. Lunde, ''Psychiatric Disorders in the Biological and Adoptive Families of Adopted Individuals With Affective Disorders,'' *Archives of General Psychiatry* 43 (1986):923–929.
47. A. L. Von Knorring, C. R. Cloninger, M. Bohman, and S. Sigvardsson, ''An Adoption Study of Depressive Disorders and Substance Abuse,'' *Archives of General Psychiatry* 40 (1983):943–950; R. J. Cadoret, T. W. O'Gorman, E. Heywood, and E. Troughton, ''Genetic and Environmental Factors in Major Depression,'' *Journal of Affective Disorders* 9 (1985):155–164.
48. Mendlewicz and Rainer, ''Adoption Study Supporting Genetic Transmission in Manic-Depressive Illness''; Wender et al., ''Psychiatric Disorders in the Biological and Adoptive Families of Adopted Individuals With Affective Disorders.''
49. Tsuang and Faraone, *The Genetics of Mood Disorders.*
50. J. P. Rice, T. Reich, N. C. Andreasen, J. Endicott, M. Van Eerdewegh, R. Fishman, R. M. A. Hirschfeld, and G. L. Klerman, ''The Familial Transmission of Biopolar Illness.'' *Archives of General Psychiatry* 44 (1987):441–447.
51. *See* previous reviews: Tsuang and Faraone, *The Genetics of Mood Disorders;* Moldin et al., ''Current Perspectives on the Genetics of Unipolar Depression''; Moldin and Reich, ''The Genetic Analysis of Depression.''
52. Moldin et al., ''Current Perspectives on the Genetics of Unipolar Depression''; McGuffin et al., *Seminars in Psychiatric Genetics.*
53. J. Mendlewicz, P. Simon, S. Sevy, F. Charon, H. Brocas, S. Legros, and G. Vassart, ''Polymorphic DNA Markers on X Chromosome and Manic Depression,'' *Lancet* 1 (1987):1230–1231.
54. *Id.*; G. Lucotte and A. Landoulski, ''Manic Depressive Illness Is Linked to Factor IX in a French Pedigree,'' *Annales de Genetique* 35 (1992):93–95; P. Pekkarinen, J. Terwilliger, P.-E. Bredbacka, J. Lonnqvist, and L. Peltonen, ''Evidence of a Predisposing Locus to Bipolar Disorder on Xq24-q27.1 in an Extended Finnish Pedigree,'' *Genome Research* 5 (1995):105–115.
55. Tsuang and Faraone, *The Genetics of Mood Disorders;* McGuffin et al., *Seminars in Psychiatric Genetics;* N. Risch and D. Botstein, ''A Manic Depressive History,'' *Nature Genetics* 12 (1996):351–353; C. Smyth, G. Kalsi, J. Brynjolfsson, J. O'Neill, D. Curtis, L. Rifkin, E. Moloney, P. Murphy, H. Petursson, and H. Gurling. ''Test of Xq26.3-28 Linkage in Bipolar and Unipolar Affective Disorder in Families Selected for Absence of Male to Male Transmission,'' *British Journal of Psychiatry* 171 (1997):578–581.
56. M. Baron, N. J. Risch, R. Hamburger, B. Mandell, S. Kushner, M. Newman, D. Drumer, and

R. H. Belmaker, ''Genetic Linkage Between X-Chromosome Markers and Biopolar Affective Illness,'' *Nature* 326 (1987):389–392.

57. M. Baron, N. F. Freimer, N. Risch, B. Lerer, J. R. Alexander, R. E. Straub, S. Asokan, K. Das, A. Peterson, and J. Amos, ''Diminished Support for Linkage Between Manic Depressive Illness and X-Chromosome Markers in Three Israeli Pedigrees,'' *Nature Genetics* 3 (1993):49–55.
58. J. A. Egeland, D. S. Gerhard, D. L. Pauls, J. N. Sussex, K. K. Kidd, C. R. Allen, A. M. Hostetter, and D. E. Housman, ''Biopolar Affective Disorders Linked to DNA Markers on Chromosome 11,'' *Nature* 325 (1987):783–787.
59. J. R. Kelsoe, E. I. Ginns, J. A. Egeland, D. S. Gerhard, A. M. Goldstein, S. J. Bale, D. L. Pauls, R. T. Long, K. K. Kidd, G. Conte, D. E. Housman, and S. M. Paul, ''Re-evaluation of the Linkage Relationship Between Chromosome 11p Loci and the Gene for Bipolar Affective Disorder in the Old Order Amish,'' *Nature* 342 (1989):238–243.
60. W. H. Berrettini, T. N. Ferraro, L. R. Goldin, D. E. Weeks, S. Detera-Wadleigh, J. I. Nurnberger, Jr., and E. S. Gershon, ''Chromosome 18 DNA Markers and Manic-Depressive Illness: Evidence for a Susceptibility Gene,'' *Proceedings of the National Academy of Sciences of the USA* 91 (1994):5918–5921; W. H. Berrettini, T. N. Ferraro, L. R. Goldin, S. D. Detera-Wadleigh, H. Choi, D. Muniec, J. J. Guroff, D. M. Kazuba, J. I. Nurnberger, W.-T. Hsieh, M. R. Hoehe, and E. S. Gershon, ''A Linkage Study of Bipolar Illness,'' *Archives of General Psychiatry* 54 (1997): 27–35; D. Blacker and M. T. Tsuang, ''Unipolar Relatives in Bipolar Pedigrees: Are They Bipolar?,'' *Psychiatric Genetics* 3 (1993):5–16.
61. O. C. Stine, J. Xu, R. Koskela, F. J. McMahon, M. Gschwend, C. Friddle, C. D. Clark, M. G. McInnis, S. G. Simpson, and T. S. Breschel, ''Evidence for Linkage of Bipolar Disorder to Chromosome 18 With a Parent-of-Origin Effect,'' *American Journal of Human Genetics* 57 (1995):1384–1394.
62. F. J. McMahon, O. C. Stine, D. A. Meyers, S. G. Simpson, J. R. DePaulo, ''Patterns of Maternal Transmission in Bipolar Affective Disorder,'' *American Journal of Human Genetics* 56 (1995):1277–1286.
63. E. S. Gershon, J. A. Badner, S. D. Detera-Wadleigh, T. N. Ferraro, and W. H. Berrettini, ''Maternal Inheritance and Chromosome 18 Allele Sharing in Unilineal Bipolar Illness Pedigrees,'' *American Journal of Medical Genetics* 67 (1996):202–207.
64. Stine et al., ''Evidence for Linkage of Bipolar Disorder to Chromosome 18 With a Parent-of-Origin Effect.''
65. M. A. Cleves, D. V. Dawson, R. C. Elston, and A. H. Schnell, ''A New Test for Linkage Applied to Bipolar Disorder and Marker D18S41,'' *Genetic Epidemiology* 14 (1997):581–586.
66. N. B. Freimer, V. I. Reus, M. A. Escamilla, A. McInnes, M. Spesny, P. Leon, S. K. Service, L. B. Smith, S. Silva, E. Rojas, A. Gallegos, L. Meza, E. Fournier, S. Baharloo, K. Blankenship, D. J. Tyler, S. Batki, S. Vinogradov, J. Weissenbach, S. H. Barondes, and L. A. Sandkuijl, ''Genetic Mapping Using Haplotype, Association and Linkage Methods Suggests a Locus for Severe Bipolar Disorder (BPI) at 18q22-q23,'' *Nature Genetics* 12 (1996):436–441.
67. Stine et al., ''Evidence for Linkage of Bipolar Disorder to Chromosome 18 With a Parent-of-Origin Effect.''
68. H. Coon, M. Hoff, J. Holik, J. Hadley, N. Fang, F. Reimherr, P. Wender, and W. Byerley, ''Analysis of Chromosome 18 DNA Markers in Multiplex Pedigrees With Manic Depression,'' *Biological Psychiatry* 39 (1996):689–696; M. C. LaBuda, M. Maldonado, D. Marshall, K. Otten, and D. S. Gerhard, ''A Follow-Up Report of a Genome Search for Affective Disorder Predisposition Loci in the Old Order Amish,'' *American Journal of Human Genetics* 59 (1996):1343–1362; S. D. Detera-Wadleigh, J. A. Badner, T. Yoshikawa, A. R. Sanders, L. R. Goldin, G. Turner, D. Y. Rollins, T. Moses, J. J. Guroff, D. Kazuba, M. E. Maxwell, H. J. Edenberg, T. Foroud, D. Lahiri, J. I. Nurnberger, O. C. Stine, F. McMahon, D. A. Meyers, D. MacKinnon, S. Simpson, M. McInnis, J. R. DePaulo, J. P. Rice, A. Goate, T. Reich, M. C. Blehar, and E. S. Gershon, ''Initial Genome Scan of the NIMH Genetics Initiative Bipolar I Pedigrees: Chromosomes 4, 7, 9, 18, 19, 20, and 21q,'' *American Journal of Medical Genetics* 74 (1997):254–262; A. De Bruyn, D. Souery, K. Mendelbaum, J. Mendlewicz, and C. Van

Broeckhoven, "Linkage Analysis of Families With Bipolar Illness and Chromosome 18 Markers," *Biological Psychiatry* 39 (1996):679–688; G. Kalsi, C. Smyth, J. Brynjolfsson, R. S. Sherrington, J. O'Neill, D. Curtis, L. Rifkin, P. Murphy, H. Petursson, and H. Gurling, "Linkage Analysis of Manic Depression (Bipolar Disorder) in Icelandic and British Kindreds Using Markers on the Short Arm of Chromosome 18," *Human Heredity* 47 (1997):268–278; J. A. Knowles, P. A. Rao, T. C. Matise, J. E. Loth, G. M. de Jesus, L. Levine, K. Das, G. K. Penchaszadeh, J. R. Alexander, B. Lerer, J. Endicott, J. Ott, T. C. Gilliam, and M. Baron, "No Evidence for Significant Linkage Between Bipolar Affective Disorder and Chromosome 18 Pericentromeric Markers in a Large Series of Multiplex Extended Pedigrees," *American Journal of Human Genetics* 62 (1998):916–924.

69. R. E. Straub, T. Lehner, Y. Luo, J. E. Loth, W. Shao, L. Sharpe, J. R. Alexander, K. Das, R. Simon, and R. R. Fieve, "A Possible Vulnerability Locus for Bipolar Affective Disorder on Chromosome 21q22.3," *Nature Genetics* 8 (1994):291–296.
70. Detera-Wadleigh et al., "Initial Genome Scan of the NIMH Genetics Initiative Bipolar I Pedigrees."
71. H. Gurling, C. Smyth, G. Kalsi, E. Moloney, L. Rifkin, J. O'Neill, P. Murphy, D. Curtis, H. Petursson, and J. Brynjolfsson, "Linkage Findings for Bipolar Disorder," *Nature Genetics* 10 (1995):8–9.
72. Straub et al., "A Possible Vulnerability Locus for Bipolar Affective Disorder on Chromosome 21q22.3."
73. W. Byerley, J. Holik, M. Hoff, and H. Coon, "Search for a Gene Predisposing to Manic-Depression on Chromosome 21," *American Journal of Medical Genetics* 60 (1995):231–233; La Bonda et al., "A Follow-Up Report of a Genome Search for Affective Disorder Predisposition Loci in the Old Order Amish"; Detera-Wadleigh et al., "Initial Genome Scan of the NIMH Genetics Initiative Bipolar I Pedigrees."
74. D. H. R. Blackwood, L. He, S. W. Morris, A. McLean, C. Whitton, M. Thomson, M. T. Walker, K. Woodburn, C. M. Sharp, A. F. Wright, Y. Shibasaki, D. M. St. Clair, D. J. Porteous, and W. J. Muir, "A Locus for Bipolar Affective Disorder on Chromosome 4p," *Nature Genetics* 12 (1996):427–430.
75. L. J. Adams, P. B. Mitchell, S. L. Fielder, A. Rosso, J. A. Donald, and P. R. Schofield, "A Susceptibility Locus for Bipolar Affective Disorder on Chromosome 4q35," *American Journal of Human Genetics* 62 (1998):1084–1091.
76. J. R. Kelsoe, A. D. Sadovnick, H. Kristbjarnarson, P. Bergesch, Z. Mroczkowski-Parker, M. Drennan, M. H. Rapaport, P. Flodman, M. A. Spence, and R. A. Remick, "Possible Locus for Bipolar Disorder Near the Dopamine Transporter on Chromosome 5," *American Journal of Medical Genetics* 67 (1996):533–540.
77. E. I. Ginns, J. Ott, J. A. Egeland, C. R. Allen, C. S. J. Fann, D. L. Pauls, J. Weissenbach, J. P. Carulli, K. M. Falls, T. P. Keith, and S. M. Paul, "A Genome-Wide Search for Chromosomal Loci Linked to Bipolar Affective Disorder in the Older Order Amish," *Nature Genetics* 12 (1996):431–435.
78. Rice et al., "Initial Genome Scan of the NIMH Genetics Initiative Bipolar I Pedigrees."
79. C. Smyth, G. Kalsi, D. Curtis, J. Brynjolfsson, J. O'Neill, L. Rifkin, E. Moloney, P. Murphy, H. Petursson, and H. Gurling, "Two-Locus Admixture Linkage Analysis of Bipolar Affective and Unipolar Affective Disorder Supports the Presence of Susceptibility Loci on Chromosomes 11p15 and 21q22," *Genomics* 39 (1997):271–278; A. Malafosse, M. Leboyer, T. d'Amato, S. Amadeo, M. Abbar, D. Campion, O. Canseil, D. Castelnau, F. Gheysen, B. Granger, B. Henrikson, M.-F. Poirier, O. Sabate, D. Samolyk, J. Feingold, and J. Mallet, "Manic Depressive Illness and Tyrosine Hydroxylase Gene: Linkage Heterogeneity and Association," *Neurobiology of Disease* 4 (1997):337–349.
80. N. Craddock, M. Owen, S. Burge, B. Kurian, P. Thomas, and P. McGuffin, "Familial Cosegregation of Major Affective Disorder and Darier's Disease (Keratosis Follicularis)," *British Journal of Psychiatry* 164 (1994):355–358.

81. H. Ewald, O. Mors, T. Flint, K. Koed, H. Eiberg, and T. A. Kruse, ''A Possible Locus for Manic Depressive Illness on Chromosome 16p13,'' *Psychiatric Genetics* 5 (1995):71–81.
82. H. M. Lachman, J. R. Kelsoe, R. A. Remick, A. D. Sadovnick, M. H. Rapaport, M. Lin, B. A. Pazur, A. M. A. Roe, T. Saito, and D. F. Papolos, ''Linkage Studies Suggest a Possible Locus for Bipolar Disorder near the Velo-Cardio-Facial Syndrome Region on Chromosome 22,'' *American Journal of Medical Genetics* 74 (1997):121–128.
83. N. S. Cotton, ''The Familial Incidence of Alcoholism: A Review,'' *Journal of Studies in Alcoholism* 40 (1979):89–116.
84. J. E. Helzer, G. J. Canino, E. K. Yeh, R. C. Bland, C. K. Lee, H. G. Hwu, and S. Newman, ''Alcoholism—North American and Asia. A Comparison of Population Surveys With the Diagnostic Interview Schedule,'' *Archives of General Psychiatry* 47 (1990):313–319.
85. W. S. Stone and I. I. Gottesman, ''A Perspective on the Search for the Causes of Alcoholism: Slow Down the Rush to Genetic Judgements,'' *Neurology, Psychiatry and Brain Research* 1 (1993):123–132.
86. T. Reich and C. R. Cloninger, ''Time-Dependent Model of the Familial Transmission of Alcoholism.'' In Anonymous, *Banbury Report 33: Genetics and Biology of Alcoholism* (Cold Spring Harbor, NY: Cold Spring Harbor Press, 1990).
87. M. A. Schuckit, D. A. Goodwin, and G. Winokur, ''A Study of Alcoholism in Half Siblings,'' *American Journal of Psychiatry* 128 (1972):1132–1136.
88. H. M. D. Gurling, R. M. Murray, and C. A. Clifford, 'Investigations into the Genetics of Alcoholism Dependence and Into Its Effect on Brain Function.'' In L. Gedda, P. Parisi, and W. E. Nance, eds., *Twin Research 3, Part C. Epidemiological and Clinical Studies* (New York: Alan Liss, 1981).
89. M. McGue, R. W. Pickens, and D. S. Svikis, ''Sex and Age Effects on the Inheritance of Alcohol Problems: A Twin Study,'' *Journal of Abnormal Psychology* 101 (1992):3–17.
90. Gurling et al., ''Investigations Into the Genetics of Alcoholism Dependence and Into Its Effects on Brain Function.''
91. K. S. Kendler, A. C. Heath, M. C. Neale, R. C. Kessler, and L. J. Eaves, ''A Population-Based Twin Study of Alcoholism in Women,'' *Journal of the American Medical Association* 268 (1992):1877–1882.
92. See a previous review: A. C. Heath, ''Genetic Influences on Drinking Behavior in Humans.'' In H. Begleiter and B. Kissin, eds., *The Genetics of Alcoholism* (New York: Oxford University Press, 1995).
93. K. S. Kendler, E. E. Walters, M. C. Neale, R. C. Kessler, A. C. Heath, and L. J. Eaves, ''The Structure of the Genetic and Environmental Risk Factors for Six Major Psychiatric Disorders in Women. Phobia, Generalized Anxiety Disorder, Panic Disorder, Bulimia, Major Depression, and Alcoholism,'' *Archives of General Psychiatry* 52 (1995):374–383.
94. McGue et al., ''Sex and Age Effects on the Inheritance of Alcohol Problems.''
95. 0.0, *Id.*, 0.1, C. B. Caldwell and I. I. Gottesman, ''Sex Differences in the Risk for Alcoholism: A Twin Study,'' *Behavior Genetics* 21 (1991):563; 0.3, R. W. Pickens, D. S. Svikis, M. McGue, D. T. Lykken, L. L. Heston, and P. J. Clayton, ''Heterogeneity in the Inheritance of Alcoholism. A Study of Male and Female Twins,'' *Archives of General Psychiatry* 48 (1991):19–28.
96. Kendler et al., ''A Population-Based Twin Study of Alcoholism in Women''; K. S. Kendler, M. C. Neale, A. C. Heath, R. C. Kessler, and L. J. Eaves, ''A Twin-Family Study of Alcoholism in Women,'' *American Journal of Psychiatry* 151 (1994):707–715.
97. A. C. Heath and N. G. Martin, ''Genetic Influences on Alcohol Consumption Patterns and Problem Drinking Results From the Australian NH and MRC Twin Panel Follow-Up Survey,'' *Annals of the New York Academy of Sciences* 708 (1994):72–85.
98. Kendler et al., ''A Twin-Family Study of Alcoholism in Women.''
99. K. S. Kendler, M. C. Neale, C. A. Prescott, R. C. Kessler, A. C. Heath, L. A. Corey, and L. A. Eaves, ''Childhood Parental Loss and Alcoholism in Women: A Causal Analysis Using a Twin-Family Design,'' *Psychological Medicine* 26 (1996):79–95.
100. Pickens et al., ''Heterogeneity in the Inheritance of Alcoholism.''

101. R. J. Cadoret, T. W. O'Gorman, E. Troughton, and E. Heywood, ''Alcoholism and Antisocial Personality. Interrelationships, Genetic and Environmental Factors,'' *Archives of General Psychiatry* 42 (1985):161–167; M. Bohman, S. Sigvardsson, and C. R. Cloninger, ''Maternal Inheritance of Alcohol Abuse. Cross-Fostering Analysis of Adopted Women,'' *Archives of General Psychiatry* 38 (1981):965–969; M. Bohman, ''Some Genetic Aspects of Alcoholism and Criminality. A Population of Adoptees,'' *Archives of General Psychiatry* 35 (1978):269–276.
102. Bohman et al., ''Maternal Inheritance of Alcohol Abuse.''
103. S. B. Gilligan, T. Reich, and C. R. Cloninger, ''Etiologic Heterogeneity in Alcoholism,'' *Genetic Epidemiology* 4 (1987):395–414; C. E. Aston and S. Y. Hill, ''Segregation Analysis of Alcoholism in Families Ascertained Through a Pair of Male Alcoholics,'' *American Journal of Human Genetics* 46 (1990):879–887.
104. S. Y. Hill, C. Aston, and B. Rabin, ''Suggestive Evidence of Genetic Linkage Between Alcoholism and the MNS Blood Group,'' *Alcoholism and Clinical Experimental Research* 12 (1988):811–814.
105. K. Blum, E. P. Noble, P. J. Sheridan, A. Montgomery, T. Ritchie, P. Jagadeeswaran, H. Nogami, A. H. Briggs, and J. B. Cohn, ''Allelic Association of Human Dopamine D2 Receptor Gene in Alcoholism,'' *Journal of the American Medical Association* 263 (1990):2055–2060.
106. J. Gelernter, D. Goldman, and N. J. Risch, ''The A1 Allele at the $D_2$ Dopamine Receptor Gene and Alcoholism: A Reappraisal,'' *Journal of the American Medical Association* 269 (1993):1673–1677.
107. T. Reich, H. J. Edenberg, A. Goate, J. T. Williams, J. P. Rice, P. Van Eerdewegh, T. Foroud, V. Hesselbrock, M. A. Schuckit, K. Bucholz, B. Porjesz, T.-K. Li, P. M. Conneally, J. I. Nurnberger, J. A. Tischfield, R. R. Crowe, C. R. Cloninger, W. Wu, S. Shears, K. Carr, C. Crose, C. Willig, and H. Begleiter, ''Genome-Wide Search for Genes Affecting the Risk for Alcohol Dependence,'' *American Journal of Medical Genetics* 81 (1998):207–215.
108. J. C. Long, W. C. Knowler, R. L. Hanson, R. W. Robin, M. Urbanek, E. Moore, P. H. Bennett, and D. Goldman, ''Evidence for Genetic Linkage to Alcohol Dependence on Chromosomes 4 and 11 From an Autosome-Wide Scan in an American Indian Population,'' *American Journal of Medical Genetics* 81 (1998):216–221.
109. H. Begleiter, B. Porjesz, T. Reich, H. J. Edenberg, A. Goate, J. Blangero, L. Almasy, T. Foroud, P. Van Eerdewegh, J. Polich, J. Rohrbaugh, S. Kuperman, L. O. Bauer, S. J. O'Connor, D. B. Chordlian, T.-K. Li, P. M. Conneally, V. Hesselbrock, J. P. Rice, M. A. Schuckit, C. R. Cloninger, J. Nurnberger, R. Crowe, and F. E. Bloom, ''Quantitative Trait Loci Analysis of Human Event-Related Brain Potentials: P3 Voltage,'' *Electroencephalography and Clinical Neurophysiology* 108 (1998):244–250.
110. L. N. Robins, J. Tipp, and T. Przybeck, ''Antisocial Personality.'' In L. N. Robins and D. A. Regier, eds., *Psychiatry Disorders in North America: the Epidemiologic Catchment Area Study* (New York: Free Press, 1991).
111. *Id.*
112. P. McGuffin and I. I. Gottesman, ''Genetic Influences on Normal and Abnormal Development.'' In M. Rutter and L. Hersov, eds., *Child Psychiatry: Modern Approaches.* 2d ed. (London, England: Blackwell, 1985); I. I. Gottesman and H. H. Goldsmith, ''Developmental Psychopathology of Antisocial Behavior Inserting Genes Into Its Ontogenesis and Epigenesis.'' In C. A. Nelson, ed., *Threats to Optimal Development: Integrating Biological, Psychological, and Social Risk Factors* (Hillsdale, NJ: Erlbaum, 1994); H. H. Goldsmith, and I. I. Gottesman, ''Heritable Variability and Variable Heritability in Developmental Psychopathology.'' In M. F. Lenzenweger and J. J. Haugaard, eds., *Frontiers of Developmental Psychopathology* (New York: Oxford University Press, 1996).
113. Goldsmith and Gottesman, ''Heritable Variability and Variable Heritability in Developmental Psychopathology.''
114. McGuffin et al., *Seminars in Psychiatric Genetics;* Goldsmith and Gottesman, ''Heritable Variability and Variable Heritability in Developmental Psychopathology.''

115. M. J. Lyons, W. R. True, S. A. Eisen, J. Goldberg, J. M. Meyer, S. V. Faraone, L. J. Eaves, and M. T. Tsuang, ''Differential Heritability of Adult and Juvenile Antisocial Traits,'' *Archives of General Psychiatry* 52 (1995):906–915.
116. K. S. Kendler, ''Genetic Epidemiology in Psychiatry: Taking Both Genes and Environment Seriously,'' *Archives of General Psychiatry* 52 (1995):895–899.
117. R. J. Cadoret, ''Psychopathology in Adopted-Away Offspring of Biologic Parents With Antisocial Behavior,'' *Archives of General Psychiatry* 35 (1978):176–184; R. J. Cadoret and C. Cain, ''Sex Differences in Predictors of Antisocial Behavior in Adoptees,'' *Archives of General Psychiatry* 37 (1980):1171–1175; Cadoret et al., ''Alcoholism and Antisocial Personality''; R. J. Cadoret, W. R. Yates, E. Troughton, G. Woodworth, and M. A. Stewart, ''Genetic-Environmental Interaction in the Genesis of Aggressivity and Conduct Disorders,'' *Archives of General Psychiatry* 52 (1995):916–924; S. Sigvardsson, C. R. Cloninger, M. Bohman, and A. L. Von Knorring, ''Predisposition to Petty Criminality in Swedish Adoptees. III. Sex Differences and Validation of the Male Typology,'' *Archives of General Psychiatry* 39 (1982):1248–1253; C. R. Cloninger, S. Sigvardsson, M. Bohman, and A. L. Von Knorring, ''Predisposition to Petty Criminality in Swedish Adoptees. II. Cross-Fostering Analysis of Gene-Environment Interaction,'' *Archives of General Psychiatry* 39 (1982):1242–1247; J. T. Nigg and H. H. Goldsmith, ''Genetics of Personality Disorders: Perspectives From Personality and Psychopathology Research,'' *Psychological Bulletin* 115 (1994):346–380; M. Bohman, ''Predisposition to Criminality: Swedish Adoption Studies in Retrospect.'' In G. R. Bock and J. A. Goode, eds., *Genetics of Criminal and Antisocial Behaviour* (Ciba Foundation Symposium 194) (Chichester, England: John Wiley, 1995).
118. M. Bohman, C. R. Cloninger, S. Sigvardsson, and A. L. Von Knorring, ''Predisposition to Petty Criminality in Swedish Adoptees. I. Genetic and Environmental Heterogeneity,'' *Archives of General Psychiatry* 39 (1982):1233–1241; S. A. Mednick, W. F. Gabrielli, Jr., and B. Hutchings, ''Genetic Influences in Criminal Convictions: Evidence From an Adoption Cohort,'' *Science* 224 (1984):891–894; Bohman, ''Predisposition to Criminality.''
119. H. G. Brunner, M. Nelen, X. O. Breakfield, H. H. Ropers, and B. A. van Oost, ''Abnormal Behavior Associated With a Point Mutation in the Structural Gene for Monoamine Oxidase A,'' *Science* 262 (1993):578–580.
120. Goldsmith and Gottesman, ''Heritable Variability and Variable Heritability in Developmental Psychopathology.''
121. Brunner et al., ''Abnormal Behavior Associated With a Point Mutation in the Structural Gene for Monoamine Oxidase A.''
122. B. K. Suarez, C. L. Hampe, and P. Van Eerdewegh, ''Problems of Replication Linkage Claims in Psychiatry.'' In E. S. Gershon, C. R. Cloninger, and J. E. Barrett, eds., *Genetic Approaches to Mental Disorders* (Washington, DC: American Psychiatric Press, 1994).
123. J. F. Gusella, N. S. Wexler, P. M. Conneally, S. L. Naylor, M. A. Anderson, R. E. Tanzi, P. C. Watkins, K. Ottina, M. R. Wallace, A. Y. Sakaguchi, A. B. Young, I. Shoulson, E. Bonilla, and J. B. Martin, ''A Polymorphic DNA Marker Genetically Linked to Huntington's Disease,'' *Nature* 306 (1983):234–238; Huntington's Disease Collaborative Research Group, ''A Novel Gene Containing a Trinucleotide Report That Is Expanded and Unstable on Huntington's Disease Chromosomes.''
124. E. S. Lander and N. J. Schork, ''Genetic Dissection of Complex Traits,'' *Science* 265 (1994):2037–2048.
125. J. L. Davies, Y. Kawaguchi, S. T. Bennett, J. B. Copeman, H. J. Cordell, L. E. Pritchard, P. W. Reed, S. C. L. Gough, S. C. Jenkins, S. M. Palmer, K. M. Balfour, B. R. Rowe, M. Farrall, A. H. Barnett, S. C. Bain, and J. A. Todd, ''A Genome-Wide Search for Human Type 1 Diabetes Susceptibility Genes,'' *Nature* 371 (1994):130–136; L. Hashimoto, C. Habita, J. P. Beressi, M. Delepine, C. Besse, A. Cambon-Thomsen, I. Deschamps, J. I. Rotter, S. Djoulah, M. R. James, P. Froguel, J. Weissenbach, G. M. Lathrop, and C. Julier, ''Genetic Mapping of a Susceptibility Locus for Insulin-Dependent Diabetes Mellitus on Chromosome 11q'' *Nature* 371 (1994):161–164; D.-F. Luo, M. M. Bui, A. Muir, N. K. Maclaren, G. Thomson, and J.-X. She, ''Affected-

Sib-Pair Mapping of a Novel Susceptibility Gene to Insulin-Dependent Diabetes Mellitus (IDDM8) on Chromosome 6q25-q27,'' *American Journal of Human Genetics* 57 (1995):911–919.

126. S. Sawcer, H. B. Jones, R. Feakes, J. Gray, N. Smaldon, J. Chataway, N. Robertson, D. Clayton, P. N. Goodfellow, and A. Compston, ''A Genome Screen in Multiple Sclerosis Reveals Susceptibility Locus on Chromosome 6p21 and 17q22,'' *Nature Genetics* 13 (1996):464–468; Multiple Sclerosis Genetics Group, ''A Complete Genomic Screen for Multiple Sclerosis Underscores a Role for the Major Histocompatibility Complex,'' *Nature Genetics* 13 (1996):469–471; G. C. Ebers, K. Kukay, D. E. Bulman, A. D. Sadovnick, G. Rice, C. Anderson, H. Armstrong, K. Cousin, R. B. Bell, W. Hader, D. W. Paty, S. Hashimoto, J. Oger, P. Duquette, S. Warren, T. Gray, P. O'Connor, A. Nath, A. Auty, L. Metz, G. Francis, J. E. Paulseth, J. Murray, W. Pryse-Phillips, R. Nelson, M. Freedman, D. Brunet, J.-P. Bouchard, D. Hinds, and N. J. Risch, ''A Full Genome Search in Multiple Sclerosis,'' *Nature Genetics* 13 (1996):472–476; S. Kuokkanen, M. Sundvall, J. D. Terwilliger, P. J. Tienari, J. Wikstrom, R. Holmdahl, U. Pettersson, and L. Peltonen, ''A Putative Vulnerability Locus to Multiple Sclerosis Maps to 5p14-p12 in a Region Syntenic to the Murine Locus *Eae2*,'' *Nature Genetics* 13 (1996):477–480.

127. D. J. Lockhart, H. Dong, M. C. Byrne, M. T. Follettie, M. V. Gallo, M. S. Chee, M. Mittmann, C. Wang, M. Kobayashi, H. Horton, and E. L. Brown, ''Expression Monitoring by Hybridization to High-Density Oligonucleotide Arrays,'' *Nature Biotechnology* 14 (1996):1675–1680; M. Chee, R. Yang, E. Hubbell, A. Berno, X. C. Huang, D. Stern, J. Winkler, D. J. Lockhart, M. S. Morris, and S. P. A. Fodor, ''Accessing Genetic Information With High-Density DNA Arrays,'' *Science* 274 (1996):610–614; S. P. A. Fodor, ''Massively Parallel Genomics,'' *Science* 277 (1997):393–395.

128. J. G. Hacia, L. C. Brody, M. S. Chee, S. P. A. Fodor, and F. S. Collins, ''Detection of Heterozygous Mutations in *BRCA1* Using High Density Oligonucleotide Arrays and Two-Color Fluorescence Analysis,'' *Nature Genetics* 14 (1996):441–447; J. DeRisi, L. Penland, P. O. Brown, M. L. Bittner, P. S. Meltzer, M. Ray, Y. Chen, Y. A. Su, and J. M. Trent, ''Use of a cDNA Microarray to Analyse Gene Expression Patterns in Human Cancer,'' *Nature Genetics* 14 (1996):457–460.

129. F. S. Collins, M. S. Guyer, and A. Chakravarti, ''Variations on a Theme Cataloging Human DNA Sequence Variation,'' *Science* 278 (1997):1580–1581.

130. D. G. Wang, J.-B. Fan, C.-J. Siao, A. Berno, P. Young, R. Sapolsky, G. Ghandour, N. Perkins, E. Winchester, J. Spencer, L. Kruglyak, L. Stein, L. Hsie, T. Topaloglou, E. Hubbell, E. Robinson, M. Mittmann, M. S. Morris, N. Shen, D. Kilburn, J. Rioux, C. Nusbaum, S. Rozen, T. J. Hudson, R. Lipshutz, M. Chee, and E. S. Lander, ''Large-Scale Identification, Mapping, and Genotyping of Single-Nucleotide Polymorphisms in the Human Genome,'' *Science* 280 (1998):1077–1082.

131. Moldin, ''Indicators of Liability to Schizophrenia''; Moldin and Van Eerdewegh, ''Multivariate Genetic Analysis of Oligogenic Disease''; ———, ''Detection and Replication of Linkage to a Complex Human Disease,'' *Genetic Epidemiology* 14 (1997):1023–1028.

132. M. A. Nance, ''Invited Editorial—Huntington Disease: Another Chapter Rewritten,'' *American Journal of Human Genetics* 59 (1996):1–6; C. Gellera, C. Meoni, B. Castellotti, B. Zappacosta, F. Girotti, F. Taroni, and S. DiDonato, ''Errors in Huntington Disease Diagnostic Test Caused by Trinucleotide Deletion in the IT15 Gene,'' *American Journal of Human Genetics* 59 (1996):475–477.

133. American College of Medical Genetics/American Society of Human Genetics Working Group on ApoE and Alzheimer Disease, ''Statement on Use of Apolipoprotein E Testing for Alzheimer Disease,'' *Journal of the American Medical Association* 274 (1995):1627–1629.

134. S. H. Friend, ''Breast Cancer Susceptibility Testing: Realities in the Post-Genomic Era,'' *Nature Genetics* 13 (1996):16–17.

135. A. Allen, B. Anderson, L. Andrews, J. Beckwith, J. Bowman, R. Cook-Deegan, D. Cox, T. Duster, R. Eisenberg, B. Fine, N. Holtzman, P. King, P. Kitcher, J. McInerney, V. McKusick, J. Mulvihill, J. Murray, R. Murray, T. Murray, D. Nelkin, R. Rapp, M. Saxton, and N. Wexler, ''*The Bell Curve:* Statement by the NIH-DOE Joint Working Group on the Ethical, Legal, and Social

Implications of Human Genome Research,'' *American Journal of Human Genetics* 59 (1996):487–488.

136. I. I. Gottesman and J. Shields, *Schizophrenia: The Epigenetic Puzzle* (New York: Cambridge University Press, 1982).
137. J. Glover, ''The Implications for Responsibility of Possible Genetic Factors in the Explanation of Violence.'' In G. R. Bock and J. A. Goode, eds., *Genetics of Criminal and Antisocial Behavior* (Ciba Foundation Symposium 194) (Chichester, England: John Wiley, 1995).
138. M. C. Weinstein, H. V. Fineberg, A. S. Elstein, H. S. Frazier, D. Neuhauser, R. R. Neutra, and B. J. McNeil, *Clinical Decision Analysis* (Philadelphia: W. B. Saunders, 1980).
139. *Id.*
140. D. W. Denno, ''Legal Implications of Genetics and Crime Research.'' In G. R. Bock and J. A. Goode, eds., *Genetics of Criminal and Antisocial Behaviour* (Ciba Foundation Symposium 194) (Chichester, England: John Wiley, 1995).
141. 426 S.E.2d 150 (Ga. 1993).
142. Brunner et al., ''Abnormal Behavior Associated With a Point Mutation in the Structural Gene for Monamine Oxidase A.''
143. A. M. Dershowitz, *The Abuse Excuse and Other Cop-Outs, Sob Stories, and Evasions of Responsibility* (Boston: Little Brown, 1994).
144. Brunner et al., ''Abnormal Behavior Associated With a Point Mutation in the Structural Gene for Monamine Oxidase A.''
145. *Id.*, 579.
146. PA-96-042; available at http:/www.nih.gov/grants/pa-files/PA-96-042.html.

# ELABORATION
## New Techniques in the Genetic Analysis of Complex Illness

Hilary Coon

Once genetic factors in a disease have been established through twin and adoption studies, the search for disease genes may begin. Unfortunately, the primary tool used to search for disease-causing mutations, linkage analysis, may be limited for the analysis of complex diseases such as schizophrenia and manic depression. A host of difficulties, including non-Mendelian inheritance, reduced penetrance, phenocopies, diagnostic uncertainty, and genetic heterogeneity (different genes that produce the same effect) plague the search for susceptibility genes. Such problems may be the cause of the failure to replicate initial reports of linkage for psychiatric disorders.[1] Because effective research on the genetic components of mental disorders depends on the successful completion of the first step in this process, the location of susceptibility genes, I will explore two additional recent gene-searching techniques. Although examples I use will focus on work that has been done on schizophrenia and manic depression, these techniques are applicable to genetic studies of any complex illness. Use of such an expanded array of tools in genetic research may speed the localization of genes influencing complex behavioral traits.

### Endophenotypes

In some circumstances, abnormal results on biologic tests are correlated with complex diseases such as mental illnesses. The use of clearly defined biological traits (endophenotypes) in genetic research instead of psychiatric diagnoses offers significant promise for identifying linked chromosomal regions.[2] Such endophenotypes can provide simpler phenotypes to measure than overt psychiatric illness, and they may have higher penetrances and show patterns of inheritance that are more likely to be a result of the effects of a detectable gene. For example, the diagnostic threshold for schizophrenia remains unclear, however unaffected relatives of schizophrenics with abnormal endophenotypes may represent carriers of a genetic liability for schizophrenia. In addition, individuals with schizophrenia with abnormal endophenotypes may be more likely to be true genetic cases rather than phenocopies, because social or other environmental forces are less likely to produce subtle biological changes than changes in behavior. Thus the endophenotypes may allow us to determine a more accurate diagnosis for use in genetic studies.

An endophenotype may not be useful for genetic analysis unless it meets the following criteria: (a) demonstrable reliability and validity within clinically affected and normal populations, (b) a low population base rate, (c) substantial heritability, (d) abnormality specific to the clinical illness in question, (e) stability across acute phases and remission phases of the illness, (f) increased risk of the abnormal phenotype among family members of clinically ill subjects, and (g) a simple inheritance pattern of the trait in clinically affected families. Several endophenotypes have been proposed, particularly in searches for susceptibility genes for schizophrenia. I will review two of these schizophrenia endophenotypes. These endophenotypes are purely physiological anomalies, having no culturally based components, and may therefore be useful in populations worldwide.

## *SPEM as an Endophenotype for Schizophrenia*

Smooth pursuit eye movement (SPEM) stands out as one of the most promising of the potential endophenotypes for schizophrenia.[3] Of primary importance, SPEM has proven to be a reliable and valid measure in its own right. Even for time periods of as long as two years, test–retest reliabilities of SPEM in individuals without schizophrenia are high, ranging from 0.54 to 0.83.[4] SPEM also is stable over time in schizophrenic patients, with correlations ranging from 0.69 to 0.78.[5] High correlations among measurement methods (0.89 to 0.99) and scoring techniques (0.82 to 0.86) demonstrate the internal consistency of the variable.[6]

Estimates of population base rates of abnormal SPEM range from 4 to 8%.[7] These rates indicate that abnormal SPEM is restricted primarily to individuals with schizophrenia and their family members. SPEM also has a strong genetic component, showing higher pooled estimates of correlations among monozygotic (MZ) twin pairs (0.69) than among dizygotic (DZ) pairs (0.36).[8]

Several population-based studies report relationships between SPEM and schizophrenic symptoms.[9] Among schizophrenic patients, from 51 to 85% are reported to have abnormal SPEM.[10] Abnormal SPEM has also been reported in 30 to 50% of patients with manic-depressive illness.[11] However, abnormal SPEM in affective illness is strongly associated with lithium treatment.[12] Further, the first-degree relatives of manic-depressive patients have rates of abnormal SPEM that do not differ significantly from the estimated population rate of 8%.[13] Abnormal SPEM appears not only during acute phases of schizophrenia, but also during periods of remission. Stabilized schizophrenic outpatients have consistently lower SPEM performance than unipolar-depressed and normal control individuals.[14] In addition, SPEM is not induced or masked by neuroleptic drug treatment.[15]

Evidence also points to a simple form of inheritance of SPEM. A dominant single-gene model for the transmission of a latent susceptibility factor contributing to both schizophrenia and SPEM fit the data from MZ and DZ twin pairs (discordant for schizophrenia) and their offspring.[16] Recessive transmission and a model considering SPEM as a phenomenon separate from schizophrenia were rejected. A recent segregation analysis has also demonstrated evidence for a single major gene for SPEM, with the single gene accounting for approximately two thirds of the observed variance in eye tracking.[17]

The reviewed population study results of an up to twenty-fold increase in abnormal SPEM among individuals with schizophrenia over normal controls or persons with other psychiatric illnesses provides the first line of evidence that schizophrenia and SPEM share underlying etiology. The clustering of abnormal SPEM in schizophrenia families (and not in families of normal controls or persons with other psychiatric illnesses) provides the second line of evidence; these data indicate not only that SPEM and schizophrenia occur *together* in the same individuals within families but also that a significant proportion of unaffected relatives of individuals with schizophrenia exhibit the abnormality.[18] Schizophrenia is known to have reduced penetrance given the evidence from monozygotic twins and from the offspring of discordant monozygotic twins;[19] therefore, we expect that a significant percentage of unaffected relatives of individuals with schizophrenia are nonpenetrant gene carriers. Given the familial association of SPEM and schizophrenia, abnormal SPEM can potentially identify these nonpenetrant gene carriers.

## *P50-Evoked Sensory Gating as an Endophenotype for Schizophrenia*

The P50-evoked auditory response also shows promise as a neurophysiological endophenotype for schizophrenia.[20] This trait is measured by presenting a sound as the stimulus and using electrodes on the scalp to measure the positive brain wave that occurs 50 milliseconds after the sound is presented. In normal persons, evoked response to an initial sound conditions neuronal pathways so that response to a repetition of the same sound is lessened. However, in most schizophrenics, evoked response is approximately equal for the two stimuli.[21] A ratio of the amplitudes of response to two sounds defines the ability or lack of ability of the brain to filter the second sound.

Evidence of sensory disturbances have been reported in individuals with schizophrenia.[22] Although SPEM may measure one manifestation of such disturbances, the evoked auditory response may provide an additional, possibly related measure. SPEM and the P50-evoked auditory response have been studied together in schizophrenic families.[23] Among individual with schizophrenia, 61.5% showed abnormal eye tracking, and all of these individuals *also* showed abnormal auditory evoked response. Relatives of individuals with schizophrenia were also more likely to have both deficits, with 23% showing both abnormal SPEM and abnormal P50. However, among the relatives, the two traits also appeared independently.

Reliability of the P50 ratio in adults has been previously documented.[24] The population base rate of abnormal P50 is low, at an estimated 5 to 8%.[25] A strong association between schizophrenia and abnormal P50 has been reported and replicated.[26] A comparison of medicated and unmedicated schizophrenic patients revealed no significant differences in P50 scores,[27] indicating minimal effects of neuroleptics on this measure. Although neuroleptics altered the amplitudes of the response to the two sounds, both the initial and the second amplitudes were higher; the ratio that defines P50 remained unaltered. It is intersting to note that recent studies indicate that clozapine may have a normalizing effect on P50.[28] Although abnormal P50 occurs among patients with manic depressive illness,[29] abnormal P50 remains stable across remission phases in schizophrenia. For bipolar patients, P50 returns to normal when the patient is not experiencing mania.[30]

The inheritance pattern of schizophrenia, together with P50 (cosegregation), gives particularly strong evidence of a shared genetic etiology.[31] Although relatives of normal controls do *not* show abnormal P50, of family members of individuals with schizophrenia, usually one parent and approximately half the siblings (including the schizophrenic proband) have abnormal P50 values.[32] More than 75 to 90% of the individuals with schizophrenia in these families have abnormal P50, demonstrating the cosegregation of the schizophrenia with abnormal P50. These results suggest that unaffected family members with abnormal P50 may be genetically at risk for schizophrenia, but do not express the illness.

Heritability of P50 has been estimated at 0.74.[33] Although the mode of inheritance for P50 has not been tested directly, data are consistent with simple autosomal dominant transmission.[34]

## *Summary*

Finding the genetic causes of schizophrenia and other psychiatric disorders using standard linkage techniques has proven to be a remarkably difficult task. One method that may help surmount some of the difficulties involved in linkage analysis of schizophrenia is the use of correlated physiological variables. Such measures can have several advantages, including higher penetrances and simpler and more clearly defined patterns of inheritance. Diagnostic uncertainty has proven to be a major stumbling block in psychiatric genetics; determining the status of an individual on a well-characterized physiological measure avoids the subjectivity and uncertainty of psychiatric diagnosis. Using endophenotypes to classify relatives of individuals with schizophrenia may allow us to delineate liability to schizophrenia more clearly by detecting unaffected carriers (unaffected or mildly affected persons with endophenotype abnormalities). This refinement in potential gene-carrying status will greatly improve the chances that susceptibility genes may be found.

Endophenotypes could potentially be applied to arguments in criminal proceedings related to diminished responsibility or increased risk for recidivism. Hypothetically, the defense could argue that the presence of some endophenotype represents proof of a biological impairment in judgement or impulse control, either during a trial or during sentencing. Conversely, the prosecution could argue that the presence of an endophenotype stigmatizes the defendant as unchangeable, and therefore in need of capital punishment or preventive detention. Because no endophenotype has been identified as clearly associated with violence or criminality, this is currently a moot point. However, even if one is discovered tomorrow, it will need to be subjected to the tests of positive predictive power and negative predictive power, as described in chapter 5 by Moldin.

## Genetically Isolated Samples

Perhaps one of the most difficult barriers to the success of the gene searches for mental illness is genetic heterogeneity. Under genetic heterogeneity, one gene (or a set of genes) may be responsible for illness in some families, and an entirely different gene (or set of genes) causes illness in other families. If a researcher unknowingly includes many "subtypes" of genetic families, the effect of one susceptibility gene in one subset of families may be obscured.

One potential answer to this problem is the use of a population that has been genetically isolated, either because of geographical barriers (as with island populations) or social constraints (as with religious communities). Within a genetic isolate, a disease gene or genes may have been introduced in the founding population, and the subsequent isolation of the individuals ensures that every case of illness is caused by the same gene or set of genes.

A special technique known as linkage disequilibrium (LD) can be undertaken in a genetically isolated population. This technique relies on the knowledge that a known sequence of DNA that shows variation from individual to individual (called a *genetic marker*) is said to be tightly linked to a disease-causing gene if the marker is located very close to the disease gene on a chromosome. In an isolated population in which each case of illness has the same underlying gene(s), affected individuals (even if they are not closely related) will also very likely share the same variant of a tightly linked genetic marker. In fact, affected individuals may share the same variants of several tightly linked genetic markers in the same region surrounding the disease gene. Again, this sharing will occur because of the rarity of recombination between markers that are so close to one another. The number of markers shared among affected individuals will depend on the number of generations since the mutation, as recombination even between very tightly linked markers becomes increasingly likely over time. This is a result of the fact that the paired parental chromosomes recombine in the process of crossing over that normally occurs every time a paternal sperm cell or maternal egg cell is formed. Researchers can look for a series of such shared marker variants (called *haplotypes*) by genotyping relatively few unrelated affected individuals in the isolated population. Once a shared haplotype is found, a disease gene is likely to reside nearby. Why? The researcher assumes the individuals all share a disease gene, but that they are unlikely to share other genes because they are only related distantly through the ancestor who introduced the disease mutation. Therefore, if they share a haplotype, it is likely to reside near the unknown disease gene.

If the mutation in an isolated population is relatively old (on the order of 100 generations) the amount of shared DNA in the common haplotype may be quite small. In this case, the population isolate may be more useful for following up initial linkage findings in a fine-scale search for the disease gene. For example, if evidence for a relatively large region of linkage were found using families from a population isolate, the next step would be to genotype unrelated affected individuals with markers very close together, a task only recently possible with the discovery and cataloging of thousands of new variable DNA markers.[37]

Some researchers have advocated the use of the LD strategy even in populations that have not been genetically isolated.[36] However, a primary underlying assumption of the technique is that all individuals share a common disease gene, so that when shared segments of DNA occur, they will correspond to that shared disease gene. In addition, several effects, including population stratification or nonindependent sampling can confound results in nonisolated populations.[37]

Genes for several single-gene diseases have been successfully mapped in isolated Finnish populations.[38] In a large bipolar family from an isolated Finnish region, a distinct haplotype on the long arm of the X chromosome (Xq24-q27.1) was found in affected members of the pedigree.[39] Work has also begun on two Finnish schizophrenia families, but with no definitive results as yet.[40]

Individuals with manic depression have also been identified in the Costa Rican Central Valley, an area that was settled by a small number of founders in the sixteenth to eighteenth centuries and has been isolated until relatively recently.[41] Linkage analysis was first used in two large families. Suggestive lod scores were found on the long arm of chromosome 18 (18q22-q23), and this region was genotyped with 14 closely spaced markers. Using LD, the affected individuals in these families were found to share haplotypes in the chromosome 18 region of linkage, even when they were only very distantly related.

The Utah schizoprenia research group has recently identified an isolated island population in Palau, Micronesia.[42] Founded by a small number of settlers in approximately A.D. 40, Palau has remained ethnically and geographically isolated for more than 2000 years, as shown by blood group clustering, linguistic analysis, and the recorded history of travel between islands.[43] Initial work has revealed several very large families of more than 100 members, each with 10 or more cases of schizophrenia.[44] This population will give us an excellent resource to search for schizophrenia susceptibility genes using linkage and LD techniques.

## *Summary*

Given that mental disorders, and especially behavioral conditions such as predisposition to criminality, are likely to be heterogeneous illnesses, affected families and individuals from genetically isolated populations offer an efficient, cost-effective, and statistically powerful resource in which to search for susceptibility genes. Linkage analysis may be more successful because a gene contributing to disease susceptibility may be more likely to reside on a single ancestral haplotype. In contrast, considerable time and resources are required to ascertain and genotype the large number of relatively small families needed to detect linkage assuming a substantial degree of genetic heterogeneity.[45] In addition, recent studies have generated a number of relatively large potential chromosomal regions of interest, especially for the major psychiatric disorders. Linkage disequilibrium techniques applied to genetic isolates may be used for fine-scale searches within these promising regions of linkage.

# Notes

1. N. Risch, "Genetic Linkage and Complex Diseases, With Special Reference to Psychiatric Disorders," *Genetic Epidemiology* 7 (1990):3–16.
2. P. S. Holzman, E. Kringlen, S. Matthysse, S. D. Flanagan, R. B. Lipton, G. Cramer, S. Levin, K. Lange, and D. L. Levy, "A Single Dominant Gene Can Account for Eye Tracking Dysfunctions and Schizophrenia in Offspring of Discordant Twins," *Archives of General Psychiatry* 45 (1988):641–647; S. Matthysse, "Genetic Linkage and Complex Diseases: A Comment," *Genetic Epidemiology* 7 (1990): 29–31; S. Matthysse, and J. Parnas, "Extending the Phenotype of Schizophrenia: Implications for Linkage Analysis," *Journal of Psychiatric Research* 26 (1992):329–344; S. O. Moldin, "Indicators of Liability Schizophrenia: Perspectives From Genetic Epidemiology," *Schizophrenia Bulletin* 20 (1994):169–184; S. O. Moldin, J. P. Rice, P. Van Eerdewegh, I. I. Gottesman, and L. Erlenmeyer-Kimling, "Estimation of Disease Risk Under Bivariate Models of Multifactorial Inheritance," *Genetic Epidemiology* 7 (1990):371–386.
3. B. A. Clementz and J. A. Sweeney, "Is Eye Movement Dysfunction a Biological Marker Variable for Schizophrenia? A Methodological Review," *Psychological Bulletin* 108 (1990):77–92; P. S. Holzman, "Recent Studies of Psychophysiology in Schizophrenia," *Schizophrenia Bulletin* 13 (1987):49–75; W. G. Iacono, "Psychophysiology and Genetics: A Key to Psychopathology Research," *Psychophysiology* 20 (1983):371–383.
4. W. G. Iacono and D. T. Lykken, "Two-Year Retest Stability of Eye Tracking Performance and a Comparison of Electro-Oculographic and Infrared Recording Techniques: Evidence of EEG in the Electro-Oculogram," *Psychophysiology* 18 (1981):49–55.
5. R. J. van den Bosch, N. Rozendaal and J. M. F. A. Mol, "Symptom Correlates of Eye Tracking Dysfunction," *Biological Psychiatry* 22 (1987):919–921.
6. Clementz and Sweeney, "Is Eye Movement Dysfunction a Biological Marker Variable for Schizophrenia?"; Iacono and Lykken, "Two-Year Retest Stability of Eye Tracking Performance and a Comparison of Electro-Oculographic and Infrared Recording Techniques."
7. P. S. Holzman, C. M. Solomon, S. Levin, and C. S. Waternaux, "Pursuit Eye Movement Dysfunctions in Schizophrenia: Family Evidence for Specificity," *Archives of General Psychiatry* 41 (1984):136–139; W. G. Iacono, M. Moreau, M. Beiser, J. A. Fleming, and T. Y. Lin, "Smooth Pursuit Eye Tracking in First-Episode Psychotic Patients and Their Relatives," *Journal of Abnormal Psychology* 101 (1992):104–116.
8. P. S. Holzman, E. Kringlen, D. L. Levy, S. Haberman, L. R. Proctor, and N. J. Yasillo, "Abnormal Pursuit Eye Movements in Schizophrenia: Evidence for a Genetic Indicator," *Archives of General Psychiatry* 34 (1977):802–805; P. S. Holzman, E. Kringlen, D. L. Levy, L. R. Proctor, and S. Haberman, "Smooth Pursuit Eye Movements in Twins Discordant for Schizophrenia," *Journal of Psychiatric Research* 14 (1978):111–122; P. S. Holzman, E. Kringlen, D. L. Levy, and S. Haberman, "Deviant Eye Tracking in Twins Discordant for Psychosis: A Replication," *Archives of General Psychiatry* 32 (1980):627–631; W. G. Iacono, "Eye Tracking in Normal Twins," *Behavior Genetics* 12 (1982):517–526.
9. L. J. Siever, R. D. Coursey, I. S. Alterman, M. S. Buchsbaum, and D. L. Murphy, "Impaired Smooth Pursuit Eye Movement: Vulnerability Marker for Schizotypal Personality Disorder in a Normal Volunteer Population," *American Journal of Psychiatry* 141 (1984):1560–1566; R. Simmons, J. Graydon, D. Gajdusek, and P. Brown, "Blood Group Genetic Variations in Natives of the Caroline Islands and in Other Parts of Micronesia," *Oceania* 36 (1965):132–170.
10. Iacono et al., "Smooth Pursuit Eye Tracking in First-Episode Psychotic Patients and Their Relatives"; Holzman et al., "Pursuit Eye Movement Dysfunctions in Schizophrenia"; P. S. Holzman, L. R. Proctor, and D. W. Hughes, "Eye Tracking Patterns in Schizophrenia," *Science* 181 (1973):179–181.
11. Iacono, "Eye Tracking in Normal Twins"; C. Shagass, R. A. Roemer, and M. Amadeo, "Eye Tracking Performance in Psychiatric Patients," *Biological Psychiatry* 14 (1974):245–261.
12. Iacono, "Eye Tracking in Normal Twins"; D. L. Levy, E. Dorus, R. Shaughnessy, N. J. Yasillo, G. N. Pandey, P. G. Janicak, R. D. Gibbons, M. Gaviria, and J. M. Davis, "Pharmacologic

Evidence for Specificity of Pursuit Dysfunction to Schizophrenia: Lithium Carbonate Associated With Abnormal Pursuit,'' *Archives of General Psychiatry* 42 (1985):335–341; P. S. Holzman, C. O'Brien, and C. S. Waternaux, ''Effects of Lithium Treatment on Eye Movements,'' *Biological Psychiatry* 29 (1991):1001–1015.

13. Holzman et al., ''Pursuit Eye Movement Dysfunctions in Schizophrenia''; D. L. Levy, N. J. Yasillo, E. Dorus, R. Shaughnessy, R. D. Gibbons, J. Peterson, P. G. Janicak, R. D. Gibbons, M. Gaviria, and J. M. Davis, Relatives of Unipolar and Bipolar Patients Have Normal Pursuit,'' *Psychiatric Research* 10 (1983):285–293.
14. Iacona and Lykken, ''Two-Year Retest Stability of Eye Tracking Performance and a Comparison of Electro-Oculographic and Infrared Recording Techniques''; W. G. Iacono and W. G. R. Koenig, ''Features That Distinguish the Smooth Pursuit Eye-Tracking Performance of Schizophrenic, Affective-Disorder, and Normal Individuals,'' *Journal of Abnormal Psychology* 92 (1983):29–41; W. G. Iacono, L. J. Peloquin, A. E. Lumry, R. H. Valentine, and V. B. Tuason, ''Eye Tracking in Patients With Unipolar and Bipolar Affective Disorders in Remission,'' *Journal of Abnormal Psychology* 91 (1982):35–44.
15. Levy et al., ''Pharmacologic Evidence for Specificity of Pursuit Dysfunction to Schizophrenia''; I. David, ''Disorders of Smooth Pursuit Eye Movements in Schizophrenics and the Effect of Neuroleptics in Therapeutic Doses,'' *Activitas Nervosa Superior* 22 (1980):115–156; P. S. Holzman, D. L. Levy, E. H. Uhlenhuth, L. R. Proctor, and D. X. Freedman, ''Smooth Pursuit Eye Movements and Diazepam, CPZ, and Secobarbital,'' *Psychopharmacologia* 44 (1975):111–114; D. L. Levy, R. B. Lipton, P. S. Holzman, and J. M. Davis, ''Eye Tracking Dysfunction Unrelated to Clinical State and Treatment With Haloperidol,'' *Biological Psychiatry* 18 (1983):813–819.
16. Holzman et al., ''A Single Dominant Gene Can Account for Eye Tracking Dysfunctions and Schizophrenia in Offspring of Discordant Twins.''
17. W. M. Grove, B. A. Clementz, W. G. Iacono, and J. Katsanis, ''Smooth Pursuit Ocular Motor Dysfunction in Schizophrenia: Evidence for a Major Gene,'' *American Journal of Psychiatry* 149 (1992):1362–1368.
18. Iacono et al., ''Smooth Pursuit Eye Tracking in First-Episode Psychotic Patients and Their Relatives''; Holzman et al., ''Pursuit Eye Movement Dysfunctions in Schizophrenia''; Holzman et al., ''Deviant Eye Tracking in Twins Discordant for Psychosis''; P. S. Holzman, L. R. Proctor, D. L. Levy, N. J. Yasillo, H. Y. Meltzer, and S. W. Hurt, Eye-Tracking Dysfunctions in Schizophrenic Patients and Their Relatives,'' *Archives of General Psychiatry* 31 (1974):143–151.
19. I. I. Gottesman, *Schizophrenia Genesis: The Origins of Madness* (New York: W. H. Freeman, 1991).
20. R. Freedman, L. E. Adler, G. A. Gerhardt, M. Waldo, N. Baker, G. M. Rose, C. Drebing, H. Nagamoto, P. Bickford-Wimer, and R. Franks, ''Neurobiological Studies of Sensory Gating in Schizophrenia,'' *Schizophrenia Bulletin* 13 (1987):669–677.
21. R. Freedman, L. E. Adler, M. C. Waldo, E. Pachtman, and R. D. Franks, ''Neurophysiological Evidence for a Defect in Inhibitory Pathways in Schizophrenia: Comparison of Medicated and Drug-Free Patients,'' *Biological Psychiatry* 18 (1983):537–551.
22. P. Venables, ''Input Dysfunction in Schiophrenia.'' In B. A. Mahler, *Progress in Experimental Personality Research* (New York: Academic Press, 1964); T. Patterson, H. E. Spohn, D. P. Bogia, and K. Hayes, ''Thought Disorder in Schizophrenia,'' *Schizophrenia Bulletin* 12 (1986):460–472.
23. C. Siegel, M. Waldo, G. Mizner, L. E. Adler, and R. Freedman, ''Deficits in Sensory Gating in Schizophrenic Patients and Their Relatives,'' *Archives of General Psychiatry* 41 (1984):607–612.
24. M. C. Waldo and R. Freedman, ''Gating of Auditory Evoked Responses in Normal College Students,'' *Psychiatric Research* 19 (1986):233–239.
25. Siegel et al., ''Deficits in Sensory Gating in Schizophrenia Patients and Their Relatives.''
26. R. Freedman, L. E. Adler, G. A. Gerhardt, M. Waldo, N. Baker, G. M. Rose, and C. Drebing, ''Neurobiological Studies of Sensory Gating in Schizophrenia,'' *Schizophrenia Bulletin* 13 (1987):669–677; L. L. Judd, L. A. McAdams, B. Budnick, and D. L. Braff, ''Sensory Grating Deficits in Schizophrenia: New Results,'' *American Journal of Psychiatry* 149 (1992):488–493.
27. R. Freedman, L. E. Adler, M. C. Waldo, E. Pachtman, and R. D. Franks. ''Neurophysiological

Evidence for a Defect in Inhibitory Pathways in Schizophrenia: Comparison of Medicated and Drug-Free Patients,'' *Biological Psychiatry* 18 (1983):537–551.

28. H. T. Nagamoto, L. E. Adler, R. A. Hea, J. M. Griffith, K. A. McRae, and R. Freedman, ''Gating of Auditory P50 in Schizophrenics: Unique Effects of Clozpaine,'' *Biological Psychiatry* (in press).
29. Freedman et al., ''Neurobiological Studies of Sensory Gating in Schizophrenia.''
30. R. D. Franks, L. E. Adler, M. C. Waldo, J. Alpert, and R. Freedman, ''Neurophysiological Studies of Sensory Gating in Mania: Comparison With Schizophrenia,'' *Biological Psychiatry* 18 (1983):989–1005.
31. M. C. Waldo, G. Carey, M. Myles-Worsley, E. Cawthra, L. E. Adler, H. T. Nagamoto, P. Wender, W. Byerly, R. Plaetke, and R. Freedman, ''Co-Distribution of a Sensory Gating Deficit and Schizophrenia in Multi-Affected Families,'' *Psychiatric Research* 39 (1991):257–268.
32. Siegel et al., ''Deficits in Sensory Gating in Schizophrenic Patients and Their Relatives.''
33. M. Myles-Worsley, H. Coon, W. Byerley, M. Waldo, D. Young, and R. Freedman, ''Developmental and Genetic Influences on the P50 Sensory Gating Phenotype,'' *Biological Psychiatry* 39 (1996):289–295.
34. Siegel et al., ''Deficits in Sensory Gating in Schizophrenic Patients and Their Relatives.''
35. T. J. Hudson, L. D. Stein, S. G. Sebastian, J. Ma, A. B. Castle, J. Silva, D. K. Slonim, R. Baptista, L. Kuglyak, and S. H. Xu, ''An STS-Based Map of the Human Genome,'' *Science* 270 (1995):1945–1954.
36. J. B. Copeman, J. Cucca, C. M. Hearne, R. J. Cornall, P. W. Reed, K. S. Ronningen, D. E. Undlien, L. Nistico, R. Buzzetti, and R. Tosi, ''Linkage Disequilibrium Mapping of a Type 1 Diabetes Susceptibility Gene (IDDM7) to Chromosome 2q31-q33,'' *Nature Genetics* 9 (1995):80–85.
37. L. B. Jorde, W. S. Watkins, D. Viskochil, P. O'Connell, and K. Ward, ''Linkage Disequilibrium in the Neurofibromatosis I (NFI) Region: Implications for Gene Mapping,'' *American Journal of Human Genetics* 53 (1993):1038–1050; L. B. Jorde, ''Invited Editorial: Linkage Disequilibrium as a Gene-Mapping Tool,'' *American Journal of Human Genetics* 56 (1995):11–14.
38. A. de la Chapelle, ''Disease Gene Mapping in Isolated Human Populations: The Example of Finland,'' *Journal of Medical Genetics* 30 (1993):857–865; J. Hästbacka, A. de la Chapelle, I. Kaitila, P. Sistonen, A. Weaver, and E. Lander, ''Linkage Disequilibrium Mapping in Isolated Founder Populations: Diastrophic Dysplasia in Finland,'' *Nature Genetics* 2 (1992):204–211; E. Tahvahainen, H. Forsius, M. Damsten, E. Karila, J. Kolehmainen, J. Weissenbach, P. Sistonen, and A. de la Chapelle, ''Linkage Disequilibrium Mapping of the Cornea Plana Congenital Gene CNA2,'' *Genomics* 30 (1995):409–414.
39. P. Pekkarinen, J. Terwilliger, P.-E. Bredbacka, J. Lonnqvist, and L. Peltonen, ''Evidence of a Predisposing Locus to Bipolar Disorder on Xq24-q27.1 in an Extended Finnish Pedigree,'' *Genome Research* 5 (1995):105–115.
40. I. Hovatta, J. Seppala, P. Pekkarinen, A. Tanskanen, J. Lonnqvist, and L. Peltonen, ''Linkage Analysis in two Schizophrenic Families Originating From a Restricted Subpopulation of Finland,'' *Psychiatric Genetics* 4 (1994):143–152.
41. N. B. Freimer, V. I. Reus, M. A. Escamilla, L. A. McInnes, M. Spesny, P. Leon, S. K. Service, L. B. Smith, S. Silva, E. Rojas, A. Gallegos, L. Meza, E. Fournier, S. Baharloo, K. Blankenship, D. J. Tyler, S. Batki, S. Vinogradov, J. Weissenbach, S. H. Barondes, and L. A. Sandkuijl, ''Genetic Mapping Using Haplotype, Association, and Linkage Methods Suggests a Locus for Severe Bipolar Disorder (BPI) at 18q22-q23,'' *Nature Genetics* 12 (1996):436–441.
42. M. Myles-Worsley, H. Coon, J. Tiobech, J Collier, P. Dale, P. Wender, P. Reimherr, A. Polloi, and W. Byerley, ''A Genetic Study of Schizophrenia in Palau, Micronesia. I. Prevalence and Familiality,'' *Neuropsychiatric Genetics* (in press); H. Coon, M. Myles-Worsley, M. Waldo, J. Holik, M. Hoff, J. Yaw, P. Dale, A. Polloi, R. Freedman, and W. Byerley, ''Segregation, Linkage, and Linkage Disequilibrium: Analysis of Micronesian Schizophrenia Families,'' *Psychiatric Genetics* 5 (1995):549.
43. R. J. Parmentier, *The Sacred Remains, Myth, History, and Polity in Belau,* (Chicago: University of

Chicago Press, 1987). J. Takayama, ''Early Pottery and Population Movement in Micronesian Prehistory,'' *Asian Perspectives* 24 (1981):1–10.

44. H. Coon, M. Myles-Worsley, J. Tiobech, M. Hoff, J. Rosenthal, P. Bennett, F. Reimherr, P. Wender, P. Dale, A. Polloi, and W. Byerley; ''Evidence for a Chromosome 2p13-14 Schizophrenia Susceptiblity Locus in Families from Palau, Micronesia,'' *Molecular Psychiatry* 3 (1998):521–527.
45. K. S. Kendler and S. R. Diehl, ''The Genetics of Schizophrenia: A Current, Genetic–Epidemiological Perspective,'' *Schizophrenia Bulletin* 19 (1993):261–285.

# Part III

# Genetic Research in Relation to Criminal and Juvenile Law

# INTRODUCTION TO PART III

This part of the book takes up difficult issues about whether any of the developments in genetic research described in this volume are relevant to criminal or juvenile law. These areas of the law have been selected for scrutiny because of the importance to them of judgments of responsibility. For the past several centuries, criminal law has struggled to delineate the role judgments of responsibility should play in decisions about fitness to stand trial, conviction, and punishment. Juvenile law has struggled in addition with the applicability of adult models to the treatment of younger offenders.

The initial contribution to the section is Rebecca Dresser's analysis of whether a ''genetics defense'' can be expected to develop in the criminal law. Dresser concludes that our increasing understanding of behavioral genetics is unlikely to cause major shifts in our understanding of criminal responsibility. Criminal law abandons ordinary notions of personal responsibility only when behavior is seen to have been compelled or otherwise outside the control of the person. Genetics will contribute to an understanding of the causal origins of behavior, but this is a quite different matter than seeing behavior as compelled. At the same time, it is important for participants in the criminal justice process—judges, juries, lawyers, and the interested public—to understand developments in genetics so that genetic information is neither misused nor ignored when it is genuinely helpful in understanding why offenders acted as they did. Criminal law, Dresser also notes, will not abandon its commitment to protecting society from dangerous offenders.

In her contribution, Mary Crossley agrees that genetic information should not generate radical changes in the paradigm of criminal responsibility. She is concerned, however, that public misunderstanding of genetics, illustrated by the tendency to assume the kind of genetic determinism warned against in earlier parts of this volume, will generate pressures for the development of a ''genetics defense.'' These pressures, she fears, will be put to the service of solving the ''crime problem'' through restraints, medication, and incapacitation, rather than through deeper critique and amelioration of social environments.

Daniel Summer is the defense attorney who raised the possibility of a genetics defense in the case of Stephen Mobley. Summer gives a sympathetic account of how the genetics defense might be raised, and why it might be relevant to judgments of responsibility. His view is that in some cases genetic characteristics should properly be viewed as causally involved in behavior in a manner that is sufficient to be regarded as exculpatory. It is fair to say, however, that in reaching this conclusion Summer probably sets a higher threshold for judgments of responsibility than would other contributors to this volume. Summer's concluding advice for the zealous defense attorney is to leave no genetics stone unturned, but for the present, evidence of genetic causation is most likely to be taken into account by courts in making sentencing decisions, particularly in mitigating against imposition of the death penalty.

Creighton Horton, a prosecutor, agrees with Dresser that genetic information is unlikely to change our fundamental paradigm of criminal responsibility. Even if genes play a clear role in causing behavior, it does not follow that they undercut the defendant's possession of the mental element of the offense, which is necessary to exculpation in some jurisdictions. Nor does understanding a genetic causal role abrogate the criminal law's traditional interest in protecting the citizenry. Disagreeing with Summer, Horton also suggests that juries are more likely to see environmental factors such as abuse as sympathetic bases for mitigation than they are genetic differences.

Juvenile law, for most of the past 50 years, has struggled with the extent to which it should follow adult paradigms of responsibility. In recent years, moved by fears of increasingly violent and young offenders, the pendulum has shifted toward the adoption of adult models of responsibility and punishment in the juvenile arena. Mark Small, in his contribution, argues that misunderstanding of genetic information will intensify these shifts. Juvenile offenders may be perceived as determined by their genes and as irredeemably "criminal." The result may be an abandonment of efforts at reform and rehabilitation. Jeff Kovnick, a psychiatrist, is also sympathetic to differentiating adult and juvenile models of responsibility. Like Small, Kovnick is concerned that misunderstanding of genetic information will harden demands for punishment of juvenile offenders. Kovnick sees an irony in this result, however, if fuller understanding of genetic and other causes of behavior suggest that punitive paradigms are more appropriate for adult than for juvenile offenders.

Two themes are paramount in this part. The first is that our developing knowledge of behavioral genetics should not undermine current paradigms of responsibility in criminal and juvenile law. The second is that if the public misunderstands the findings of genetics, inflating them to demonstrate genetic determinism, resulting pressure may do great damage both to individual offenders and to our public policy with respect to crime. These themes urge great caution in the face of burgeoning genetic knowledge. The concerns raised in the final part of this volume intensify the call for caution.

# Chapter 6
# CRIMINAL RESPONSIBILITY AND THE "GENETICS DEFENSE"

Rebecca Dresser

Principles of criminal responsibility and punishment assume that human beings for the most part are capable of controlling their behavior. Criminal liability rules presuppose that the ordinary person has a reasonable opportunity to conform to legal standards delineating permissible conduct. According to this vision of human behavior, it is morally fair for society to impose restrictions and burdens on the individual who chooses not to comply with these standards.

The existing criminal justice system has evolved over hundreds of years. It primarily reflects common-sense metaphysical and moral judgments about human nature and free will. In the past century, however, more deterministic models of human action have intruded on the law's accepted wisdom. The current speculation about genetic predispositions to violent and other forms of criminal behavior is simply the latest of these potential intrusions.

Could behavioral genetics research pose a serious challenge to the folk psychological model of human behavior implicit in the criminal law? Could this emerging body of scientific work succeed in altering principles that have so far resisted the deterministic influence? What would be required for behavioral genetics research to alter existing notions of criminal responsibility?

In this chapter, I address these and other questions raised by scientific investigations in behavioral genetics. I first will review the traditional justifications of punishment and their reliance on the presumption of individual choice and control. I will then examine three instances in which the presumption of individual responsibility has been challenged in the courts. Next I will consider three impediments to the criminal law's acceptance of a genetics excuse. Finally, I will address potential alternative forms of social control for persons deemed unable to refrain from criminal conduct. I conclude that serious disruptions in legal notions of criminal responsibility are unlikely, but that courts will need guidance on the proper response to claims of a genetics defense.

## Principles of Punishment and Criminal Responsibility

Although the civil law frequently imposes financial and other burdens on persons for their failure to fulfill legal standards and responsibilities, punishment is society's most severe response to individual misbehavior. Through punishment, society deprives persons of physical freedom, contact with loved ones, opportunities for employment and recreation, and myriad other activities important to quality of life. Moreover, punishment is stigmatizing and damaging to the convicted person's reputation; for the most serious offenses, punishment can yield loss of life or of basic civil rights such as voting. Persuasive justification is required for the state to take such oppressive action against an individual.

Traditionally, four major justifications for punishment are presented. According to *retributive theory,* persons violating society's criminal prohibitions deserve punishment. Explanations of retributive punishment vary. Most primitive is the retaliatory "eye for an eye" account, in which society is given moral warrant to (and by some interpretation, is morally required to) impose on an offender the same harm the offender imposed on members of society.[1]

A more recent retributive account portrays society as a system in which each of us is granted freedom to pursue our activities and projects as long as we respect the rights of others to do the same. When someone fails to meet this condition and interferes with the protected freedom of others, society may rectify the moral imbalance by restricting the intruder's own freedom.[2] A third variation sees punishment as promoting respect for autonomous persons. Holding individuals responsible for their bad acts acknowledges their status as moral agents and gives them an opportunity to recognize and do penance for their wrongdoing.[3]

As this brief description indicates, retributive theory is morally defensible only if individuals are able to choose and control what they do. Retributivists characterize punishment as deserved when persons make autonomous decisions to advance their own interests by taking unfair advantage of others.

H. L. A. Hart's discussion of the fair distribution of punishment further highlights the importance of individual freedom and autonomy to retributive accounts. Although retributive theory is commonly perceived as harsher than the alternative punishment justifications, Hart pointed out its crucial protective dimension. The retributive constraint on punishment prevents society from imposing punishment on individuals purely to advance the interests of others. Persons may be punished only if they *deserve* it—that is, only if they have chosen to engage in criminal conduct.[4]

Consistent with this view, retributive theory excuses from punishment persons incapable of choosing to behave according to standards set by the criminal law. On the retributive model, such persons cannot fairly be blamed for their actions. George Fletcher described the retributive inquiry as essentially normative: Would it "be fair under the circumstances to expect the actor to resist the pressures of the situation and abstain from the criminal act"?[5] If legal decision makers conclude that this expectation would be unfair, retributive theory would label punishment morally inappropriate.

The significance of the retributive constraint on punishment becomes evident when one examines the three remaining punishment justifications: deterrence, rehabilitation, and incapacitation. All have their foundation in utilitarian ethical theory, and all are targeted at reducing future crime, in contrast to the retributive focus on past wrongdoing. According to utilitarian theory, punishment is justified when the burdens punishment inflicts on an individual are outweighed by the benefits society obtains in the form of crime reduction.[6]

Deterrence operates at two levels. Punishment advances specific deterrence aims if its burdens are sufficiently heavy to discourage a convicted offender from committing future crimes. Punishment is justified as promoting general deterrence when its negative effects convey a message discouraging potential offenders in the general population from violating the criminal law. Deterrence relies on the assumption that the individual criminal and would-be offenders in society can shape their future conduct to avoid the burdens of punishment. Without individual control over behavior, deterrence would be impossible to achieve and thus would be unacceptable as justification for punishment.

Rehabilitative theories of punishment also are premised on the existence of human choice and control over behavior. The *rehabilitative approach* sees punishment as justified if it makes persons aware of and inclined to choose law-abiding conduct when their freedom is restored. Offenders are exposed to new ways of promoting their interests, through education or job training, in hopes that they will make different behavioral choices in the future.

Incapacitation is the sole approach that justifies punishment without relying on the concept of individual control. On this model, restraining dangerous people is a morally defensible practice if it protects society from future crime. Putting offenders in prison is analogous to caging dangerous animals. The convicted person is seen as unable to behave in a law-abiding manner, and external controls are the appropriate way to prevent future harm to others. In selective incapacitation, individuals identified as probable recidivists are confined for a lengthier period than others convicted of the same offense.

In sum, if offenders cannot adequately control their future conduct, the utilitarian cannot cite deterrence and rehabilitation as morally acceptable reasons for imposing the pain of punishment. Without individual control, neither deterrence nor rehabilitation will effectively reduce future crime; thus the burdens of punishment will not be outweighed by its benefit to society.

In contrast, incapacitation remains acceptable in the absence of individual control. Incarceration is permitted as long as an individual remains a threat to society. Moreover, because utilitarian theory omits the retributive constraint on punishment, the theory permits incarceration of dangerous persons even when they have done nothing to deserve punishment. If incapacitation alone is deemed an adequate justification for punishment, persons who are not responsible for their criminal behavior may be imprisoned simply to protect society.

Currently, however, the U.S. criminal justice system accepts the retributive constraint on punishment. The formal position is that persons should be punished only when they can fairly be held responsible for their criminal acts. As long as this stance persists, individual responsibility will continue to be a precondition to justified punishment. Unless the criminal justice system adopts incapacitation as its primary moral justification, the system's moral warrant will depend strongly on the presumption of individual control. In this situation, evidence that certain individuals are genetically predisposed to antisocial behavior could raise questions about the fairness of punishing such persons.

## Excusing Conditions

Existing legal doctrines recognize certain instances in which accused persons ought not to be held responsible and, accordingly, ought not be punished for committing a criminal offense. The organizing theme for legal doctrines of excuse is a judgment that the criminal conduct at issue was in some sense involuntary. As Michael Moore noted, "When an agent is caused to act by a factor outside his control, he is excused; . . . those acts not caused by some factor external to his will are unexcused."[7] I will focus on three situations in which individual responsibility has been questioned on grounds that the wrongful act was caused by "some factor external to [the defendant's] will."

### *The Voluntary Act Requirement*

A longstanding criminal law rule holds that punishment is morally appropriate solely for persons who voluntarily commit criminal acts. According to this rule, it is unfair to punish someone who lacked the ability to refrain from the harmful conduct. The Model Penal Code (MPC), a respected and influential model criminal code, labels involuntary the following behavior: reflexes, convulsions, conduct during hypnosis or as a result of hypnotic suggestion, and "bodily movement that otherwise is not a product of the effort or determination of the actor, either conscious or habitual."[8]

According to the MPC drafters, the voluntary act requirement exists because "the law cannot hope to deter involuntary movement or to stimulate action that cannot be physically performed; the sense of personal security would be undermined in a society where such movement or inactivity could lead to formal social condemnation of the sort that a conviction necessarily entails.''[9] The rule is interpreted narrowly, however, and does not absolve criminal actors who cannot remember their conduct, behaved impulsively, or acted without subjective awareness that they were engaging in criminal conduct.

In the 1960s, the U.S. Supreme Court issued two important opinions interpreting the voluntary act requirement. In these cases, the Court considered whether punishment of drug addicts and alcoholics should be prohibited as cruel and unusual punishment in violation of the Eighth Amendment of the U.S. Constitution. Their discussions addressed the question of whether the criminal conduct of addicts and alcoholics is sufficiently voluntary to justify punishment.

In *Robinson v. California,*[10] the Supreme Court invalidated a statute that made it a criminal offense ''to be addicted to the use of narcotics.'' The Court's majority opinion referred to addiction as an illness, and asserted that treating addiction as a criminal offense would ''doubtless be universally thought to be an infliction of cruel and unusual punishment.'' The opinion contained language suggesting that addicts should not be held responsible for their conditions. For six years, commentators debated *Robinson's* implications, wondering whether the decision meant that states could no longer criminalize narcotics use, possession, or other behavior, such as theft or burglary, associated with an individual's drug addiction.

*Powell v. Texas*[11] clarified the Court's position. In this case, the Court considered the constitutionality of punishing an alcoholic for public intoxication. Powell claimed that he was unable to stop drinking and that while intoxicated he lacked control over his behavior. A psychiatrist testified that Powell was a chronic alcoholic, an ''involuntary drinker,'' and someone who could not control his behavior once he became intoxicated. Both admitted, however, that Powell retained some control during the initial stage of a drinking episode.

The Supreme Court upheld Powell's conviction. According to the plurality opinion, Powell's situation could be distinguished from Robinson's because the Texas statute punished ''public behavior which may create substantial health and safety hazards.'' In contrast, the California law at issue in *Robinson* punished the ''mere status'' of *being* an addict, without requiring any evidence of actual criminal conduct. According to the plurality, *Robinson* did not intend to label behavior associated with drug or alcohol addiction involuntary, compelled, or otherwise beyond the control of the individual addict. Members of the plurality labeled existing medical knowledge inadequate to support a finding that chronic alcoholics ''suffer from such an irresistible compulsion to drink and to get drunk in public that they are utterly unable to control their performance of either or both of these acts.''[12] Four Justices disagreed with this conclusion, however, and argued that Powell's conviction violated the Eighth Amendment prohibition on punishing someone ''for being in a condition he is powerless to change.''[13]

*Powell* held that the Constitution permits states to label conduct associated with alcoholism and drug addiction as voluntary behavior justifying punishment. The decision leaves states free to classify such behavior as involuntary if they so choose, however. The Supreme Court did express concern regarding the consequences of such a policy choice:

> If Leroy Powell cannot be convicted of public intoxication, it is difficult to see how a State can convict an individual for murder, if that individual, while exhibiting normal behavior in all other respects, suffers from a ''compulsion'' to kill, which is an ''exceedingly strong influence,'' but ''not completely overpowering.'' Even if we limit our consideration to chronic alcoholics, it would seem impossible to confine the principle. . . .[14]

Since these cases were decided, some state courts have considered whether certain behavior of alcoholics and addicts should be excused from criminal punishment. Courts generally have been unwilling to consider the actions of addicts and alcoholics sufficiently compelled to be excluded from the realm of personal responsibility. As one commentator remarked, "The relative paucity of cases in which [alcoholic or drug-addicted] defendants have pressed a loss-of-control defense, together with the near universal hostility accorded such arguments by the few courts to reach the issue, is significant evidence that a judicially created involu[ntariness] doctrine is unlikely to emerge in the foreseeable future."[15]

## *The Insanity Defense*

The evolution of the modern insanity defense is another illustration of the law's reluctance to depart from the presumption that persons ordinarily exercise choice and control over their actions. Traditionally, the insanity defense applied only when it could be "clearly proved that, at the time of the committing of the act, the party accused was labouring under such a defect of reason, from a disease of the mind, as not to know the nature and quality of the act he was doing; or, if he did know it, that he did not know he was doing what was wrong."[16] Somewhat later, many jurisdictions added what is known as the "irresistible impulse" formulation, which excuses someone whose mental disease or defect "completely deprives the person of the power of choice or volition."[17]

Except for a short-lived departure during the 1950s, the insanity test has been considered primarily a moral and social decision for legal fact-finders, based on their common sense ideas of justice and fairness. In 1954, the infamous *Durham* decision sought to transform the insanity test into a matter of scientific judgment, by adopting an insanity standard that excused persons whose criminal acts were "the product of a mental disease or defect."[18] Judges soon realized, however, that psychiatrists tended to adopt a deterministic view of human behavior that severely conflicted with the legal presumption of individual control.

Under the *Durham* test, legal insanity was largely resolved by psychiatrists testifying whether the defendant met the profession's criteria for mental illness. Their training led most psychiatrists to attribute all behavior of a mentally ill defendant to the illness. It took just a few years for courts to reject *Durham* and return to a version of the traditional tests. As one court put it, "The determination whether a man is or is not held responsible is not a medical but a legal, social, or moral judgment."[19] Though the expert witness could supply legal fact-finders with information relevant to their decision, the ultimate decision belonged to those representing ordinary members of society.

Contrary to popular perceptions, the insanity defense is infrequently raised and rarely successful. One court reviewing the data estimated the insanity acquittal rate at no more than 0.2 percent of felony cases.[20] Moreover, social science studies suggest that juries reach insanity verdicts at similar rates regardless of the particular insanity test they are instructed to apply.[21] According to one commentator, juries tend to acquit as insane only those defendants who appear, based on the evidence presented, to have been clearly irrational when the crime was committed.[22]

## *The XYY Defense*

The courts have already encountered one attempt to attribute criminal behavior to a person's genetic characteristics. During the 1960s, research results suggested that males having an extra Y chromosome were more likely to engage in aggressive and violent behavior. Studies of prisoners revealed a higher incidence of males with the XYY characteristic than existed in the general population. By the 1970s, some defendants who had the extra Y chromosome claimed that their behavior should be excused on grounds that this empirical evidence was sufficient to establish legal insanity.

These efforts were unsuccessful. Once again, the courts exhibited extreme reluctance to relinquish the presumption of individual control. One court described the standard for mounting a successful XYY defense:

> An insanity defense based on chromosomal abnormality should be possible only if one establishes with a high degree of medical certainty an etiological relationship between the defendant's mental capacity and the genetic syndrome. Further, the genetic imbalance must have so affected the thought processes as to interfere substantially with the defendant's cognitive capacity or with his ability to understand or appreciate the basic moral code of his society.[23]

A demanding causal standard underlies the law's conservative approach toward exculpation. To date, the law has required persuasive evidence of psychological incapacity or compulsion to excuse an actor from personal responsibility for criminal conduct. Could scientific data on genetic predispositions to addictive or violent behavior ever comprise such evidence? In the next section, I consider three significant impediments to legal acceptance of a genetics defense.

## Problems in Establishing a Genetics Defense to Criminal Liability

Those seeking legal recognition of an excusing genetic condition face three major hurdles. These include (a) meeting standards for admissibility of scientific evidence, (b) meeting standards for establishing that a defendant was not personally accountable for an offense, and (c) avoiding rules that would permit liability for defendants failing to take reasonable precautions to avoid genetically created risks to others.

### *Standards for Admission of Scientific Evidence*

Until recently, courts determined the admissibility of scientific evidence according to a standard set forth in *Frye v. United States.*[24] The *Frye* test allowed judges to admit a scientific principle or technique that had gained ''general acceptance in the field to which it belongs.'' The rule was criticized as overly stringent in that it excluded valid data that were too new or detailed to meet the general acceptance standard.[25]

In 1993 the Supreme Court adopted new criteria for admissibility. In *Daubert v. Merrell Dow,*[26] the Court adopted what appears to be a more flexible approach. Courts are now to admit evidence on scientific knowledge if the evidence is relevant, valid, and reliable. The Supreme Court suggested that judges consider a variety of factors in making the decision on admission. These include whether the proffered theory or technique can be (and has been) tested, whether it has been subjected to peer review and publication, a technique's known or potential error rate, and a technique's consistency with professional standards for its use. The Court stated that general acceptance among the relevant scientific community should also be considered but should no longer be determinative.

Courts applying the *Frye* test have excluded medical and scientific evidence purporting to establish an individual's lack of responsibility. This occurred in some cases in which defendants sought to raise an XYY defense.[27] A more recent example involves defendants attempting to offer psychiatric testimony that they suffer from ''pathological gambling,'' a disorder now recognized in the American Psychiatric Association's *Diagnostic and Statistical Manual.*[28] These individuals claim that their disorder prevents them from refraining from gambling and related criminal activities. Most courts, however, have refused to admit the testimony, ruling that the necessary expert consensus that the disorder significantly deprives persons of control is absent.[29]

Commentators predict that courts will be more open to scientific evidence now that the *Frye* rule has been replaced. Defendants seeking to establish a genetics defense will still have to meet the *Daubert* standard, however. In light of the current state of behavioral genetics research, it is likely that courts will be reluctant to admit such evidence in the foreseeable future. Because judges are assigned the gatekeeper role, however, admission remains a possibility in individual cases.

### *Antideterminism in the Law*

If a defendant succeeded in introducing scientific data on a genetic predisposition to criminal behavior, what sort of evidence could defeat the law's free will presumption? First, the defendant would be required to link the evidence of genetic abnormality to an impairment relevant to the defendant's capacity to behave in a law-abiding manner. As one legal commentator noted, "Even if someone's physiological system looks odd or abnormal in a statistical sense, this in itself is not a ground for exculpation: What must be determined is whether the person's forms of thought and feeling reflect relevant forms of deficit—in reasoning, self-control, and the like."[30]

Second, the defendant must establish the necessary causal link. Precedent suggests that demonstrating a correlation between a particular genetic condition and the behavior at issue will be insufficient. Evidence of other environmental, psychological, and biological correlations with criminal conduct has generally failed to persuade courts that specific defendants should be excused. Moreover, at this point it appears unlikely that scientific evidence addressing a defendant's conduct will be able to separate the contribution of any genetics-based deficit from other causal influences originating in biology or environment.[31]

The attempt to present a genetics defense will encounter resistance at the theoretical level as well. Legal commentators generally favor a theory of human behavior that preserves a role for individual choice and control. Although most modern authorities admit that all behavior is in some sense caused, they distinguish causation from compulsion. In philosophical terms, they are compatibilists—that is, they hold that even if all behavior is causally determined, persons may exercise choice and control within the causal network.[32]

According to the compatibilist model, free will does not depend on the complete absence of causal restrictions on behavior but instead on an absence of compulsion. Thus for a defendant to be excused, it must be shown that because of her genetic condition, she could not fairly be expected to avoid the criminal act. Legal decision makers must be persuaded that the genetic predisposition is so powerful as "to establish the breakdown of rationality and judgment that is incompatible with moral agency."[33] Experience with the internally-based defenses discussed previously suggests that this will be a difficult task.

### *Liability for Failure to Avoid a Known Dangerous Condition*

A third challenge for those endorsing a genetics defense is posed by legal doctrines attributing criminal liability to involuntarily dangerous persons who fail to take reasonable precautions to avoid harm to others. The approach is most commonly applied when intoxicated defendants argue that because of their intoxication they lacked the requisite control and awareness to be held responsible for their criminal acts. States vary in their willingness to allow defendants to present such evidence. The MPC takes a middle ground, allowing such evidence to negate purpose or knowledge, which are highly culpable mental states required to establish liability for certain offenses.

The MPC excludes evidence of intoxication, however, when defendants are charged with crimes requiring a showing of culpable recklessness or negligence. Defendants act recklessly when they are aware of the risk their conduct creates, and negligently when they are unaware but should be aware of such risk. (To convict for most crimes of violence against the person, proof of recklessness or negligence is sufficient.) Though acknowledging that intoxication can in fact deprive a person of subjective awareness of risk, the MPC drafters referred to the culpable awareness that ordinarily exists for persons at some point in the drinking process:

> Awareness of the potential consequences of excessive drinking on the capacity of human beings to guage the risks incident to their conduct is by now so dispersed in our culture that it is not unfair to postulate a general equivalence between the risks created by the conduct of becoming a drunken actor and the risks created by his conduct in becoming drunk.[34]

A similar approach is adopted in assessing the liability of persons with epilepsy or other medical conditions who engage in driving or other potentially risky activities with awareness that they could become impaired and thus endanger others.[35]

These precedents suggest that criminal defendants with reason to know or actual awareness of their predispositions to antisocial behavior would be found similarly culpable for failing to take reasonable precautions to avoid their harmful behavior. They also raise the possibility that certain persons who suspect that they or their children possess such a genetic profile might be held responsible for seeking testing to determine whether they are dangerous to society.[36]

## Alternative Social Controls

The previous discussion reveals the law's general unwillingness to give up the presumption of individual control over behavior. Critics argue that this attitude leads in at least some cases to unjust imposition of punishment on persons who in fact lacked the necessary freedom to be held personally accountable.

Defenders of the law's conservatism rely on a variety of arguments. Some are utilitarian, focusing on the threat to society that a more liberal attitude would yield. There is concern that a greater willingness to attribute criminal behavior to biological or environmental conditions ''is bound to take us down a slippery slope, at the end of which we would have nullified the entire criminal law.''[37] The feared outcome would be a system in which no one could be held responsible for any of the crimes they commit.

Other commentators offer philosophical support for a system that continues to assess criminal responsibility according to ordinary common sense intuitions of the defendants' peers.[38] Finally, at this point the scientific evidence itself provides sufficient reason to reject the claims of biological or genetic compulsion. No respected scientist has offered evidence that antisocial behavior is genetically compelled; the general expectation is that all behavior is affected by a complex array of biological and environmental factors.[39] One should also consider the likely outcome of a move toward excusing persons from criminal responsibility because of a genetic predisposition. Freedom is not the inevitable outcome when a defendant is excused from punishment. Insanity acquittees and other excused defendants may still be restrained through a civil commitment or related procedure. The retributive constraint applies only to criminal punishment; it represents a judgment that the stigma and blame associated with punishment should be reserved for moral agents who chose to engage in criminal wrongdoing. In the civil system, however, incapacitation can stand alone as moral justification for restraint and other forms of coercion.

It is likely that defendants avoiding criminal liability through a genetics defense will face other forms of coercive state action. Existing legal mechanisms will be adjusted or new ones created to protect society from individuals who have shown themselves dangerous to society and unable to control their dangerous propensities. This occurred in California after *Robinson,* when the state supreme court upheld a statute permitting involuntary civil commitment for up to five years for persons found to be addicts.[40] Similarly, in 1983 the U.S. Supreme Court upheld a commitment system for insanity acquittees authorizing indeterminate confinement for such individuals as long as they were deemed mentally ill and dangerous to others.[41] Excused defendants subjected to these systems could be confined for substantially longer periods than was possible if they had been convicted and sentenced in the criminal system.

In sum, defendants who successfully argue that their genetic makeup produces an uncontrollable predisposition to antisocial behavior are unlikely to benefit from this achievement. It will not take long for authorities to produce alternative restraint mechanisms for this population. Indeed, as one author has noted, "If a defense against criminal punishment based on genotype becomes widely accepted, pressure to restrain even people of that genotype who have not yet committed any antisocial acts will be strong."[42]

## Conclusion

Criminal law in the United States is determined primarily by state legislatures and state courts. In the absence of federal legislation or federal court decisions applying the U.S. Constitution to this area, the fate of a genetics defense lies with the states. This situation creates the possibility that some jurisdictions may admit evidence on and accept the defense.[43]

The initial decision to raise a genetics defense will be made by individual attorneys and their clients. State trial judges will decide whether the defense may be asserted, and judges and juries will decide whether the evidence is sufficient to excuse. Appellate judges will then evaluate any decision in which review is sought. As in many areas of criminal law, approaches may differ substantially among the states, and development of general rules will be slow.

In this framework, judges and attorneys would benefit from scientific and philosophic guidance. Experts in these areas should consider preparing educational materials and guidelines for the busy judges and lawyers who will encounter genetics excuse claims. Without such materials, legal arguments and decisions may rest on unwarranted assumptions about the quality of the scientific evidence put forward. Behavioral genetics researchers could also benefit from materials and guidance on presenting their data. Too often, such data are publicly announced as supporting practical and policy judgments extending far beyond what should be inferred from the findings.[44]

If past is prologue, genetic determinism is unlikely to modify the rules governing criminal liability. Unless a particular genetic condition is so strongly linked to criminal behavior that juries and judges are persuaded that the condition compels the behavior, the ordinary presumption of personal control will prevail. In the event that such a link is established, individuals unaware of their genetic compulsion may be excused from criminal liability. Their escape from blame will be short-lived, however, because once they are on notice of their dangerous condition, they will be held responsible for adopting a lifestyle that prevents them from endangering others. If they are deemed incapable of making such arrangements, the state will impose some form of control. In a legal system controlled and operated by ordinary persons, common sense perceptions and morality are likely to retain their dominant role in attributions of criminal responsibility.

## Notes

1. *See, e.g.,* Richard W. Burgh, "Do the Guilty Deserve Punishment?," *Journal of Philosophy* 79 (1982):193–213.
2. Herbert Morris, "Persons and Punishment," *Monist* 52 (1968):475–501.
3. J. E. McTaggert, "Hegel's Theory of Punishment." In G. Ezorsky, ed., *Philosophical Perspectives on Punishment* (Albany: State University of New York Press, 1972).
4. H. L. A. Hart, "Prolegomenon to the Principles of Punishment." In *Punishment and Responsibility* (Oxford: Oxford University Press, 1968).
5. George Fletcher, "Excuse: Theory." In S. h. Kadish, ed., *Encyclopedia of Crime and Justice* (New York: Free Press, 1983).
6. *See* Richard B. Brandt, *Ethical Theory: The Problems of Normative and Critical Ethics* (Englewood Cliffs, NJ: Prentice-Hall, 1959).
7. Michael S. Moore, "Causation and the Excuses," *California Law Review* 73 (1985):1091–1149.
8. Model Penal Code § 2.01(2) (1985).
9. Model Penal Code and Commentaries, cmt to § 2.01, 214–215 (1985).
10. 370 U.S. 660 (1962).
11. 392 U.S. 514 (1968).
12. 392 U.S. at 535.
13. 392 U.S. at 567.
14. 392 U.S. at 534.
15. *See* Richard C. Boldt, "The Construction of Responsibility in the Criminal Law," *University of Pennsylvania Law Review* 140 (1992):2245–2332.
16. M'Naghten's Case, 8 ENG. REP. 718, 722 (1843).
17. United States v. Kunak, 17 C.M.R. 346 (Ct. Mil. App. 1954).
18. Durham v. United States, 214 F.2d 862 (D.C. Cir. 1954).
19. United States v. Freeman, 357 F.2d 606, 619 (2d Cir. 1966).
20. United States v. Lyons, 739 F.2d 994, 996 n.8 (5th Cir. 1984).
21. Rita J. Simon, *The Jury and the Defense of Insanity* (Boston: Little, Brown, 1967).
22. Michael S. Moore, *Law and Psychiatry: Rethinking the Relationship* (Cambridge: Cambridge University Press, 1984).
23. People v. Yukl, 372 N.Y.S.2d 313, 319 (N.Y. Sup. Ct. 1975). Evidence of a genetic predisposition to impulsive and violent acts might also be offered to reduce a defendant's level of culpability. For example, a defendant facing murder charges could seek mitigation based on the provocation doctrine. Defendants qualifying for this doctrine are convicted of manslaughter rather than murder. The doctrine applies to defendants who can establish that they were "disturbed or obscured by passion to an extent which might render ordinary men, of fair average disposition, liable to act rashly or without due deliberation or reflection." *Maher v. People,* 10 Mich. 212 (Sup. Ct. 1862). A defendant's chance of successfully invoking this doctrine would be reduced, however, by the requirement to establish that the situation would have provoked an ordinary person to act rashly.
24. 293 F.2d 1013 (D.C. Cir. 1923).
25. Jay A. Gold, Miles J. Zaremski, Elaine R. Lev, and Deborah H. Shefrin, "*Daubert v. Merrell Dow:* The Supreme Court Tackles Scientific Evidence in the Courtroom," JAMA 270 (1993):2964–2967.
26. 509 U.S. 579 (1993).
27. *See* Maureen P. Coffey, "The Genetic Defense: Excuse or Explanation?," *William and Mary Law Review* 35 (1993):353–399.
28. *DSM-IV.* (American Psychiatric Association, 1994).

29. Sanford H. Kadish and Stephen J. Schulhofer, *Criminal Law and Its Processes,* 5th ed. (Boston: Little, Brown, 1989).
30. Michael H. Shapiro, ''Law, Culpability, and the Neural Sciences.'' In R. Masters and M. McGuire, eds., *The Neurotransmitter Revolution: Serotonin, Social Behavior, and the Law* (Carbondale: Southern Illinois University, 1994).
31. Coffey, ''The Genetic Defense,'' 391.
32. *See* Shapiro, ''Law, Culpability, and the Neural Sciences,'' 183 (citing Flew).
33. Sanford H. Kadish, ''Excusing Crime,'' *California Law Review* 75 (1987):257–289.
34. Model Penal Code and Commentaries, cmt. to § 2.08, 359 (1985).
35. *See, e.g.,* People v. Decina, 138 N.E.2d 799 (1956).
36. Dennis S. Karjala, ''A Legal Research Agenda for the Human Genome Initiative,'' *Jurimetrics Journal* 32 (1992):121–203.
37. Meir Dan-Cohen, ''Actus Reus.'' In S. H. Kadish, ed., *Encyclopedia of Crime and Justice* (New York: Free Press, 1983).
38. *E.g.,* Moore, ''Causation and the Excuses.''
39. David Wasserman, ''Science and Social Harm: Genetic Research Into Crime and Violence,'' *Philosophy and Public Policy* (Winter 1995):14–19.
40. In re De La O, 378 P.2d 793 (1963).
41. Jones v. United States, 463 U.S. 354 (1983).
42. Karjala, ''A Legal Research Agenda for the Human Genome Initiative,'' 162.
43. *Id.,* 161.
44. Wasserman, ''Science and Social Harm,'' 18.

# Elaboration
# The ''Genetics Defense'': Hurdles and Pressures

Mary Crossley

Dresser's chapter 6 lucidly considers whether evidence of genetic influences on criminality will alter existing notions of criminal responsibility. I fundamentally agree with her ultimate conclusion that ''genetic determinism is unlikely to modify the rules governing criminal liability.'' This commentary will build on her broad analysis to provide more focused attention to the difficulties inherent in attempting to raise a genetic condition as the basis for excusing a defendant of legal responsibility for criminal behavior. In doing so, I will consider both a series of evidentiary hurdles and the moral and legal philosophical bases for blaming or assessing culpability. I will close by suggesting that, despite any conviction that foreseeable advances in behavioral genetics should not provide a basis for excusing criminal behavior, the public's seeming embrace of genetic determinism, combined with political forces, could create pressures favoring the recognition of the so-called genetics defense.

To delve more deeply into the evidentiary and theoretical barriers to establishing a genetic condition as a basis for excusing criminal behavior, let us consider a hypothetical case in which a defendant charged with violent criminal behavior seeks to exculpate himself by introducing expert testimony to establish that he is afflicted with what has been called the ''aggression gene.''[1] In 1993 a team led by Han Brunner reported a link between a gene and males in a single Dutch family who exhibited borderline mental retardation and impulsive aggressive behavior.[2] The gene isolated is believed to code for monoamine oxidase A (MAOA), a compound that metabolizes dopamine, serotonin, and noradrenaline, which in turn are involved in the transmission of neural signals in the brain.[3] Under the existing framework for excuses of criminal liability, the accused raising an excuse based on this study would have to prove a series of five factual and legal propositions to avoid criminal responsibility for his actions.

First, the accused would have to prove that he in fact possessed the recessive gene on his X chromosome that the researchers had linked to a predisposition to aggressive behavior. This mutated gene has been identified only in the Dutch kindred studied by Brunner and is believed to be extremely rare. Thus the study may have uncovered what might be called a ''private family gene''—a mutation isolated in a single family group—and would thus seem to be of little help for criminal defendants outside this group. Let us assume, however, that our hypothetical defendant is able to supply expert testimony that an analysis of his DNA indicates the presence of the mutated MAOA gene on his X chromosome. If he can supply this testimony, this initial hurdle will probably be the most straightforward for the defendant to clear, as courts are increasingly willing to accept evidence of DNA analysis offered for various purposes.

Second, the accused will have to prove that possession of the gene creates in its bearers a statistical predisposition to excessive impulsive aggression. This is where things start to get sticky. Under Federal Rule of Evidence 702, expert testimony is admissible ''if scientific . . . knowledge will assist the trier of fact to understand the evidence or to determine a fact in issue. . . .'' Thus in federal courts and those state courts that have adopted federal evidentiary standards, the defendant's proffered evidence must both rise to the level of ''scientific knowledge'' *and* ''assist the trier of fact to . . . determine a fact in issue.''

In *Daubert v. Merrill Dow Pharmaceuticals,*[4] the U.S. Supreme Court recently spoke to the question of what kinds of evidence qualify as ''scientific knowledge.'' Without creating a bright-line test, the Court indicated that a trial judge must first assess whether the reasoning or methodology underlying the testimony is scientifically valid.[5] Thus the judge, before deciding whether to admit the evidence, will consider both arguments asserting the scientific validity of the conclusions drawn by the MAOA study and arguments criticizing that study. For example, critics skeptical of the study's conclusions have pointed out that because the suspect gene was studied in only one family, the study's findings may not be generalizable to the larger population, even if the mutated gene is found in the larger population.[6]

In the broader realm of behavioral genetics in general, scientists continue to debate the validity of studies claiming to link genes to human behavior. Criticisms fall into four general areas: misuse of statistical methods, vague definitions of the trait being studied, bias in the selection of cases studied, and inadequate sample sizes.[7] In addition, attempts to use twin studies and adoption studies to define the heritability of behavioral traits have also come under fire as failing to account sufficiently for predictable environmental influences on behavior.[8] And at the most fundamental level, skeptics decry the tendency on the part of researchers, the media, and the public to imply factual cause from statistical correlation.[9] Simply because variable A and B are statistically correlated does not prove that A caused B. Instead, B may have caused A, or a third variable, C, may have produced both A and B.

Whether criticisms of a particular study's methodology cast doubt on its scientific validity, and hence admissibility at trial, is a question for the trial judge to weigh. The mere existence of questions regarding a conclusion's validity will not suffice to exclude the evidence from the jury's consideration. The Supreme Court rejected the idea that ''the subject of scientific testimony must be 'known' to a certainty; arguably there are no certainties in science.''[10] Moreover, a threshold finding of validity and consequent admission of a controversial study does not guarantee its *use* by the jury; even scientific testimony that passes the validity threshold for admissibility will still be subject to vigorous cross-examination and the presentation of contrary evidence.[11] David Faigman has suggested that factual questions regarding human behavior, like the questions raised by the genetics defense, do not easily lend themselves to the scientific method, but that the proper response by the trial judge is not simply to exclude scientific evidence that is less than totally conclusive but instead to demand the best science available and to be aware of its limitations.[12] Nonetheless, a given study into behavioral genetics may be so discredited that it cannot be seen as qualifying as ''scientific knowledge.'' For example, it seems clear that the original study linking hyperaggressiveness to XYY males was flawed in its design[13] and would be properly excludable on grounds of lack of validity.

Assuming, however, that the trial judge could be convinced of the validity of the MAOA gene study, our hypothetical accused would still have to show that the genotype he has been shown to carry is phenotypically expressed in him. In other words, he must show that he has the ''disorder'' that the study linked to the MAOA mutation. This showing is not assured by the mere presence of the mutation, for most genetic disorders are not characterized by complete penetrance. Mark Leppert has told us that for most behavioral traits the accepted range of penetrances for postulated major genes is approximately 30 to 70%.[14] Moreover, genotypes that are phenotypically expressed may fall along a potentially wide spectrum of expression, with either relatively mild or more marked effects.

Even if the accused in our hypothetical case can convince the trial judge that the study regarding the linkage between a genetic defect and a predisposition to excessive aggression is scientifically valid *and* that the defective genotype is phenotypically expressed in the accused, he faces yet a fourth hurdle: He must present evidence that the defective gene predisposing him to aggressive behavior, and not some other influence or personal motive, in fact led to his violent offense. This feat may prove difficult. First, studies claiming to show statistical correlations between a genetic condition and certain behaviors may have little relevance to an individual case. As Dorothy Nelkin and Susan Lindee explained: "The scientific concept of genetic predisposition assumes the existence of a biological condition signaling that an individual may suffer a future disease or behavioral aberration. But predisposition in the clinical sense is a statistical risk calculation, not a prediction."[15]

In other words, some persons genetically predisposed (according to the MAOA study) do not act violently. More important, perhaps, even of those genetically predisposed to violence, it may be impossible to link violent conduct to a genetic, rather than environmental, influence. All that a gene does is to influence biochemical processes within the body; the translation of those processes into the complexities of human behavior is inevitably modulated by social, economic, and cultural factors.[16] Erik Parens stresses this point by focusing on the limited role that genes play both within a person's internal environment and in relation to the external environment:

> It is enormously important to remember that genes are but one component of fabulously complex biological, and ultimately biopsychosocial systems. . . . [A] fuller account [of complex behaviors] will require thinking in terms of nonlinear and dynamic interactions among many genes, hormones, nutrients, and other biological factors in the "internal" environment that is the body. And as genetics always will be only one important part of biology, biology always will be only one important part of any richer account of human behavior. Such an account will have to consider, not only interactions *among* social and biological factors in the "external" environment, but will have to consider the complex interactions *between* the internal and external environments.[17]

Clearly, violent behavior is a "multiply determined phenomenon" that is influenced by external factors (including characteristics of the victim and the setting), as well as internal factors (including recent alcohol intake, current emotional status, and neurological factors).[18] Even a researcher on the Brunner team emphasized the influence of the environment on the MAOA gene's expression: "Even in this particular type of syndrome . . . we have people who are happily married with children. They had the right type of support."[19]

In light of the foregoing, it seems that our hypothetical accused faces significant hurdles to tracing a complete chain of causation from the presence of the mutated MAOA gene to a specific act of criminal behavior. The statistical correlations uncovered by researchers at best indicate only probabilities that "ultimately *some* antisocial or criminal behavior is likely to occur, not that any specified act [was] somehow genetically foreordained."[20]

But even if our hypothetical defendant could clear all four of the foregoing evidentiary hurdles by convincing the trial judge that the proffered evidence qualifies as "scientific knowledge" regarding a causal linkage between his genetic makeup and his criminal conduct, a fifth and final hurdle awaits him: He must still convince the judge that, if the jury believes this evidence, it would provide a legal basis, consistent with existing approaches to excuse, for excusing his conduct. In other words, the court must be persuaded that evidence showing that a person's genetic makeup causally influenced his criminal conduct effectively renders the person nonculpable. Unless the judge accepts this conclusion, the genetic evidence described earlier is simply irrelevant to the question of guilt or innocence, and therefore should be ruled inadmissible because it will not "assist the trier of fact . . . to determine a fact in issue."[21] Thus this last hurdle is essential for the defendant to clear.

In contrast to the initial hurdles discussed, which raised questions about the validity and power of probabilistic genetic information, this final hurdle raises questions that are essentially philosophical in nature, and, I agree with Dresser that the law's antideterministic stance will make the defendant's task difficult. This conclusion, however, does not simply reflect that the genetics defense's novelty runs afoul of the law's inherent conservatism but that the defense, even when cast in the strongest terms plausible, is unlikely to lead us to find a defendant blameless and undeserving of some punishment, in either legal or moral terms. Fundamentally, the decision to punish, rather than excuse, criminal behavior is founded on a determination of culpability, or blameworthiness.[22] And many commentators agree that it is only when some mental defect significantly impairs an individual's capacity for rationality or when some external force compels criminal conduct (and thus overrides the effectiveness of an individual's rational will) that the law has been willing, and then only grudgingly, to find blameless an individual who has engaged in criminal behavior.[23]

In this regard, two concepts touched on by Dresser bear reemphasis. First, cause is not compulsion, and mere but-for causation of criminal activity by a force beyond an individual's control does not in and of itself excuse the individual. Surely we do not excuse the individual who engages in securities fraud *because* he is excessively acquisitive *because* his parents, in raising him, overemphasized the value of money and material possessions. Yet this individual's criminal conduct is in part caused (though not compelled) by external forces. This illustration suggests how accepting a purely causal basis for excusing criminal conduct would support a deterministic stance under which no person could be blamed for wrongful conduct as long as some underlying cause could be identified.[24]

Second, the true basis for legal and moral excuse is not simply causation by external forces but such a great breakdown in an individual's capacity for practical reasoning that we no longer attribute moral agency to the individual.[25] Commentators use the term "practical reasoning" to refer to the process of rationally formulating goals, selecting means to obtain those goals, and acting to further those goals by the means selected.[26] Persons who lack the capacity to use reason to guide their actions in this manner may be excused because they seem fundamentally different from our understanding of persons as rational, moral agents.[27]

Stephen Morse explained how these two concepts interact in the context of the insanity defense:

> Claims of legal insanity are usually supported and explained by using mental disorder as a variable that at least in part caused the defendant's offense[, . . . but] the criteria for legal insanity primarily address the defendant's reasoning, rather than mechanistic causes. For example, in addition to a finding of mental disorder, acquittal by reason of insanity requires that the defendant did not know right from wrong or was unable to appreciate the wrongfulness of her act. . . . The law excuses a legally insane defendant . . . because her practical reasoning was nonculpably irrational, not because her behavior was caused by abnormal psychological or biological variables.[28]

Correspondingly, for a genetics defense to excuse a defendant it seems that the defendant will have to show that his genetic condition influenced his behavior either by significantly impairing his capacity for rationality or by compelling him to act as he did. Simply arguing that the condition caused his criminal behavior by predisposing him to aggression will not suffice; nor will evidence that the defendant faced more difficulty in choosing to act lawfully than do most persons.[29] We are all expected to use our practical reasoning skills to compensate for whatever antisocial influences on our actions we may experience. It is only when a defendant can prove that an abnormal genetic condition so utterly robbed him of rationality or choice that it also stripped him of moral agency that he should be excused. If this threshold—loss of moral agency—is proved, however, excuse should be available regardless of whether the causative agent is mental illness (the conventional basis of the insanity defense) or a genetic condition.

For the foreseeable future, it seems unlikely that defense attorneys will be able to prove that a genetic condition renders an individual so incapable of practical reasoning or void of rationality that he cannot be blamed for his behavior. Most past studies in the field of behavioral genetics have focused on the question of a trait's heritability, without focusing on the genome itself and trying to identify the responsible genetic material and how it affects behavior. The simple statistical linkage of inheritance and behavioral traits does nothing to advance the proposition that an affected person lacks the ability to engage in practical reasoning. The MAOA gene study, by contrast, claims to have identified the responsible genetic defect and has suggested a plausible scenario for how the defect affects behavior. There has been no suggestion, however, that a biochemical interruption in neural transmissions, by itself, causes a total breakdown in the capacity for rational self-direction.[30]

Based on this analysis, I support a conclusion that research regarding genetic influences on behavior should not engender an expansion of existing excuses available under the criminal law. With that said, a cautionary note should nonetheless be sounded, suggesting two reasons why, despite the law's traditionally conservative approach toward excusing criminal behavior, some pressures toward the legal recognition of a genetics defense may develop.

First, the American public appears increasingly willing to attribute a high level of deterministic power to genetics. Nelkin and Lindee recently have chronicled in detail the myriad ways in which genetics figures in popular culture and have shown that media ranging from comic strips to movies to advertisements reflect how beliefs about genetic determinism permeate our culture.[31] An attempt to establish a genetically based excuse to criminal liability seems most likely to take root in a culture that understands (though perhaps mistakenly) genetics to determine not only humans' physical structure and functioning but also their mental and behavioral functioning. One explanation for the paucity of cases in which criminal defendants are actually excused based on an insanity defense showing mental defect or disorder is that juries view the mental health sciences as "soft," imprecise, and subjective.[32] By contrast, if Americans view genetics as a "hard" science subject to objective proof, they may be willing to cede more authority for granting excuse to genetics than they do to psychiatry or psychology. Deborah Denno suggested that concern regarding this "aura of truth" associated with genetic evidence is one reason that has prompted most courts to date to shun genetic evidence proffered to prove excuse.[33] As suggested next, however, courts may not be the only lawmakers determining the future of a genetics defense.

The potential for a popular acceptance of the genetics defense should be unsurprising in light of the limited understanding of genetics that most members of the American public possess. For a person whose primary instruction in genetics involved Gregor Mendel and his pea plants and who sees the genetic determination of eye color as a model for how genetics "works," it may not seem farfetched to conclude that behavior is similarly determined. The error, of course, is that no scientific evidence indicates that complex human behaviors like aggression are produced inevitably by the presence of a single gene.[34] But the seeming fit between the a-responsibility of genetic makeup and a-responsibility for the "resulting" criminal conduct may be persuasive for many persons whose understanding of genetics is unsophisticated. As Dan Brock has written, "One's specific and unique genetic structure is a paradigm of what is viewed as beyond one's control and for which one cannot be held responsible.''[35] Thus a lack of sophisticated public understanding regarding genetics may enhance the chances that efforts to establish a genetics defense will succeed.

The straightforward response to this line of reasoning is to call, appropriately in my opinion, for public educational efforts regarding the meaning and limitations of genetic information. These efforts would strive to correct the often misleading impressions created by both journalistic reporting and popular culture. Any call to public education, however, would be naive to ignore the risk that genetic information could be oversimplified and misappropriated by political or ideological forces involved in the process of presenting information regarding genetics to the public.[36] In other words, we must recognize that some individuals or groups could, for political purposes, potentially seek to foster the public's acceptance of genetic behavioral determinism.

To give an example, it stands to reason that groups taking a strong ''law and order'' stance while simultaneously decrying the expansive social programs of ''big government'' might find the legislative establishment of a genetics defense useful. Persons using the genetics defense to avoid criminal punishment are likely to be subject to some form of nonpunitive restraint on their freedom, such as involuntary civil commitment or involuntary pharmacological interventions.[37] And because the cause of their criminal behavior would be identified as genetic and thus commonly understood to be static and unchanging,[38] one could argue that permanent restraint is appropriate to protect society.[39] This approach—combining recognition of a genetics defense with long-term restraints—could be touted as addressing the crime problem by keeping dangerous people off the streets.

As numerous commentators have suggested, recognizing a new class of the ''genetically criminal'' could also serve a second ideological purpose: Internalizing the source of criminality to the individual's genetic makeup relieves society of any responsibility for influencing individuals' actions. And if social environment is understood to play no role in fostering or discouraging criminal behavior, then citing crime prevention begins to ring hollow as a justification for social programs aimed at ameliorating destructive social environments.[40]

This commentary closes with this admittedly speculative line of thought simply to suggest what *could* happen as research in behavioral genetics continues to pursue links between genetics and antisocial behavior. My purpose in sounding this cautionary note is not to suggest that advances in behavioral genetics will inevitably be used in an ideologically driven or abusive fashion or that the research should itself be limited or abandoned.[41] The purpose, instead is to highlight the need for scientists, scholars, and policy makers to be conscious of the ideological forces that may be drawn into the fray over the social meaning of genetic information.

## Notes

1. Virginia Morrell, ''Evidence Found for a Possible 'Aggression' Gene,'' *Science* 260 (1993):1722–1723.
2. H. G. Brunner, M. Nelen, X. O. Breakefield, H. H. Ropers, B. A. van Oost, ''Abnormal Behavior Associated With a Point Mutation in the Structural Gene for Monoamine Oxidase A,'' *Science* 262 (1993):578–580.
3. Han G. Brunner, ''MAOA Deficiency and Abnormal Behavior: Perspectives on an Association.'' In *Genetics of Criminal and Antisocial Behavior* (Ciba Foundation Symposium 194) (Chichester, England: John Wiley, 1995).
4. 509 U.S. 579 (1993).
5. *Id.,* 2796.
6. Morrell, ''Evidence Found for a Possible 'Aggression' Gene.''

7. Charles C. Mann, ''Behavioral Genetics in Transition,'' *Science* 264 (1994):1686–1689.
8. Stanton Peele and Richard DeGrandpre, ''My Genes Made Me Do It.'' *Psychology Today* 28 (1995):50–53, 62, 64, 66, 68.
9. Rochelle Cooper Dreyfuss and Dorothy Nelkin, ''The Jurisprudence of Genetics,'' *Vanderbilt Law Review* 45 (1992):313–348; Patricia J. Falk, ''Novel Theories of Criminal Defense Based Upon the Toxicity of the Social Environment: Urban Psychosis, Television Intoxication, and Black Rage,'' *North Carolina Law Review* 74 (1996):731–811.
10. *Daubert,* 113 Ct. at 2795. *See* Edward J. Imwinkelried, ''Evidence Law Visits Jurassic Park: The Far-Reaching Implications of the *Daubert* Court's Recognition of the Uncertainty of the Scientific Enterprise,'' *Iowa Law Review* 81 (1995):55–78.
11. *Daubert,* 113 S. Ct. at 2798.
12. David Faigman, ''Mapping the Labyrinth of Scientific Evidence,'' *Hastings Law Journal* 46 (1995):555–579.
13. Ruth Hubbard and Elijah Wald, *Exploding the Gene Myth* (Boston: Beacon Press, 1993), 105 (1993); Deborah W. Denno, ''Legal Implications of Genetics and Crime Research.'' In *Genetics of Criminal and Antisocial Behavior* (Ciba Foundation Symposium 194) (Chichester, England: John Wiley, 1995).
14. *See* Mark Leppert, elaboration on chapter 4, this volume.
15. Dorothy Nelkin and M. Susan Lindee, *The DNA Mystique: The Gene as a Cultural Icon* (New York: W.H. Freeman, 1995).
16. Stanton and DeGrandpre, ''My Genes Made Me Do It.''
17. Erik Parens, ''Taking Behavioral Genetics Seriously,'' *Hastings Center Report* 24 (1996):13–18.
18. Stephen H. Dinwiddie, ''Genetics, Antisocial Personality, and Criminal Responsibility,'' *Bulletin of the American Academy of Psychiatry & Law* 24 (1996):95–108.
19. Mann, ''Behavioral Genetics in Transition'' (quoting Xandra O. Breakefield).
20. Dinwiddie, ''Genetics, Antisocial Personality, and Criminal Responsibility,'' 101.
21. FED. R. EVID. 702.
22. Sanford H. Kadish, ''Excusing Crime,'' *California Law Review* 75 (1987):257–289.
23. Stephen J. Morse, ''Brain and Blame,'' *Georgetown Law Journal* 84 (1996):527–549; Joshua Dressler, *Understanding Criminal Law* (New York: Matthew Bender, 1987), § 17.03, discussing underlying theories of excuse.
24. Michael S. Moore, ''Causation and the Excuses,'' *California Law Review* 73 (1985):1091–1149.
25. *Id.*
26. Dressler, *Understanding Criminal Law,* at § 17.03[E], 186.
27. Moore, ''Causation and the Excuses.''
28. Morse, ''Brain and Blame,'' 529.
29. *Id.,* 541–542.
30. Brunner et al., ''Abnormal Behavior Associated with a Point Mutation in the Structural Gene for Monoamine Oxidase A.''
31. Dorothy Nelkin and M. Susan Lindee, *The DNA Mystique.*
32. Michael L. Perlin, ''Unpacking the Myths: The Symbolism Mythology of Insanity Defense Jurisprudence,'' *Case Western Reserve Law Review* 40 (1989–90):599–731.
33. Denno, ''Legal Implications of Genetics and Crime Research.''
34. David Wasserman, ''Science and Social Harm: Genetic Research into Crime and Violence,'' *Report From the Institute for Philosophy and Public Policy* (Winter 1995):14–19.
35. Dan W. Brock, ''The Human Genome Project and Human Identity,'' *Houston Law Review* 29 (1992):7–22.
36. Dinwiddie, ''Genetics, Antisocial Personality, and Criminal Responsibility.''
37. Wasserman, ''Science and Social Harm.''
38. Denno, ''Legal Implications of Genetics and Crime Research.''
39. California's recent adoption of mandatory postrelease chemical castration for persons twice convicted of child molestation provides an example of this type of approach to protecting society from persons who are believed to be dangerous but who are not subject to criminal confinement (in

this case because their sentences have already been served). Dave Lesher, "Molester Castration Measure Signed," *Los Angeles Times,* September 18, 1996, A3.

40. Dreyfuss and Nelkin, "The Jurisprudence of Genetics."
41. Parens, "Taking Behavioral Genetics Seriously."

# ELABORATION
# The Use of Human Genome Research in Criminal Defense and Mitigation of Punishment

Daniel A. Summer

> If justice and humanity shall ever have to do with the treatment of the criminal, and if science shall ever be called upon in this, one of the most serious and painful questions of the ages, it is necessary, first, that the public shall have a better understanding of crime and criminals.[1]

In the sensational 1924 murder trial of Richard Loeb and Nathan Leopold, Jr., accused of kidnaping and bludgeoning to death a young schoolmate, famed criminal defense attorney Clarence Darrow implored the trial judge to spare the lives of his clients because the boys may have been corrupted by the seed of remote ancestors. Darrow passionately argued to the court:

> I know that they cannot feel what you feel and what I feel; that they cannot feel the moral shocks which come to men that are educated and who have not been deprived of an emotional system or emotional feelings. I know it and every person who has honestly studied this subject knows it as well. Is Dickie Loeb to blame because out of the infinite forces that conspired to form him, the infinite forces that were at work producing him ages before he was born, that because out of this infinite combinations he was born without it? If he is, then there should be a new definition for justice.[2]

On the conclusion of the proceedings, and after days of judicial deliberation, the trial judge spared the boys from the gallows and instead imposed a sentence of life imprisonment and 99 years.

Darrow's description of his clients' inability to internalize a value system without any hint of insanity characterizes the modern definition of antisocial personality disorder. Yet Darrow would offer this disorder as mitigating evidence and claim his clients were helpless in its grip. Darrow's successful use of this theory of the origin of his clients' deviant behavior heralded the genetic-based or genetic-defect criminal defense. However, it would not be until the late twentieth century that modern molecular genetics would confirm the scientific validity of Darrow's radical defense.

The revolution in molecular genetics and human genome research offers the prospect of a new class of defenses to criminal conduct and mitigation of punishment. Unlike traditional affirmative defenses such as alibi, self-defense, and accident or the legal concept of insanity as a defense, a new forensic argument can be empirically posited based on advances in human genome research and its role in human behavior. In essence, a person who has inherited, through no fault of his or her own, a particular genetic trait or mutant gene that precludes or to some extent interferes with the ability to form the legal concept of criminal intent either (a) should not be held legally accountable for the criminal act; or (b) should have his or her punishment mitigated in some form.

It is fundamental in the criminal law that the state prove the existence of both the criminal act and intention to act by the defendant beyond a reasonable doubt before a conviction may be legally obtained. Thus even if the person actually committed the criminal act, the state must still prove the person acted with criminal intent. Criminal intent has been defined in several ways such as *scienter,* which means guilty knowledge, and *mens rea,* which means a guilty mind and guilty willfulness. Therefore, any credible scientific evidence that calls into question the ability of persons accused of a crime to form *scienter* or *mens rea* or intent creates the basis for a new affirmative defense or evidence in mitigation of punishment.

Recent advances in neuroimaging techniques and neurobiological research, along with classic twin, adoption, and family studies, have provided compelling circumstantial evidence for a genetic etiology of some forms of criminal behavior. Circumstantial evidence is the proof of facts and circumstances, by direct evidence, from which one may infer other related or connected facts that are reasonable deductions from the direct evidence. For example, a SPECT image demonstrating that psychopaths process emotion-laden words differently from normal research participants circumstantially offers proof that psychopaths are suffering from some underlying disorder with a biologic or genetic etiology.[3] Similarly, decreased serotonin turnover, as indicated by low CSF 5-HIAA, has been definitively linked to deficient impulse control and impulsive violence, offering additional circumstantial evidence that impulsive aggression may be under genetic control.[4]

It is, however, the recent breakthroughs in molecular neurogenetic research that offers new and powerful direct evidence of a genetic basis for antisocial and criminal behavior. Direct evidence is evidence that points directly and immediately to the question at issue. It is the eyewitness to the event. For example, the research linking high novelty-seeking behavior and its association with the 7 repeat allele in the locus for the D4 dopamine receptor gene (D4DR) provides direct evidence for a specific genetic locus involved in neurotransmission and a specific personality trait.[5]

Human genome research may provide evidence that creates an affirmative defense for one accused of a criminal act or evidence that may only mitigate punishment or a combination of both. Evidence that rises to the level of an affirmative or actual defense allows the accused to rely on the evidence to establish that he or she is not guilty of crime even though the accused may have actually committed the act. Evidence that mitigates punishment does not preclude a finding of legal guilt but serves only to lessen the harshness of punishment for the deed.

## Legal Standard for Admissibility of Human Genome Research as Evidence in a Court of Law

The U.S. Supreme Court has articulated a new and clear standard for the admissibility of novel scientific evidence into evidence. In *Daubert v. Merrell Dow Pharmaceuticals,*[6] the Supreme Court reexamined Rule 702 of the Federal Rules of Evidence and the case of *Frye v. United States,*[7] which held that expert opinion based on a scientific technique was inadmissible unless the technique was "generally accepted" as reliable in the relevant scientific community. In explicitly rejecting the *Frye* "general acceptance" test for admissibility of scientific evidence, the Court held that there are no certainties in science, and to qualify as scientific knowledge an inference or assumption must only be derived by the scientific method.

The Court in *Daubert* declined to establish a definitive checklist for the admissibility of novel scientific evidence; however, the court's opinion does suggest the following questions be satisfactorily answered in the affirmative:

1. Is the reasoning and methodology underlying the proposed expert testimony scientifically valid and is the reasoning and methodology relevant to the facts in issue?

2. Has the proposed scientific theory been tested or is it capable of being tested? Can the theory be subject to empirical testing to determine its falsifiability, refutability, or testability?
3. Has the scientific theory been subjected to peer review and publication?
4. Does the theory enjoy a degree of general acceptance in the scientific community?

If these questions can be answered in the affirmative, then the trial courts should, but are not required to, allow into evidence the scientific testimony regarding the theory.

It would appear that human genome research and its underlying methodology and reasoning linking genetic variables to impulsive, antisocial, and some forms of criminal behavior would easily satisfy the *Daubert* requirements for admission into evidence in a court of law. Although human genome research remains controversial when applied to human behavior, the underlying scientific methodology employed is well-accepted in the scientific community. Indeed as the Supreme Court held, "The focus, of course, must be solely on principles and methodology, not on the conclusions that they generate."[8]

## Presenting a "Genetic-Defect" Defense

For the purpose of this discussion, the term *genetic defect* can refer to a mutant form of a gene or to an otherwise nonmutated gene that encodes for the creation of biologic product that ultimately influences some form of human behavior.

In practice only rarely will a criminal defendant be in a position to use a "genetic-defect" type defense in a court of law. In general the criminal act must be extreme in nature but not be motivated by typical factors such as revenge, profit, hate, ulterior purpose, and so on. In other words, the criminal act must not have any particular rationale, yet not be a product of true psychosis or insanity.

The defendant should also have a personal or family history of behavior in some way linked to the criminal act. A psychiatric, medical, or clinical diagnosis of a disease or disorder should have been made or suggested before the criminal act in question. Genetic counseling should unearth a prevalence of a disease or disorder in the defendant's family's history.

Defendants who have a family or personal history of a disease or disorder, and who have been diagnosed as suffering from the disorder and who commit a criminal act of an extreme nature that is not the product of insanity yet defies any rational explanation might avail themselves of following recent advances in human genome research to present a "genetic-defect" defense as an affirmative defense or as evidence in mitigation of punishment.

## Mood Disorders

Classic twin, adoption, and family studies have consistently demonstrated strong circumstantial evidence for the heritability of the mood disorders. These disorders have been clearly shown to occur at much higher rates in first-degree biologic relatives than in the general population.

Human genome research has provided direct evidence for a susceptibility gene for one form of mood disorder—bipolar affective disorder (BAFD)—on chromosome 18 with a parent-of-origin effect. In a study of the linkage of bipolar disorder to chromosome 18, researchers have provided the first compelling direct evidence of a linkage of BAFD to chromosome 18 and the first molecular evidence of a parent-of-origin effect to the disorder.[9]

Persons suffering from mood disorders often demonstrate severe episodes of mania or major depression. One who commits a criminal act while suffering from one of the extreme mood swings of major depression may use human genome research for a "genetic-defect" defense if the criminal act appears related directly or indirectly to the disorder. The theoretical basis of this defense as an affirmative defense would be that extreme depression or BAFD precludes the formation of criminal intent. Alternatively, one could argue that the extreme moods "clouded judgment" and thus impaired the formation of intent that would mitigate the punishment for committing the offense.

To establish this defense, as a practical matter, it is first necessary to obtain genetic counseling on the defendant and family to ascertain a history of BAFD or severe depression or mania. Evidence that biological relatives have suffered from a mood disorder should be presented to the court. Thereafter, a clinical diagnosis for a mood disorder like BAFD must be offered by the defendant along with expert testimony about the disorder and its sequelae. Finally, either by way of affidavit or expert testimony, human genome research linking BAFD to chromosome 18 must be introduced along with the actual genotyping of the defendant, if possible.

## Case History: *South Carolina v. Susan Smith:* Major Depression–Homicide–Mitigation of Punishment

Susan Smith, the South Carolina mother of two young sons, was convicted of the horrific murder of her children by drowning them in a lake while they were strapped in their car seats. The prosecution contended the murders were performed by Smith to enable her to marry her wealthy lover who purportedly wanted no children. Her defense lawyer maintained that the murders were actually an aborted suicide attempt by Smith in which she was unable to save her children once her efforts at suicide were abandoned. During the trial, Smith's lawyer was able to successfully demonstrate to the jury that she was likely suffering from major depression, a disorder that had been inherited from her father or mother or both.

During the penalty phase of the trial, in which life or death was the sole issue for the jury's resolution, Smith's attorney called on experts to testify about the family history of depression. Testimony included Smith's diagnosis of major depression, Smith's father's suicide, and other acts relating to severe depression involving Smith, including her suicide attempts and molestation as a youth. Her attorney told the jury that the Smith family was "genetically loaded" for depression. He argued, "You saw the history of depression in her bloodlines going back three generations on both sides of her family—the alcoholism, the depression, the attempted suicides . . . was that a choice she made? No."[10]

Smith's counsel was ultimately successful in convincing the jury to impose a sentence of life, rather than death, in large part because of the argument, supported by compelling circumstantial evidence, that Smith had inherited a form of major depression that impaired her ability to form criminal intent, which was a mitigating factor.

## Case History: *State of Georgia v. Anabelle S:* Bipolar Disorder, Mania–Compulsive Shopping/Shoplifting–Mitigation of Punishment

Anabelle S. is a middle-aged affluent White female with no criminal history. At age 45 she begins to suffer cyclical mood swings and, during episodes of mania, she begins to compulsively shop, purchasing the same items numerous times. During one shopping episode, she becomes manic and walks out of the establishment without paying for the items and with no effort at concealment. She is apprehended and convicted of shoplifting and placed on probation. Anabelle is placed under the care of a family physician who prescribes Prozac. While under the influence of Prozac and on probation she commits three new shoplifting offenses during manic episodes. All of the offenses involve the theft of numerous items of the same type of clothing, with no effort at concealment.

Anabelle is referred for psychiatric treatment and is diagnosed with BAFD. A family history reveals one of Anabelle's daughters shows signs of BAFD. Anabelle, who was adopted, has no pedigree to review. Trial counsel provided to prosecutors Anabelle's psychiatric records, family history, and research linking genetics and BAFD. After reviewing the case history and human genome research, the prosecutor recommended that Anabelle receive a concurrent probated sentence with a requirement for continued psychiatric treatment.

## Huntington's Disease

Huntington's disease is an inherited chronic progressive neurodegenerative disorder characterized by mid-life onset, involuntary movements, cognitive decline, and emotional disturbance. The disease is caused by a mutant autosomal dominant gene localized to the short arm of chromosome 4. The gene was cloned in 1993 and codes for a normal protein called *huntingtin.* The mutant gene codes for an abnormal protein that leads to selective neuronal death. A definitive diagnosis of Huntington's disease can be achieved by confirmatory genetic testing. Genetic testing for Huntington's disease is almost 100% sensitive and specific. A clinical diagnosis of Huntington's disease is necessary to establish that the individual is symptomatic from the disease. Clinical manifestations often include depression, episodic violence, irritability, emotional liability, and impulsiveness.

One who engages in a criminal act and who suffers from Huntington's disease may raise the "genetic-defect" as an affirmative defense to the crime. The basis for this defense is that the disease causes certain clinical symptoms that preclude or drastically interfere with the creation of criminal intent or *scienter.*

To successfully present a defense to a crime based on the presence of the mutant Huntington gene, two factors must be established in court. First, the mutant gene must be confirmed by genetic testing; second, the defendant must be symptomatic from the disease and the symptoms must be related to the criminal act. Expert testimony must be introduced about the presence of the mutant gene in the person and its sequelae. In addition, a family history should be offered into evidence as well as testimony relating to the defendant's behavior.

### Case History: *State of Georgia v. Glenda Sue Caldwell:* Huntington's Disease–Homicide–Affirmative Defense

In 1985 Glenda Sue Caldwell waited up for her son to arrive home. When he walked through the door, she shot and killed him for no apparent reason. Caldwell then went to her daughter's room and shot at her sleeping daughter, missing her by only inches. She was convicted of her son's murder and was sentenced to life in prison.[11]

While in prison Caldwell began to develop numerous symptoms of Huntington's disease, a disease that had taken the life of her father and brother. Caldwell was given the confirmatory genetic test establishing that indeed she had the mutant Huntington's gene. A follow-up clinical diagnosis confirmed that she was symptomatic from the disease at the time of the homicide.

Based on the evidence that she had Huntington's disease as determined through genetic testing, Caldwell won a new trial. At the new trial, she was found not guilty by reason of insanity after offering into evidence the results of her genetic test for Huntington's disease and additional expert testimony. Caldwell was freed from prison.

## Antisocial Personality Disorder/Sociopathy/Psychopathy

Antisocial personality disorder, sociopathy, or psychopathy is a personality disorder characterized by a pattern of irresponsible and antisocial behavior beginning in childhood or early adolescence and continuing into adulthood. Individuals with this disorder engage in behavior that repeatedly brings them into conflict with society. They appear to be incapable of significant loyalty to any individual or group. They are usually grossly selfish, callous, irresponsible, impulsive, and are unable to feel remorse or to learn from experience. They are unable to internalize any kind of moral value system. They often engage in repeated criminal conduct with punishment offering no deterrent whatsoever. It is likely that the 6 to 7% of criminals who commit 85 to 90% of repeated violent criminal acts are disproportionately persons suffering from this disorder.

Family, twin, and adoption studies provide some circumstantial evidence for the heritability of this disorder. Neuroimaging and neuroendocrine studies, especially involving disturbances of CNS 5 HT function in humans with aggressive and impulsive behavior, provide compelling circumstantial evidence for a biologic and genetic etiology for this disorder.

Over the past four years human genome research has produced several ground-breaking studies identifying normal and mutant genes directly related to antisocial, high-risk-taking, and impulsive criminal behavior. These studies offer compelling direct evidence for the argument that a criminal defendant's conduct, in certain limited circumstances, may have been influenced or dictated by genetic factors and not necessarily criminal intent or the exercise of pure free will.

In the earliest research linking a mutated gene with impulsive aggression and antisocial behavior, Brunner and colleagues published a study titled *Abnormal Behavior Associated With a Point Mutation in the Structural Gene for Monoamine Oxidase A.* According to the study, the researchers identified a Dutch family in which many of the males engaged in repeated aggressive, impulsive, and antisocial behavior. Genetic testing of the males revealed a point mutation in the MAOA gene resulting in severe impairment in neurotransmitter metabolism affecting the breakdown of 5-HT, noradrenaline, and dopamine.[12] The mutated MAOA gene, although not directly coding for aggressive–impulsive behavior, was clearly demonstrated to be directly involved in the consequent aggressive–impulsive behavior.

In yet another MAOA study, providing replication of the Brunner findings, a line of transgenic mice was isolated in which transgene integration caused a deletion of the gene encoding MAOA. In this study, researchers demonstrated that MAOA-deficient males showed increased aggressiveness.[13] The researchers concluded that the study supported the contention that particularly aggressive behavior in humans known to have a mutated MAOA gene was a direct consequence of MAOA metabolic deficiency.

Another "knock-out" study involved a line of transgenic mice in which transgene integration caused a deletion of the gene encoding for nitric oxide (NO). Researchers accidentally discovered that the mutant male mice lacking the neurotransmitter NO engaged in extraordinarily violent behavior and excessive sexual aggression.[14] The researchers theorized that "nitric oxide may be the neurotransmitter that puts a brake on some behaviors." One of the studies' authors concluded, "What we might have here is an example of serious criminal behavior which can be explained by a single gene defect."

Perhaps the most powerful evidence yet for a deterministic genetic etiology of a human behavioral trait comes from a study titled "Dopamine D4 Receptor (D4DR) Exon 111 Polymorphism Associated With the Human Personality Trait of Novelty Seeking."[15] In this study researchers concluded that individuals whose personality could be measured and described as impulsive, excitable, quick-tempered, and extravagant (all traits similar to those of antisocial personality) had a significant association with a particular exonic polymorphism, the 7 repeat allele in the locus for the D4 dopamine receptor gene (D4DR). As a consequence of the findings of this research the authors concluded that, "Given the significant heritability of many human behaviors and the rapid progress of the human genome project, it is likely that additional genes that influence normal and abnormal psychological characteristics will be found in the future."[16]

A psychopathic or antisocial personality criminal defendant who wishes to use this type of human genome research as an affirmative defense or as evidence in mitigation of punishment faces a difficult task and numerous obstacles. However, a "genetic-defect" defense based on this research is indeed possible.

First, the defendant should have a prior or current diagnosis for antisocial personality disorder as defined in the *DSM-IV*.[17] Many attorneys feel that such a diagnosis is itself an aggravating, rather than mitigating, factor. However, if such a clinical diagnosis is offered to a jury or judge, along with human genome research and other scientific research establishing the likelihood of a biologic and genetic etiology for the disorder, such a diagnosis could be, as Dr. Robert Hare contended, "The kiss of life rather than a kiss of death."[18]

Second, a family historian or clinical psychologist should offer a family history for evidence of the disorder. A pedigree that presents other similarly affected individuals provides compelling evidence for the contention that the trait has a genetic etiology. Genetic counseling is critical in the effort to present this evidence.

Third, expert testimony should be introduced to establish that antisocial personality disorder or psychopathy is a recognized personality disorder in the medical and psychiatric literature. Testimony should be elicited regarding the diagnostic criteria for this disorder and its sequelae.

Fourth, the defendant should undergo neurochemical testing to detect identifiable disturbances in various neurotransmitter systems that have been implicated in impulsive aggression such as 5-HT, dopamine, epinephrine, norepinephrine, and MHPG. Neuroimaging using PET or SPECT to identify orbitofrontal lobe impairment would also be useful.

Fifth, if the defendant has a clinical diagnosis of antisocial personality disorder or psychopathy, a family history of the disorder, abnormal PET or SPECT, or disturbances in serotenergic or dopamergic systems, then he or she would be a candidate for genotyping for mutations or a specific gene that encodes for high-risk and novelty-seeking behavior.

## Case History: *State of Georgia v. Stephen A. Mobley:* Psychopathy, Antisocial Personality Disorder–Homicide–Death Penalty–Mitigating Evidence

On February 17, 1991, Stephen A. Mobley walked into a Domino's Pizza Store in Hall County, Georgia, robbed the store manager, and shot him in the back of the head as he was fleeing. Several weeks later Mobley was apprehended after a high-speed car chase through Atlanta. Mobley was questioned by the police and he confessed to numerous armed robberies and the murder and armed robbery at Domino's.

Once he was indicted for the crimes, Mobley offered to plead guilty to the charges in exchange for a sentence of life in prison without parole. The state refused the offer, insisting on the death penalty. Mobley was tried and convicted for the murder and armed robbery and the jury sentenced him to death by electrocution.

As is discussed in the introduction to this volume, Mobley's history of antisocial behavior began in early childhood and persisted throughout adolescence and adulthood. As a young child, Mobley repeatedly lied, cheated, and stole property from his parents and sister. Mobley's parents made numerous efforts to correct their son's deviant behavior by traditional forms of punishment, reform school, and intensive psychological counseling. All efforts to modify Mobley's behavior were unsuccessful, and his misbehavior grew progressively worse, culminating in prison sentences for forgery and theft.

At age 16 Mobley was diagnosed at a psychiatric facility near Atlanta as having an antisocial personality disorder. Mobley was discharged from the facility and Mobley's parents were advised that nothing more could be done for their son.

In his late teens and early twenties, Mobley engaged in numerous petty crimes and promiscuous bisexual activity. Mobley also changed jobs frequently (usually after being fired for stealing cash receipts). In his mid-twenties, Mobley embarked on his crime spree after stealing a gun in Macon, Georgia. Mobley began committing numerous armed robberies, culminating in the armed robbery and murder of the Domino's manager.

While Mobley was in jail awaiting his trial for the armed robbery and murder, he engaged in numerous fights with other inmates, anally sodomized a cellmate on two occasions, had the word *Domino* tattooed on his back, and made various comments to the jail guards such as, "When I beat this case, I plan to apply for the opening at Domino's that I heard about," and "You should've seen that fat slob beg for his life before I shot him." When Mobley was refused an extra aspirin during sick call, he told a guard, "You are looking more like a Domino's employee every day."[19]

Notwithstanding these transgressions, Mobley also presented himself as an extremely intelligent and charming person with a quick wit and a gift for creative writing. Mobley was a voracious reader and was conversant in almost any subject. Even though Mobley was quite sane and understood both the charges against him and the high likelihood of receiving the death penalty, he seemed almost powerless to control his impulses.

Several months before his trial was to begin, the results of the Brunner MAOA study were published. Mobley and his attorney realized that they had no legal defense to the armed robbery and murder charges because of Mobley's numerous confessions, and they also recognized that they had no traditional "mitigating" evidence that they could offer the jury to convince them to spare his life. The Brunner MAOA deficiency study offered the only hope of developing mitigating evidence for Mobley if it could be shown that his antisocial personality disorder and his impulsive and aggressive behavior was under genetic control rather than a product of Mobley's own choice or free-will.

In an effort to obtain funding to determine if Mobley was suffering from MAOA metabolic deficiency, or some other genetic mutation implicated in his antisocial personality disorder, Mobley's lawyers assembled a substantial evidentiary showing for the trial judge to convince him that Mobley might indeed be suffering from a biologic or genetic disorder that could be associated with his antisocial and criminal conduct. The lawyers hoped to present evidence of a biologic or genetic etiology for antisocial personality, not as an affirmative defense but to use only as possible mitigating evidence to sway the jury to spare Mobley from execution by electrocution.

In addition to introducing the Brunner MAOA study, Mobley introduced evidence demonstrating an extensive family history of antisocial and criminal conduct going back several generations; Mobley's own clinical diagnosis for antisocial personality disorder at age 16; expert affidavits from Dr. Xandra O. Breakfield (molecular neurogenetics), Dr. Bahjat A. Faraj (neurochemistry), and Dr. J. Stephen Ziegler (clinical psychology). After several hearings the trial judge refused to give Mobley funding to do any preliminary testing, concluding that "the genetic connection is not at a level of scientific acceptance that would justify its admission into court." Mobley was convicted and sentenced to death without the jury ever having considered the research associating genetics and antisocial behavior.[20]

Mobley's death sentence was affirmed on direct appeal to the Georgia Supreme Court.[21] Mobley remains on death row in Georgia.

## Conclusion

The revolution in molecular genetics and human genome research will continue to provide information about the origins of human behavior. Human genome research that demonstrates genetic variables control or influence antisocial, impulsive, and criminal behavior mandates that lawyers exploit this research in defense of their client's accused of criminal conduct. The complexity of the subject matter and research should not be a deterrent to lawyers charged with the duty of zealously representing their clients. Understanding the advances in human genome research and using this information, especially in defense of persons accused of capital offenses, may make the difference between a life or death sentence.

## Notes

1. Clarence Darrow, *Crime: Its Cause and Treatment* (New York: Thomas Y. Crowell, 1922).
2. K. Tierney, *Darrow, A Biography* (New York: Thomas Y. Crowell, 1979).
3. S. Boschert, "Differently Abled: Studies Suggest Psychopaths Lack Standard Affective Brain Function," *Clinical Psychiatry News* (December 1994):1–22.
4. D. Nielson, D. Goldman, M. Virkkunen, R. Tokola, R. Rawlings, and M. Linnoila, "Suicidality and 5-Hydroxyindoleacetic Acid Concentration Associated With a Tryptophan Hydroxylase Polymorphism," *Archives of General Psychiatry* 51 (January 1994):34–38.
5. R. Ebstein, O. Novick, and R. Umansky, B. Priel, Y. Osher, D. Blaine, E. R. Bennett, L. Nemanov, M. Katz, and R. H. Belmaker, "Dopamine D4 Receptor (D4DR) Exon III Polymorphism Associated With the Human Personality Trait of Novelty Seeking," *Nature Genetics* 12 (1996): 78–80.
6. 113 S. Ct. 2786 (1993).
7. 293 F. 1013 (1923).
8. *Daubert,* 113 S. Ct. 2786.
9. O. Stine, J. Xu, R. Koskela, F. J. McMahon, M. Gschwend, C. Friddle, C. D. Clark, M. G. McInnis, S. G. Simpson, T. S. Breschel, E. Vishio, K. Riskin, H. Feilotter, E. Chen, S. Shen, S. Folstein, D.A. Meyers, D. Botstein, T. G. Marr, and J. R. DePaulo, "Evidence for Linkage of Bipolar Disorder to Chromosome 18 With a Parent-of-Origin Effect," *American Journal of Human Genetics* 57 (1995):1384–1394.
10. South Carolina v. Susan Smith, Docket Nos. 94-GS-44-906 and 94-GS-44-907.
11. R. Ellis, "She's Not a Cold Blooded Killer: Unique Defense Frees Mom Convicted of Killing Son," *Atlanta Journal and Constitution,* September 28, 1994.
12. H. G. Brunner, M. Nelen, X. O. Breakfield, H. H. Ropers, and B. A. Van Oost, "Abnormal

Behavior Associated With a Point Mutation in the Structural Gene for Monoamine Oxidase A,'' *Science* 262 (1993):578–580.

13. O. Cases, I. Sief, J. Grimsky, P. Gaspar, K. Chen, S. Pournin, V. Muller, M. Agnet, C. Babinet, J. C. Shih, and E. De Maeyer, ''Aggressive Behavior and Altered Amounts of Brain Serotonin and Norepinephrine in Mice Lacking MAOA,'' *Science* 268 (1995):1763–1766.
14. S. Begley, ''The Mice that Roared . . . and Killed and Raped, Due to a Genetic Defect,'' *Newsweek,* December 1995.
15. Ebstein, ''Dopamine D4 Receptor (D4DR) Exon III Polymorphism Associated With the Human Personality Trait of Novelty Seeking.''
16. (Washington, DC: American Psychiatric Association, 1994).
17. Boschert, ''Differently Abled.''
18. Mobley v. Georgia, 455 S.E.2d 61 (Ga. 1995).

# ELABORATION
## The "Defective Gene" Defense in Criminal Cases

Creighton C. Horton II

Recent advances in scientific knowledge and technology regarding DNA raise interesting questions for practitioners of criminal law. As knowledge of the relationship between certain genetic characteristics and behavior increases, defense attorneys are increasingly likely to look to genetics as an avenue to excuse or at least mitigate their clients' criminal conduct. The notion would be that society ought not treat harshly those who, through no fault of their own, inherited a particular genetic trait or mutant gene that substantially contributed to their errant behavior.

In this elaboration, I will consider whether those who have a genetic defect, which in some manner causes or contributes to their antisocial conduct, should be excused from legal accountability based on that defect. I will also focus on the specific issue of whether a genetic link to antisocial personality disorder would or could serve as a basis for a defense to criminal conduct. For the purpose of this commentary, I assume that such a link can be definitively established scientifically, which, given the complexity of factors affecting human behavior, is assuming a great deal. I also assume that it can be further established that a particular criminal defendant actually possesses the gene or genes that cause the criminal behavior.

Antisocial personality disorder is a diagnosis in the *Diagnostic and Statistical Manual of Mental Disorders,* 4th Edition, known as *DSM-IV.*[1] *DSM-IV* is the standard diagnostic volume used by both psychiatrists and psychologists in their evaluations. A diagnosis of antisocial personality disorder describes a pervasive behavioral pattern of disregard for the rights of others, involving such things as: (a) failure to conform to social norms; (b) engaging in unlawful behavior; (c) deceitfulness; (d) impulsivity; (e) aggressiveness; (f) reckless disregard for the safety of others; (g) consistent irresponsibility; and (h) lack of remorse for the consequences of one's conduct. It is a descriptive diagnosis involving enduring patterns of behavior that lead to significant distress or impairment in social, occupational, or other important areas of functioning. If the character traits evident in the individual's pattern of behavior do not rise to the level of *significantly* impairing functioning, then the diagnosis is not antisocial personality disorder but simply that the person exhibits antisocial personality traits that do not rise to the level of a mental disorder. (See the diagnostic criteria for antisocial personality disorder in *DSM-IV.*[2])

Notably, antisocial personality disorder does not involve features associated with major mental illnesses that involve a cognitive break with reality. Schizophrenia, for example, includes symptoms such as delusions, hallucinations, disorganized speech, grossly disorganized or catatonic behavior, and so forth. Traditionally, defendants diagnosed with schizophrenia are likely candidates for the insanity defense. Those with antisocial personality disorder have traditionally not had a viable defense based on that "diagnosis." They have been viewed as either not mentally ill at all or, at least, not mentally ill *enough* to qualify for the defense.

Between 1979 and 1983 in Salt Lake County, Utah, Arthur Gary Bishop kidnaped, sexually molested, and killed five young boys, ranging from 4 to 13 years of age. At trial, Bishop's attorneys advanced the theory that he had a mental disorder in that he was a homosexual pedophile and also suffered from antisocial personality disorder. The defense attempted to use this theory to reduce the level of Bishop's culpability from first-degree murder, carrying the possibility of a death sentence, down to manslaughter, a less serious offense. The prosecution agreed with the diagnosis but took the position that Bishop's homosexual pedophilia and antisocial personality provided an explanation for why he did what he did but in no way constituted a defense to his crimes. The prosecutor argued that the defense counsel was trying to turn a motive into a defense.

The jury rejected the argument that antisocial personality disorder, coupled with the defendant's homosexual pedophilia, in any way excused or mitigated Bishop's conduct. He was convicted of five counts of capital murder and sentenced to death.

Would the result have been different had Bishop's attorney been able to make a definitive scientific link establishing that his antisocial personality disorder was the result of a genetic defect? The central question raised is whether a defective gene would preclude the formation of criminal intent.

What exactly, then, is criminal intent? In any criminal case, the prosecution must prove not only that the defendant engaged in conduct prohibited by law but that he or she did so with a certain state of mind. Summers describes criminal intent in terms of guilty knowledge (the legal concept of *scienter*) or a guilty mind/guilty willfullness (the concept of *mens rea*). Later in this discussion I will address the issue of whether a defective gene defense might abrogate criminal liability on the theory that it blocks criminal intent.

The closest parallel in existing law to the proposed genetic defect defense is the insanity defense. Historically, it has been viewed as unfair to punish those who are so mentally ill that they commit acts without appreciating the wrongfulness of what they are doing because of disordered thinking. Thus those who are unable to appreciate the wrongfulness of their conduct because of mental illness might be found "not guilty by reason of insanity" rather than convicted of a criminal offense. The insanity defense sometimes also encompasses what is known as the *irresistible impulse doctrine,* the notion that because of the mental illness a defendant is not able to conform his or her conduct to the requirements of law. This concept is similar to Summer's idea that a criminal defendant's conduct, if influenced or dictated by genetic factors, might not be criminal because it would not be the exercise of "pure free will."

Arguably, then, a defective gene defense could incorporate both whether a defendant has the capacity to appreciate the wrongfulness of his or her conduct and whether he or she has the ability to conform the conduct to the requirements of law. Although undoubtedly there will be significant obstacles to the scientific establishment of a genetic link to antisocial conduct, equally difficult may be linking such a genetic predisposition to a viable legal defense, because there are a number of public policy issues that come into play.

For brevity I will refer to the gene or combination of genes that may theoretically result in a diagnosis of antisocial personality disorder as the *antisocial gene.* In what way would the presence of such a gene preclude the formation of criminal intent? And if society recognizes that a person with a genetic predisposition toward aggression and antisocial behavior cannot legitimately be convicted or sentenced for criminal acts, would the same rationale extend to exonerate a criminal defendant based on environmental deficits? Should a defendant be able to demonstrate that he or she came from a deprived and disadvantaged background and because of that upbringing was predisposed to act in antisocial ways that he or she might not otherwise have engaged in?

Underlying this analysis is the notion that the criminal law's legitimate reach extends only to those who are "morally blameworthy." Thus the law cannot hold legally accountable those who are not morally blameworthy. The purpose of the criminal law under this model is to identify ''bad'' people who do bad things and punish them for their misdeeds. (This view seems to have roots in religious ideology of good against evil, and the labels we use in the criminal justice system reinforce the concept. For example, when people are convicted, we say they are ''guilty'' rather than, say, ''accountable'' or ''responsible.'') Under this model, anyone who can demonstrate that they are not fundamentally ''bad'' cannot legitimately be brought under the jurisdiction of the criminal law.

There are, however, purposes other than punishing the blameworthy served by the criminal justice system. Most notably, the system is designed to protect public safety by incapacitating those who intentionally or knowingly commit antisocial acts, particularly violent acts. As a consequence, it is not presently a defense to a criminal act that a defendant was raised in disadvantaged circumstances, although that fact might be considered a relevant sentencing consideration. Even for sentencing purposes, however, those who commit acts of violence, particularly extreme acts of violence against other members of society, are generally not given much, if any, consideration based on their disadvantaged circumstances.

In other words, the criminal law generally deals with people on the basis of what they have done and, in a sense, what they have become, not what caused them to become that way. Had Arthur Gary Bishop presented evidence of a genetic link, it is unlikely that his attorneys would have been any more successful in arguing that his offenses should be mitigated down to something less than capital murder. I do think, however, that if a genetic link can be established, such evidence might become admissible in a capital sentencing proceeding, because the issue of whether a defendant should receive a death sentence is one that allows the broadest introduction of potentially mitigating evidence capable of influencing a judge or jury to not impose the ultimate sanction.

Even in the context of a capital sentencing proceeding, however, the genetic defect argument would likely be the argument of last resort. As between arguing that a defendant was born with good qualities and then subjected to negative environmental influences and arguing that the defendant was antisocial from the time he was born, most defense attorneys would rather argue environmental rather than genetic influences. Juries are more likely to identify with and show leniency toward defendants whom they see as having positive human attributes (i.e., remorse for one's misdeeds) than they are toward defendants whom they see as being ''bad apples'' since birth, even if they are that way because they inherited ''bad genes.'' The genetic argument ultimately may be viewed as advancing the rather unpersuasive proposition that the defendant should be shown leniency because ''this whole thing (the double axe murders) would never have happened if only my client (the defendant) had never been born.''

## Criminal Intent

What must be shown to establish criminal intent? Utah's experience with its insanity defense in recent years is of interest. In 1983 the Utah legislature abolished the insanity defense as it previously existed in Utah. (About the same time, Idaho and Montana abolished their insanity defenses entirely.) The previous formulation of Utah's law was similar to the Model Penal Code formulation, providing that a person could not be convicted of a criminal offense if, as a result of mental disease or defect, he or she lacked substantial capacity to appreciate the wrongfulness of his or her conduct or to conform his conduct to the requirements of the law.

The Utah legislature replaced the Model Penal Code standard with what has been called a *mens rea* standard. That is, a defendant can be convicted of a criminal offense, even if mentally ill at the time, if he or she was not so mentally ill that the mental illness prevented him or her from acting intentionally, or with whatever state of mind provided by statute for the crime. For example, in a murder case, the prosecution generally has to prove beyond a reasonable doubt that the defendant acted *intentionally or knowingly* when he caused the death of the victim. Under the new statute, the *mens rea*, or criminal intent that must be proved is simply that the defendant acted intentionally or knowingly. There is no other mental state required, such as whether the defendant thought his or her conduct was wrong.

As a consequence, mentally ill defendants may be convicted of murder in Utah if they acted intentionally or knowingly, notwithstanding the fact they may not have appreciated the wrongfulness of what they were doing. The only place in Utah law where Model Penal Code concepts (appreciating wrongfulness and ability to conform conduct to law) are retained are as potential mitigating factors at a capital sentencing hearing, after one has been convicted of aggravated murder.[3]

Not surprisingly, Utah's new law[4] was challenged for doing away with the traditional insanity defense and allowing for the conviction of those who, because of mental illness, were unable to appreciate the wrongfulness of their actions or to control their conduct (but who could still act with intent). The Utah Supreme Court, in *State v. Herrera,*[5] upheld Utah's statutory scheme as constitutional under both the state and federal constitutions. The court noted the delicate balancing of public policy in determining accountability for criminal acts, stating that "government must balance society's interest in order, protection, punishment, and deterrence with the particularly arduous responsibility for caring for the insane and mentally deficient." The court also acknowledged that "in a very real sense, the confinement of the insane is the punishment of the innocent; the release of the insane is the punishment of society.[6] The court found that Utah's *mens rea* statute was a permissible approach to assessing criminal responsibility, particularly in light of the fact that the 1983 legislation also provided for the alternative verdict of "guilty and mentally ill," a verdict that could for sentencing purposes buffer some of the harsher consequences of eliminating the more traditional insanity defense.[7]

It appears then that it would be far from automatic that the identification of an antisocial gene would result in elimination of criminal responsibility. Some states specifically eliminate personality disorders from their definitions of "mental illness" to preclude those with antisocial personality disorders from using that "diagnosis" as a defense. For example, the Utah statute provides that "mental illness does not mean a personality or character disorder or abnormality manifested only be repeated criminal conduct."[8]

If people with the antisocial gene can rob, maim, and murder, and intend to do so, they may not be able to avoid criminal liability on a theory that they lack sufficient criminal intent. Let us look at another theory of nonresponsibility—that of an "irresistible impulse" defense.

## Irresistible Impulse

Could the establishment of scientific proof of an antisocial gene lead to a defense based on the notion that those with the gene should not be punished for their antisocial acts because they could not help themselves? Many state legislatures as well as the U.S. Congress have abolished the "irresistible impulse" defense in insanity cases. This occurred in the wake of the attempted assassination of President Ronald Reagan in 1981. The American Psychiatric Association issued a position paper in December 1982 stating that it had no opposition to legislatures abolishing that section of the insanity statute having to do with irresistible impulse. The association said that "the line between an irresistible impulse and an impulse not resisted is probably no sharper than that between twilight and dusk." It also advanced the notion that persons with antisocial personality disorder *should* be held accountable for their behavior.[9]

The U.S. Supreme Court has ruled that the irresistible impulse component of an insanity defense is not constitutionally required.[10] So it seems unlikely that a successful defense could be mounted on the basis that an antisocial gene is responsible for someone acting out in an antisocial manner because they "lost control" to the extent that they *could not* resist the impulse to do what they did, rather than they *did not* resist the impulse, at least in those jurisdictions that have statutorily abolished irresistible impulse as a defense.

Another question that would be raised if this defense were advanced is whether the presence of such a gene *necessarily* would cause a person to act out in an antisocial manner. If others with the gene are able to control themselves and conform their conduct to the law, the premise that the gene controls the behavior may be flawed. It may, indeed, indicate a predisposition toward violent or antisocial behavior, but not necessarily a guarantee of antisocial conduct.

On the other hand, if science were to be able to identify specifically an antisocial gene that would cause all persons who possess it to act out in an antisocial manner, then the outcry for public safety might result in legislation to preventatively detain such persons, similar to statutes that exist in some states to confine those who suffer from a mental abnormality or personality disorder that makes them likely to engage in predatory acts of sexual violence.[11]

## Conclusion

Significant public policy considerations are implicated in the proposed genetic defect defense. Even if such a link could be established scientifically, which is by no means clear given the complexity of causal factors likely to influence human behavior, that link would just be the first step in the analysis. In today's climate of increasing concern about crime and the safety of citizens, legislatures might move to specifically restrict this type of defense, particularly in cases involving acts of violence. In evaluating the policy considerations, it is noteworthy and not surprising that the largest concentration of those who have antisocial personality disorders is in our prisons. As noted by Summer, it is likely that the 6 to 7% of criminals who commit 85 to 90% of repeated violent criminal acts are composed of persons with antisocial personality disorder. It seems likely that any shift in criminal justice philosophy that could result in such persons no longer being held criminally responsible for their acts based on the notion that their moral culpability is blocked by a genetic defect would be approached very cautiously by courts, juries, and legislatures out of a concern for public safety.

As the U.S. Supreme Court has stated, "The doctrines of *actus reus, mens rea,* insanity, mistake, justification, and duress have historically provided the tools for a constantly shifting adjustment of the tension between the evolving aims of the criminal law and changing religious, moral, philosophical, and medical views of the nature of man. This process of adjustment has always been thought to be the province of the States.[12]

In summary, the criminal justice system is not likely to embrace the notion that if antisocial or violent behavior can be understood through genetic linkage, that linkage becomes a defense rather than an explanation for conduct. Such a "genetic-defect defense" would encounter significant obstacles in the U.S. system, which is firmly grounded in the philosophy of personal responsibility for one's actions, except for those who suffer severe impairment in their cognitive thinking because of major mental illness.

On the other hand, if identifying genes that lead to antisocial aggressive tendencies is the first step in finding some way to counteract those tendencies, all of society would be benefited. For instance, if it were discovered that a gene or combination of genes produced antisocial conduct as a result of a chemical imbalance in the brain, perhaps researchers could develop ways to mitigate the effects of the genes, similar to the way antipsychotic medication may alleviate some of the disordered thinking associated with schizophrenia. With the rise of violent crime across the country and the resulting public safety concerns, anything that might serve to reduce antisocial tendencies, violence, and crime would be most welcome.

## Notes

1. *DSM-IV* (Washington, DC: American Psychiatric Association, 1994).
2. *Id.,* 649–650.
3. UTAH CODE ANN. § 76-3-207 (3)(d) (1995).
4. UTAH CODE ANN. § 76-2-305 (1995).
5. 895 P.2d 359 (Utah 1995).
6. *Id.,* 362.
7. *Id.,* 367.
8. UTAH CODE ANN. § 76-2-305(4) (1995).
9. "American Psychiatric Association Statement on the Insanity Defense" (Washington, DC: Author, December 1982).
10. Leland v. Oregon, 72 S. Ct. 1002 (1952). 343 U.S. 790
11. *See* Kansas v. Hendricks, 117 S. Ct. 2072 (1997). 521 U.S. 346
12. Powell v. Texas, 392 U.S. 514, 535–536 (1968), as cited in Montana v. Egelhoff, 518 U.S. 37, 56 (1966).

# Chapter 7
# JUVENILE LAW AND GENETICS

Mark A. Small

The section describing the juvenile court is based on an earlier article by Mark Small titled, "Introduction to Juvenile Justice: *Comments and Trends*" *Behavioral Sciences and Law* 15 (1997):119–124. Copyright John Wiley & Sons, Ltd. Adapted with permission.

The social construction of childhood is culturally and historically dependent. At any given time, legal policies affecting children rest on particular images of childhood. At the turn of the century, the image of vulnerable children in need of guidance prompted the establishment of legal institutions that functioned primarily to civilize and socially control youth. Child labor laws were enacted to protect youth from harmful working conditions, and compulsory education was established to inculcate proper values. Perhaps the best indication of the prevailing view of childhood was the founding of the juvenile court. Designed to rehabilitate wayward youth, the juvenile court operated on the assumption that children were worthy of redemption and capable of change.

Advances in genetic research may one day alter the way in which youthful behavior is viewed. If the image of children shifts from pliable beings whose ultimate personality is dependent on the environment to persons whose behavior is largely genetically programmed, corresponding shifts in juvenile law may also take place. Because of the flexibility in the construction of childhood and the uncertain issue of who has control over children, children's law presents a unique stage on which to play out social issues. For example, the primary legal arena for issues of segregation (*Brown v. Board of Education*),[1] abortion (*Planned Parenthood of Southeastern Pa. v. Casey*),[2] obscenity (*Bethel v. Fraser*),[3] and religion (*Wisconsin v. Yoder*)[4] have all been children's law. Given this background, the first frontier for issues related to genetic research very likely will be children's law and, if the research is targeted at crime, juvenile law.

The purpose of this chapter is to explore how discoveries of genetic factors related to antisocial behaviors may influence juvenile law. First, I will present a historic overview of the development of juvenile law. Next, I will discuss current trends in juvenile crime and corresponding developments in juvenile law. Finally, I will discuss the potential impact of genetic research on juvenile law.

## History of Juvenile Law in the United States

Before 1899 deviant acts by juveniles were considered within the general body of criminal law. Age, however, could reduce or even exculpate the blameworthiness of juveniles. The defense of "infancy" could be used by those under the age of 7 as a complete defense to crime. For those children aged 7 to 14, the presumption of the capacity to form criminal intent was rebuttable by a showing of maturity (and the concomitant ability to form intent) by the state. For example, the state could show that a 12-year-old's attempt to cover up evidence of a crime indicated a knowledge of wrongdoing, thus rebutting the presumption of immaturity. Children older than 14 were treated as adults by the criminal law.

Around the turn of the century several juvenile courts were created, the first in Chicago, Illinois, in 1899. They were founded on the assumption that the interests of the state and the juvenile were the same. The state, as *parens patriae,* would act on behalf of youths and provide them with the treatment needed to ensure that they overcame their youthful mistakes and adopted proper cultural values. The overriding rationale for the juvenile court was rehabilitation, and the paramount question was a youth's amenability to treatment.

Today, children typically fall under the jurisdiction of juvenile courts in one of several ways. First, a juvenile court may obtain jurisdiction if a child is accused of violating a section of the criminal code that applies to all citizens (e.g., theft, robbery). Second, a child might fall under the jurisdiction of the juvenile court by committing a ''status'' offense, an offense peculiar to childhood status (e.g., truancy, violation of curfew). When offenses require court-ordered services, a child might be classified as, a ''child in need of supervision,'' another grounds for jurisdiction. Finally, some juvenile–family courts also have jurisdiction over domestic adjudications such as abuse, neglect, and custody determinations, as well as power to civilly commit juveniles.

Because the juvenile court was not instituted to punish juveniles, there was no need for formal protections of due process. As initially designed, there was nothing adversary about juvenile court proceedings, and what traditionally counted as ''due process'' (e.g., presence of attorney, notice of charges, compulsory witnesses on one's behalf) might interfere with the therapy that the court would initiate. Indeed, appearances by lawyers were rare and most juvenile judges had no formal legal training. To avoid potentially stigmatizing youth, proceedings were closed and records sealed.

To further highlight differences with the adult criminal system, rather than a determination of guilt the fact-finding stage of juvenile proceedings was referred to as an ''adjudication'' of delinquency. The juvenile counterpart to the sentencing phase of the trial was ''disposition.'' Consistent with the rehabilitative ideal, at the disposition phase of the proceedings the focus was on treating the offender rather than meting out punishment for the offense. When possible, juveniles could be diverted to receive appropriate treatment during any stage of the proceedings.

Because the central concern was treatment and rehabilitation, the key actors in the juvenile court were mental health professionals. Perhaps the best description of the early juvenile court was written by Judge Julian Mack in 1909:

> The child who must be brought into court should, of course, be made to know that he is face to face with the power of the state, but he should at the same time, and more emphatically, be made to feel that he is the object of its care and solicitude. The ordinary trappings of the courtroom are out of place in such hearing. The judge on a bench, looking down upon the boy standing at the bar, can never evoke a proper sympathetic spirit. Seated at a desk, with the child at his side, where he can on occasion put his arm around his shoulder and draw the lad to him, the judge, while losing none of his judicial dignity, will gain immensely in the effectiveness of his work.[5]

## The Fall of the Rehabilitative Ideal

In the mid-1960s the U.S. Supreme Court began a series of opinions that changed the nature of juvenile law. These opinions directly challenged the assumptions on which the juvenile court was founded. As Justice Fortas wrote in the majority opinion in *Kent v. United States,* ''There may be grounds of concerns that the child gets the worst of both worlds: that he gets neither the protections accorded to adults nor the solicitous care and regenerative treatment postulated for children.''[6] In *Kent,* the Court held that juveniles were entitled to a hearing before being waived to adult court to face criminal charges.

A year later in *In re Gault,*[7] the Supreme Court described juvenile courts as ''kangaroo courts'' characterized by arbitrariness, ineffectiveness, and the appearance of injustice. The appellant, Gerald Gault, had been committed at age 15 to the Arizona State Industrial School ''for the period of his minority [to age 21], unless sooner discharged by due process of law.'' His behavior leading to this loss in liberty was making an obscene telephone call. Had he been an adult convicted of the same offense, he could have been sentenced only to a fine of $5 to $50 or imprisonment of up to two months. Other facts surrounding the incident highlighted deficiencies in the juvenile justice system: Gerald's parents were not notified of his arrest; neither Gerald nor his parents were notified on the charge; Gerald was not provided access to counsel or the opportunity to summon and cross-examine witnesses; and the judge interrogated Gerald during the hearing and compelled him to testify against himself.

*Gault* and its progeny have resulted in more rights being afforded juveniles, though not to the extent that they are afforded adults in criminal court. For example, *Gault* required states to provide the right to counsel, to written and timely notice of the charges, and to the privilege against self-incrimination. Since *Gault,* the Supreme Court has held that protections of the double jeopardy clause apply to juveniles,[8] and that the state's burden of proof is ''beyond a reasonable doubt'' at the adjudicatory phase of a hearing.[9]

However, the Court has not yet abandoned the rehabilitative promise of the juvenile courts. In denying juveniles a right to a jury trial, the Court expressed some optimism regarding rehabilitation.

> The juvenile court held high promise. We are reluctant to say that, despite disappointments of grave dimensions, it still does not hold promise, and we are particularly reluctant to say . . . that the system cannot accomplish its rehabilitation goals. So much depends on the availability of resources, on the interest and commitment of the public, on the willingness to learn and on understanding as cause and effect and cure. In this field, as in so many others, one perhaps learns by doing. We are reluctant to disallow the States to experiment further and to seek in new and different ways the elusive answers to the problems of the young, and we feel that we would be impeding that experimentation by imposing the jury trial. . . .
>
> If the formalities of the criminal adjudicative process are to be superimposed upon the juvenile court system, there is little need of its separate existence. Perhaps that ultimate disillusionment will come one day, but for the moment we are disinclined to impetus to it.[10]

The juvenile justice system originally came under attack because there was little proof that the courts were providing effective treatment. Specifically, the assumption that juveniles are especially amenable to treatment was challenged. As a panel of the National Academy of Sciences has concluded, the assertion that ''nothing works'' in juvenile corrections has yet to be refuted.[11] The most well-validated treatment for delinquent behavior remains getting older. To be sure, the appropriate response to ''nothing works'' is that ''nothing has been tried.'' Many evaluation studies focused on programs that were poorly conceived, inadequately staffed, and seriously underfunded.

Evaluation studies are also complicated by the fact that most juveniles have multiple problems; educational delays, family disorganization, a lack of community support, economic poverty, and poor social skills.[12] Some small experimental programs that have an intensive, integrated response to such problems have shown success, but they have yet to be tried on a large scale. However, studies show clearly that most juvenile offenders do not recidivate, no matter what intervention is provided. For juveniles who do recidivate, there is little reason to expect special amenability to treatment, relative to adult offenders.[13]

## Juvenile Crime

Previous changes to the juvenile justice system were driven by judicial efforts to protect juveniles. More recent changes are being driven by legislative efforts to protect society from juveniles. Current changes in juvenile codes have resulted from the changing nature of juvenile crime, which has worsened, most noticeably since 1985. From various governmental agencies charged with collecting statistics, the following data present a current snapshot of juvenile crime. Although many of the trends in adult crime have remained flat in recent years, there has been some change in violent crimes committed by youth. Notably,

- Based on total arrest statistics for 1992, juveniles were responsible for 13% of all violent crimes and 23% of all property crimes. Fifteen percent of all persons entering the justice system on a murder charge were juveniles.[14]
- In 1991, according to the National Crime Victimization Survey, juveniles were responsible for nearly one in five (19%) of all violent crimes (i.e., rape, personal robbery, and simple and aggravated assault).[15]
- The proportion of violent crimes committed by juveniles is disproportionately high compared with their share of the U.S. population, and the number of these crimes is growing.[16]
- "The increase in murder by very young people after 1985 has not at all been matched by increases among the older groups (ages 24 and over). Among them murder rates have even declined. Thus, much of the general increase in the aggregate homicide rate (accounting for all ages) in the late 1980's is attributable to the spurt in the murder rate by young people that began in 1985."[17]
- In a 1990 nationally representative survey of students in Grades 9 to 12, one in five reported carrying a weapon (gun, knife, club) at least once in the previous 30 days.[18]
- The current population of teenagers responsible for much of the recent crime is relatively small compared to the cohort of children ages 5 to 15 who will be moving into crime-prone ages (16–24) in the near future.[19]
- If juvenile arrest rates remain constant through the year 2010, the number of juvenile arrests for violent crime will increase by one fifth; if rates increase as they have in recent history, juvenile violent crime arrests will double.[20]
- Since 1985, among African American males ages 14 to 17, murder rates have been about four to five times higher than among White males of the same age group.[21]

## Current Trends in Juvenile Law

The current trends in juvenile crime have not gone unnoticed by scholars or legislatures. Several visions of reform are currently being offered to either modify or eliminate the juvenile justice system. Potentially the most influential academic scholarship on the debate of the nature of the juvenile court is contained in the *Juvenile Justice Standards,* a 23-volume set of standards and commentary on various topics of juvenile court administration, procedure, and substance.[22] The volumes were sponsored by the American Bar Association (ABA) and the Institute of Judicial Administration, and many of these volumes have been adopted as official ABA policy. The *Standards* reject rehabilitation as a primary basis for a juvenile justice system, instead emphasizing due process, just desserts, diversion, deinstitutionalization, and decriminalization.

The *Standards* would replace offender-based dispositions with determinate sentences proportionate to the offenses in question. On the theory that the juvenile justice system causes more harm than good, a presumption at intake would favor referrals to community agencies in place of filing charges. Moreover, postadjudicative dispositions would be to the least restrictive alternatives. Status offenses would be "decriminalized" and removed from the jurisdiction of the juvenile court.

Other proposed reforms to the juvenile justice system vary from suggestions to eliminate the court altogether to creating some hybrid, where juveniles enjoy both the full panoply of rights afforded adults and have available appropriate treatment options in dispositions.[23] Future reforms in juvenile law are largely a political question, though there is little consensus among political parties as to the direction future changes should take. Liberals are disenchanted with the lack of efficacy in treatment, and conservatives are frustrated with the perceived kid-glove treatment of serious offenders. A fair reading of possible changes would suggest that there likely will be more formal moves toward legalizing the juvenile justice system and an increased focus on retribution and due process.[24]

Despite pressure from the outside, there is still much ambivalence on the part of those who work within the juvenile justice system about the system's proper goals. Although it is indisputable that the decision in *Gault* led to significant change, it is also clear that many juvenile courts have failed to implement its mandate fully. Many juvenile courts still neglect the due process rights basic to an adversary system, and studies show that both judges and lawyers still believe in a juvenile justice system that is based on rehabilitation.[25] Thus any reform efforts should take into account the potential resistance of actors within the system.

## A Medical Model of Juvenile Crime

In a bold prediction, Adrian Raine asserted that "a future generation will reconceptualize non-trivial recidivistic crime as a disorder."[26] The thought is premised on the notion that future researchers will be able to construct models of human behavior, reliant in part on genetic research, that will allow for screening, diagnostic prediction, and treatment of disorders related to crime. Should such models prove influential in the understanding of criminal behavior, juvenile law might change dramatically—or perhaps not at all. In a speculative spirit, the following are issues that might be considered in light of potential advances in genetic research relating crime to identifiable disorders.

### *Indirect Impact: Acceptance of Genetic Explanations of Criminality*

The first and perhaps most important issue is the public acceptance of theories of criminality that are grounded more in nature than nurture. There are no known genetic causes of criminal behavior, and the presence of genetic markers directly related to crime have yet to be proven. Yet history reveals that even tentative theories proposing a biological basis of criminal behavior have been readily accepted and quickly translated into criminal justice policies; the most notorious example being the eugenics movement.[27] Although eugenics has been discredited, alternate biological theories stand ready to take its place. For example, recently the presence of serotonin has been correlated with criminality, and calls are being made for changes in public policy to reflect this new discovery.[28]

One reason for the ready acceptance of biologically based explanations of crime may be that the alternative explanation (that crime is principally the product of the environment) has proven so unmanageable. Although we know that it is possible to change the environment, in terms of reducing crime efforts have been largely unsuccessful. A biological basis of criminal behavior eases social consciousness and commitment. Taken to the extreme, politicians would not need to spend money on social programs, nor would citizens need to feel guilty about failing to fund efforts aimed at changing the environment (e.g., education, housing, poverty). Very simply, to the extent people accept theories of biological determinism, they absolve themselves from changing environmental conditions that may contribute to crime.

Moreover, an understanding of human behavior that stresses biological determinism fits well within many of the current blueprints for juvenile reform. If youthful criminal behavior is perceived more as the result of biological than environmental influences, arguments can be made to abandon the rehabilitative ideal. Portraying youth not as vulnerable innocents but as predisposed victimizers justifies transferring juveniles to adult court and promotes reforms that make the juvenile system more punitive. With a public already dubious of rehabilitative efforts and more determined to punish wrongdoers, theories that attribute individual responsibility for criminal behavior are likely to be popular.

## *Direct Impact*

Pragmatically, the impact of medical models of juvenile crime likely would differentially affect the juvenile justice system depending on the stage of the process one examines: prevention, intervention, waiver, adjudication, or disposition. Assuming the juvenile justice system remains in the same form as today and assuming relevant genetic discoveries occur, more questions are raised than answers provided when considering how genetic discoveries might alter the juvenile justice system.

### Screening for Prevention

Assume that you have been chosen to pick out future delinquents: to predict those individuals who will commit violent criminal acts in the future. The purpose is to determine eligibility for entrance into a violence prevention program. A complicated model has been developed that takes into account a host of factors, including genetic markers. You are to use this model as a profile to predict future delinquents. Let us further make some lavish assumptions to aid you in this task. First, the profile is 90% accurate, meaning that for every 10 true delinquents profiled, 9 will be correctly identified as true delinquents. Second, although fewer than 0.5% of all juveniles were arrested for a violent offense in 1992,[29] we will assume the base rate to be 1%. Thus for every 1000 juveniles chosen at random, 10 are presumed to be true delinquents.

The typical table portrayal would look something like what is presented in Figure 7-1.

For every 108 individuals picked out to be delinquent by the profile, in fact only 9 turn out to be true delinquents. Prediction models always overpredict rare events, and despite the seriousness of juvenile violent crimes, the base rate is low. Even with generous assumptions (e.g., no profile comes close to being 90% accurate), the "success" rate is only 8.33 percent (9/108).

The 99 false positives are troublesome. Their entry into the violence prevention program is unnecessary, expensive, stigmatizing, and degrading. Also troublesome is the cost of applying the profile. Given the seriousness of juvenile crime, perhaps such a profiling

**Figure 7-1**
Delinquent Profile

| | | Truth: Delinquent | Truth: Nondelinquent | |
|---|---|---|---|---|
| Profile | Delinquent | 9 (Hit, true positive) | 99 (Alarm, false positive) | 108 |
| | Nondelinquent | 1 (Miss, false negative) | 891 (True negative) | 892 |

program could be justified if the costs were low. Would society be willing to fund the profiling of 1000 14-year-olds and the subsequent treatment of 108 to potentially "cure" the 9 who will commit future violent criminal acts? If so, this would be an unprecedented level of funding for this type of prevention program.

An argument could be made that this scenario is misleading. Profiling 1000 juveniles at random is inefficient. Only those juveniles who can be identified as already being at risk would be further tested. However, selectively profiling juveniles raises problems of its own. The first is that the success rate will likely remain low. Even if the base rate of some preselected targeted group of juveniles were 5% (very high), the corresponding success rate would be 32 percent (45/140), still yielding a considerable high number of false positives. The second problem is the selection of the risk factors to narrow the population. Depending on which risk factors are identified to target this subpopulation of juveniles, there may be persuasive arguments that the targeting violates the due process clause or the equal protection clause of the Fourteenth Amendment.[30]

## Intervention

The next issue to be considered is the effect of a discovery of genetic markers associated with criminal behavior on the types of intervention programs implemented. Would current programs be sufficient, or should new programs be developed to account for these genetic discoveries? If prevention were the goal, options might range from counseling to compulsory sterilization.[31]

The timing of intervention is also an issue. One of the most well-known findings in criminal justice is the age–crime curve; the consistent finding that the age of offense typically peaks in the late teenage years and significantly declines as age advances. As noted earlier, most juveniles either desist or mature out of criminal careers. This suggests that intervention efforts be targeted at the young, perhaps the very young. Finally, unless the programs were voluntary, constitutional challenges would likely be made.

*Waiver*

The legislative response to "get tough" with juveniles chiefly has been accomplished by making it easier for juveniles to be tried as adults. Indeed, the major legislative trend in state juvenile codes has been the amendment of juvenile statutes to allow easier transfer to adult courts for juveniles accused of committing serious offenses. Currently, all states allow juveniles to be tried as adults in criminal courts when certain criteria are met. There are three principal means of transfer: judicial waiver, prosecutorial discretion, and statutory exclusion. Traditionally, juvenile court judges would assess the particular characteristics of the offender and offense and make a determination to transfer the youth to adult court. Such transfers usually occurred after a finding that the youth was not amenable to treatment.

Reflecting a growing concern for juvenile crime, over the past 20 years states have begun enacting legislation to automatically transfer juveniles based on age and offense seriousness, thus removing judicial discretion. "In half the states, laws have been enacted that exclude some offenses from juvenile court and a number of states have also expanded the range of excluded offenses. One quarter of the states have given prosecutors the discretion to charge certain offenses either in juvenile or criminal court."[32] Between 1988 and 1992, the number of cases judicially waived to criminal court increased by 68%. The types of cases being waived may seem surprising. In 1992 a national survey of criminal court transfers found that 32% of judicial waivers involved violent offenses, whereas 62% involved either property charges or public order offenses.[33]

The presence of genetic markers could influence waiver as a factor to be considered in determining whether the child is amenable to treatment. At this stage in the juvenile justice system, amenability to treatment is typically interpreted as asking whether the juvenile is treatable at all.[34] The focus is more on the juvenile's criminal history than on treatment options per se, and presence of a genetic disorder may influence the decision to transfer. Juveniles might counter the influence of genetic markers associated with antisocial behavior by showing the presence of genetic markers associated with prosocial behavior.

As an alternative possibility, if anatomical destiny prevails as the dominant psychology, legislatures may establish evidence of a genetic predisposition to crime as an additional criteria listed by statute to automatically transfer juveniles to adult court. This also raises constitutional questions and may violate constitutional rights of juveniles.

Of potential relevance is the case of *Robinson v. California*,[35] where the U.S. Supreme Court held that a California statute that made it a criminal offense for a person to be addicted to narcotics violates the Eighth and Fourteenth Amendments. Repulsive to the Court was the notion that one could be punished for having the "status" of narcotics addiction. Consider how the Court might deliberate about the status of having a genetic marker associated with crime, given the following language from the *Robinson* opinion:

> It is unlikely that any State at this moment in history would attempt to make it a criminal offense for a person to be mentally ill, or a leper, or to be afflicted with a venereal disease. *A State might determine that the general health and welfare require that the victims of these and other human afflictions be dealt with by compulsory treatment, involving quarantine, confinement or sequestration.* But, in light of contemporary human knowledge, a law which made a criminal offense of such a disease would doubtless be universally thought to be an infliction of cruel and unusual punishment in violation of the Eighth and Fourteenth Amendments.[36]

Clearly, punishing one simply for having a genetic predisposition would be unconstitutional, but one could argue that this language supports the civil commitment of juveniles to treatment facilities and the possible use of genetic markers in decisions to transfer.

### *Adjudication*

The typical fact-finding phase of juvenile proceedings provides the state the opportunity to prove that the juvenile possessed the proper intent to commit an offense and in fact did commit the offense. As possible defenses, juveniles may present evidence of alibis, justifications, excuses, or alternatively show that the state failed to prove elements of the crime charged beyond a reasonable doubt. The presence of genetic markers might be offered to "excuse" an offense based on some genetic predisposition of the juvenile; though absent a direct causal link between the genetic marker and criminal behavior, the "excuse" will likely be unsuccessful.

The defense of "excuse" is seldom used in either the adult or juvenile criminal justice systems (e.g., the insanity defense is used in less than 1% of trials), but evidence of a genetic predisposition to crime might be used as a mitigating factor during the disposition phase if the offense is not serious. Serious violent offenses will be treated accordingly, but the presence of a genetic marker may alter the course of proceedings for a nonserious offense. Assuming some appropriate genetic-based treatment is available, diversion out of the system may be possible.[37]

### *Disposition (Treatment)*

Would evidence of genetic markers related to antisocial behavior alter the types of dispositional alternatives available? For example, what types of rehabilitation programs could be started, or how would existing treatment alternatives be altered? New types of treatment programs might attempt to perform some genetic correction or devise new chemical or surgical procedures; old programs might be modified to take into account genetic explanations of crime. Like current treatment programs, the efficacy of any new or modified program will depend on adequate funding, implementation, and demonstrated effectiveness.

There is little reason to believe new programs will come into existence in any large scale. Even the acceptance of genetic explanations of criminal behavior and development of corresponding treatment programs would not likely increase public funding. Information is available that demonstrates the high correlation with alcoholism and drug addiction to crime, yet prevention and treatment programs are not funded at meaningful levels nor are they popularly supported.

## Conclusion

Who controls the social construction of childhood controls the policies that affect children. The extent to which genetic discoveries will influence juvenile law will depend on their ability to define the nature of childhood. Currently, competing visions of childhood struggle to influence legal policy, particularly in the area of juvenile law. Often perceived as a wayward youth in need of guidance by those within the system and as an incorrigible "gangsta" by those outside the system, the picture of the juvenile delinquent is currently in flux. Academic writings contain even more variations of the nature of childhood, with corresponding recommendations for policy changes.[38] Although various social constructions of childhood may compete in the academic marketplace for prominence, the courts and the legislatures are the battlegrounds with direct consequences for child policy.

A medical model of crime has potential to influence the social and legal construction of childhood, thus impacting the juvenile justice system, directly and indirectly. Indirectly, the influence of genetic research likely will have the greatest impact in reinforcing attitudes that childhood behavior, including delinquent behavior, is biologically determined. Thus current prevention and treatment efforts based on an environmental understanding of behavior need not be taken seriously. Absent the development of corresponding treatments that follow these genetic discoveries, rehabilitation may be abandoned altogether. This may fuel current reform efforts that seek to eliminate the juvenile justice system, or at least reduce funding for prevention and treatment programs.

Should a medical model of juvenile crime dominate our understanding of behavior, specific components of the juvenile justice system may be directly impacted, though the nature of the impact is unclear. Prevention and treatment efforts, decisions to waive, and adjudication and disposition may all be affected, though these will likely be subject to constitutional challenges.

## Notes

1. 347 U.S. 483 (1954).
2. 505 U.S. 833 (1992).
3. 478 U.S. 675 (1986).
4. 478 U.S. 675 (1986).
5. Julian Mack, ''The Juvenile Court,'' *Harvard Law Review* 23 (1909):104–120.
6. 383 U.S. 541, 556 (1966).
7. 387 U.S. 1 (1967).
8. Breed v. Jones, 421 U.S. 519 (1975).
9. In re Winship, 397 U.S. 538 (1970).
10. McKeiver v. Pennsylvania, 403 U.S. 528 (1971).
11. Lee Sechrest, Susan White, and Elizabeth Brown, *The Rehabilitation of Criminal Offenders: Problems and Prospects* (Washington, D.C.: National Academy of Sciences, 1979).
12. S. W. Henggeler, ''Delinquency in Adolescence.'' In Michael Rutter and Henri Giller, eds., *Juvenile Delinquency: Trends and Perspectives* (New York: Guildford Press, 1984).
13. G. B. Melton, J. Petrila, N. G. Poythress, and C. Slobogin, *Psychological Evaluations for the Courts: A Handbook for Mental Health Professionals and Lawyers* (New York: Guilford Press, 1987).
14. Federal Bureau of Investigation, *Crime in the United States* (Washington, DC: Government Printing Office, 1984–1993).
15. Bureau of Justice Statistics, *Criminal Victimization in the United States* (Washington, DC: Aurthor, 1992).
16. Office of Juvenile Justice and Delinquency Prevention, *Juvenile Offenders and Victims: A Focus on Violence* (Pittsburgh, PA: National Center for Juvenile Justice, 1995).
17. A. Blumstein, ''Violence by Young People: Why the Deadly Nexus?'' *National Institute of Justice Journal* (1995):2–9.
18. Centers for Disease Control, ''Weapon-Carrying Among High School Students—United States, 1990,'' *1991 Morbidity and Mortality Weekly Report:* 40.
19. Blumstein, ''Violence by Young People.''
20. Office of Juvenile Justice and Delinquency Prevention, *Juvenile Offenders and Victims.*
21. Blumstein, ''Violence by Young People.''
22. American Bar Association, *Juvenile Justice Standards* (Chicago: Author, 1980).
23. B. C. Feld, ''The Transformation of the Juvenile Court,'' *Minnesota Law Review* 75 (1991):691–725; G. B. Melton, ''Taking Gault Seriously: Toward a New Juvenile Court,'' *Nebraska Law Review* 68 (1989):146–181; I. M. Rosenberg, ''Leaving Bad Enough Alone: A Response to the Juvenile Court Abolitionists,'' *Wisconsin Law Review* (1993):164–185.
24. G. B. Melton, ''The Clashing of Symbols: Prelude to Child and Family Policy,'' *American Psychologist* 42 (1987):345–354; S. Morse and C. H. Whitebread, ''Mental Health Implications

of the Juvenile Justice Standards.'' In G. B. Melton, ed., *Legal Reforms Affecting Child and Youth Services* (New York: Hawthorn Press, 1982).
25. Melton et al., *Psychological Evaluations for the Courts.*
26. Adrian Raine, *The Psychopathology of Crime* (New York: Academic Press, 1993).
27. Daniel J. Kevles, ''Vital Essences and Human Wholeness: The Social Readings of Biological Information,'' *Southern California Law Review* 65 (1991):255–278.
28. W. W. Gibbs, ''Seeking the Criminal Element,'' *Scientific American* (March 1995):101–106.
29. Office of Juvenile Justice and Delinquency Prevention, *Juvenile Violence by Young People.*
30. J. B. Attanasio, ''The Constitutionality of Regulating Human Genetic Engineering: Where Procreative Liberty and Equal Opportunity Collide,'' *Utah Chi. Law Review* 53 (1986):1274–1303; G. P. Smith and T. J. Burns, ''Genetic Determinism or Genetic Discrimination?'' *Journal of Contemporary Health Law, and Policy* 11 (1994):23–61.
31. Lawrence Taylor, *Born to Crime: The Genetic Causes of Criminal Behavior* (Westport, Conn.: Greenwood Press, 1984).
32. Office of Juvenile Justice and Delinquency Prevention, *Juvenile Violence by Young People.*
33. *Id.*
34. E. Mulvey, ''Judging Amenability to Treatment in Juvenile Offenders.'' In N. D. Reppucci, L. Weithorn, E. Mulvey, and J. Monahan, eds., *Children, Mental Health, and the Law* (Los Angeles: Sage, 1984).
35. 370 U.S. 660 (1962).
36. *Id.,* 667.
37. D. W. Denno, ''Human Biology and Criminal Responsibility: Free Will or Free Ride?'' *University of Pennsylvania Law Review* 137 (1988):615–671.
38. Melton, ''The Clashing of Symbols''; M. A. Small, ''Policy Implications for Children's Law in the Aftermath of *Maryland v. Craig,*'' *Seton Hall Constitutional Law Journal* 1 (1990):109–130.

# Chapter 8
# JUVENILE CULPABILITY AND GENETICS

Jeffrey A. Kovnick

Over the past three decades the U.S. legal system has undergone a significant transformation in the way it treats youths who are thought to have committed a crime. Juvenile courts have moved away from the rehabilitative model and toward one geared more to retribution and incapacitation. Paternalism has been replaced with standards of due process; the flexibility previously accorded to juvenile court judges has been replaced with determinate sentencing and more frequent transfer to certain classes of juveniles to adult court. The presumption that juveniles were incapable of forming criminal intent has been replaced with a fear of youth and a desire to impose harsh sentences.

This move toward a more punitive theory of punishment is based on a policy decision and reflects current social attitudes not empirical scientific information that children's cognitive or moral developmental capacities are adult-like. As a result of this attitudinal shift, juveniles are being seen as increasingly blameworthy for their delinquent actions. On the other hand, the historical foundation of juvenile court is based on trying to understand why a particular youth is turning toward delinquency and trying, if possible, to reverse the individual's course. Advances in scientific understanding may aid the courts in this process. Science may be able to inform the courts as to the etiology of the juvenile's actions and suggest interventions to help rehabilitate the individual or perhaps even prevent future delinquency. Whether the juvenile courts are interested in such information is another matter. Although evidence regarding a person's genetic predisposition to crime is generally not accepted in court, if there were a basis to show that doing so would be in society's best interest, that situation might change. In the end, courts will balance scientific information with other concerns such as public safety or retributive sentiment in deciding the position it will take in the area of juvenile culpability. This chapter uses Mark Small's analysis in chapter 7, this volume, of the history of the juvenile justice system and his speculation on matters related to genetic etiologies and crime as a springboard to explore issues related to culpability, particularly as it relates to juveniles and genetics.

In the first section I will address the issue of culpability in general. Law should be viewed in the context of the particular group of people being governed because our decisions of who we punish and why we do so are socially and culturally imbedded. Furthermore, to be just law must address the extent to which our behavior is predetermined or a result of free will and conscious choice. These notions are critical when deciding who to hold morally blameworthy for their actions. In the second section I address the issue of courtroom testimony of psychiatrists and psychologists. In adult court, specific defenses or sentencing decisions may rely heavily on mental health issues, which clinicians are uniquely qualified to address. Courts have occasionally dealt with issues relating to biological and genetic factors. However, in adult court the admissibility of testimony related to such factors is often constricted. In the third section I will discuss juvenile culpability from a legal and psychological perspective. As juvenile crime rates increase, jurisdictions across the country are acting in a manner that places more emphasis on retribution and incapacitation. This is occurring despite U.S. Supreme Court decisions recognizing that juveniles should be viewed as less culpable because of their relative developmental immaturity. Current psychological theory supports that recognition. In an effort to address issues relevant to culpability, mental health testimony is frequently elicited. This testimony may be similar to that proffered in adult court but may go further. Because of their unique nature, juvenile courts may become a forum for addressing issues related to causes of behavior that adult courts will not or cannot address. In the final section I will look at the issue of genetics as it may relate to culpability of youth in juvenile court. If a genetic component is found to be related in some direct way to criminal or violent behavior there may be broad implications in several areas. For example, culpability may be enhanced or mitigated. Also, if such behavioral predispositions can be identified early in life, society in general will need to address issues that have potential implications for all of us.

## Law and Culpability

The U.S. legal system punishes criminal acts based on one or more goals. In this section I will describe the goals or theories of punishment and how a person's mental state may affect culpability. A discussion of culpability requires explaining the determinants of behavior. I will review the notion of free will versus determinism, particularly as it relates to genetic factors.

### *Theories of Punishment*

Most commentators recognize the four main goals of punishment as retribution, deterrence, incapacitation, and rehabilitation.[1] Retribution and deterrence are directly relevant to the ultimate issue of criminal liability. Deterrence is based on the utilitarian notion that "punishment is threatened and imposed in order to achieve beneficial social consequences. In more direct terms, the ultimate purpose of punishment is to prevent or minimize criminal behavior."[2] A rational person is deterred, in theory, from committing a crime "if the costs of crime are set high enough to assure that the gains to be derived from it are not profitable."[3] Retribution, on the other hand, regards punishment as deserved "because the offender has engaged in a wrongful act."[4] Furthermore, "the premise of retribution is that individuals are responsible moral agents capable of making free choices between right and wrong. Punishment is appropriate for wrong choices."[5] Incapacitation and rehabilitation are utilitarian notions that are each concerned with the effect of punishment on the individual offender rather than the general population and that are primarily relevant to decisions about sentencing.

Our society in general is increasingly in favor of harsher punishments. Recent moves by various legislatures impose a "three strikes and you're out" rule or to institute "truth in sentencing" reforms that mandate that criminals serve their entire sentences without eligibility for early parole. The juvenile justice system, although historically based on a system that strived for rehabilitation, is in the midst of a rapid shift toward a more retributive stance, particularly for violent or recidivistic individuals. If U.S. society chooses to heed a retributive model of punishment and assume that "individuals are responsible moral agents capable of free choice . . . ," it follows that only blameworthy individuals should be punished.[6]

In the process of determining guilt or innocence, our legal system weighs many factors. The law recognizes that the nature of the individual's behavior and mental state is critical when assessing culpability. This is reflected in several areas. For example, an otherwise criminal action would be considered justified if it was done in self-defense. Also, a person must have a particular level of intent, or *mens rea,* to be found guilty of a crime.[7] Each crime has attached a specific *mens rea* that identifies the particular mental state or level of intent necessary to be found guilty of that crime. Purposefulness, knowledge, recklessness, and negligence are the Model Penal Code levels of *mens rea* for most crimes.[8] If the requisite level of intent is not proven, the person may either be found not guilty for the specific crime or, alternately, guilty of a lesser crime. Culpability may also be addressed by using an insanity defense. The federal courts and all but three states have such a defense, although jurisdictions differ on the standard that they apply. States generally have a ''cognitive prong'' as part of the defense, which permits exculpation on the basis of either (a) knowledge of the nature and quality of the criminal act or (b) lack of knowledge that the act was wrong. Many states also permit a defense of insanity based on a ''volitional prong,'' which states that as a result of their mental disorder, they were not able to control their behavior (i.e., they had an irresistible impulse). Most legal systems have determined that legally insane people are so severely impaired that, morally, punishment would be unjust. Culpability is also implicitly addressed at the time of sentencing. In general, a person found guilty of a crime may introduce clinical evidence to address his or her potential for rehabilitation or risk of future offending. Finally, culpability may be diminished as a result of the persons age.[9]

## *Free Will Versus Determinism*

When discussing culpability for a violent act, one inevitably needs to examine the extent to which our behavior is a product of our free will and to what extent it is determined by influences beyond our control. The concept of free will proposes that all human behavior is mediated by the intent and agency of the individual actor. Determinism posits that our behavior is the product of a broad array of causal factors that govern the choices we make.[10] Our legal system presumes that adults act on their own volition—in other words, with free will—unless it can be proven otherwise. Scientific thought, on the other hand, tends to be more deterministic, continually seeking explanatory mechanisms for worldly phenomenon. Psychiatrists and other mental health professionals look for causal explanations for a person's violent behavior. Depending on their individual theoretical orientation, those explanations may be more or less intrapsychically or biologically based, but the underlying presumption is that there is an explanatory mechanism for the way people act.[11]

Although the discipline of law proceeds on the assumption that behavior is freely chosen and psychiatry proceeds on the assumption that there is a causal explanation for any action, it is generally conceded that such dichotomous thinking is insufficient. Various scholars have tried to reconcile these issues. For example, *degree determinism* suggests that all behavior occurs at some point on a continuum between the hypothetical conditions of completely free choice and completely determined action.[12] A different concept, *conditional free will,* posits that individuals are free to choose their behaviors within a predetermined range of options.[13] Each provides alternate ways of recognizing influences of both free will and determinism for any action.

If one proceeds on the assumption that behavior is derived at least in part by determined factors, the logical question is how to decide the extent to which our behavior is determined and what factors are responsible for our actions. Many things may influence our choices, including our family influences, education, peers, socioeconomic variables, and biologic and genetic makeup. The discipline of human behavioral geneticists look at heritable influences on a person's behavior. When these scientists address the issue of violent criminal acts there is general agreement that finding one particular gene for crime is unthinkable. Any genetic influences on behavior clearly manifest as a result of continuous interaction with environmental, social, and cultural influences. Furthermore, any genetic influences are likely mediated by neurochemical or physiological factors that in turn affect things such as impulsivity and fear avoidance. As Gottesman and Goldsmith noted,

> Notions such as "genes for crime" are nonsense, but the following kind of notion is reasonable: There may be partially genetically influenced predispositions for certain behavioral tendencies, such as impulsivity, that in certain experiential contexts, make the probability of committing certain kinds of crimes higher than for individuals who possess lesser degrees of such behavioral tendencies.[14]

Thus the relevant genetic question is not whether behavior is freely chosen or determined. Rather, we acknowledge the complexity of human development and are concerned with understanding more about the critical periods of epigenesis, how specific genes are responsible for defining probabilities and predispositions for behavior, and how variables interact to produce or modify a particular behavior.

## Clinical and Scientific Testimony in Court

Courts occasionally look to experts for guidance on how certain factors affect the particular issue at hand. This section will explore how certain psychological, biological, and genetic factors might influence courtroom proceedings.

### *Considerations of Mental State or Psychological Factors*

Psychiatric experts frequently testify in criminal proceedings at the guilt and sentencing phases of a trial. Because a person's mental state is directly relevant to culpability and potentially relevant at sentencing, such testimony is often critical to the outcome of a case. At the guilt phase, clinical testimony most frequently addresses issues relevant to sanity and the defendant's *mens rea.* A plea of not guilty by reason of insanity (NGRI) is available in federal court as well as all but three state courts. In nearly all cases, if a person is found NGRI for a specific crime, the person is committed to a mental hospital until such time that he or she is determined to be no longer mentally ill or dangerous.[15]

Although the law sees insanity as a full excuse for behavior and deems such persons not blameworthy for the crime, insanity acquittees are often incapacitated in a hospital for longer than if they were found guilty and had served their sentence. On the other hand, in cases in which the *mens rea* was not proven, the defendant would either be found not guilty or, alternately, found guilty of a less serious crime. Mental illness is one factor that may be taken into account when mounting a defense based on *mens rea.* Clinical testimony is particularly relevant at the sentencing phase of a trial. Although a convicted person is already adjudged to be blameworthy, such testimony frequently is introduced to address issues of rehabilitation potential or risk of future offending.

Clinical input regarding issues of insanity, *mens rea,* or sentencing is relevant when the individual has a severe mental illness or when emotional problems were deemed particularly relevant. Psychopathology may be manifested differently in adults and youth because of the developmental complexity of individuals. In their effort to understand patterns of behavior and seek ways to treat individuals, psychiatrists have long recognized that certain adults interact with others in a manner characterized by "a pervasive pattern of disregard for, and violation of, the rights of others." They may be said to have an antisocial personality disorder (ASPD). This cannot be diagnosed before the person is 18 years old.[16] Although persons with this diagnosis frequently break the law, "criminality is neither necessary nor sufficient to establish its presence."[17] In youth, several diagnostic categories are associated with behavioral problems and subsequently a relatively high referral to juvenile courts. These disorders include attention deficit hyperactivity disorder (ADHD), oppositional defiant disorder (ODD), learning disabilities (LD), substance abuse, and conduct disorder (CD). Furthermore, some disorders that are seen primarily in adults may develop in childhood or adolescence—depression, schizophrenia, and bipolar disorder (manic depression). Adolescents suffering from these disorders tend to have a higher rate of behavioral disturbance than those who do not.

The disorder that is seen with the greatest frequency in juvenile court is conduct disorder. A diagnosis of conduct disorder is given to those who have a pattern of aggression, property destruction, deceitfulness, theft, and rule violations.[18] Because most teenagers engage in isolated or infrequent delinquent acts, it is important to emphasize the repetitive pattern of acts that typify those few youths who commit the vast majority of serious offenses. One facet of these disorders is oftentimes a lack of appreciation of wrongfulness of one's actions because of a lack of moral inhibition. Although evidence of mental illness or mental retardation is necessary for an insanity defense, a disorder such as ASPD or conduct disorder is not considered potentially exculpatory because courts assume that persons break the law of their own free will. However, because juvenile courts traditionally have looked at the youth's background history when deciding on disposition, factors that may have been critical in developing such a pattern of behavior might be evaluated at disposition and be a basis for a leniency and attempted rehabilitation. Research will be valuable to help courts decide on the most appropriate disposition for a particular youth with conduct disorder. This might be accomplished, in part, by being able to more accurately delineate the degree that any one particular factor is responsible for development of conduct problems and to more fully understand the natural history of the disorder. Likewise, it will be important to gain a deeper understanding of which youth are likely to be more aggressive or dangerous and to become more adept at recognizing the relationship of the disorder to other psychopathology.

### *Biological Factors*

Advances in science have increasingly shown that biology has a significant impact on our behavior. Despite this, courts have been reluctant to negate individual responsibility and culpability. Denno noted that U.S. courts have rarely found hormonal or neurophysiological abnormalities of an individual to be a basis for exculpation or mitigation of his or her criminal actions. Examples of biological conditions that have been linked to nonpurposeful or violent behavior and that defendants have attempted to use to excuse criminal responsibility include premenstrual syndrome, testosterone excess, hypoglycemia, epilepsy, and serotonin imbalance.[19] Any defendant attempting to excuse his or her behavior based on such conditions would have to provide evidence that the abnormality was directly linked to a specific violent action.

### *Genetic Factors*

Numerous studies have shown that there is a significant heritable component to criminal behavior.[20] Although these studies provide the most evidence for nonviolent offenses, recent adoption research points out that antisocial behavior in a biological parent predisposes his or her adolescent offspring to increased aggression and conduct disorder and that an adverse adoptive home environment magnifies those negative outcomes.[21] Courts in the United States have rarely dealt with genetics as a defense for criminal behavior. The most well-known occasions have been several cases in the 1970s that included males with the XYY chromosomal abnormality.[22] Researchers at the time noted that persons with such a genotype were overrepresented in prisons. Subsequently, there was considerable scientific speculation that the XYY genotype might be associated with aggressive or criminal behavior. A more recent and sophisticated analysis of that research has revealed significant methodological limitations.[23] Nonetheless, based on the best scientific data available at the time, the defense lawyers asserted that those individuals might not be considered responsible for their conduct and therefore might qualify for the insanity defense.[24] However, absent a direct causal link between the genetic anomaly and one's behaviors, there is general agreement in legal circles that "the XYY syndrome" will not meet the standard for exculpation.[25] Although persons with the XYY genotype "may be weakly disposed toward criminal activity . . . a host of environmental factors are implicated in the development of criminal activities by them."[26]

Alcoholism is another area in which it has been proposed that one's genetic makeup might in some way be linked to criminal behavior. However, in the 1968 case of *Powell v. Texas,* the U.S. Supreme Court concluded that chronic alcoholism was not a defense for criminal behavior.[27] The Court noted the lack of agreement in the medical community for defining alcoholism as a disease. They also stressed the difference (based on genetic predisposition) between being strongly influenced to act and between being compelled to do so. Furthermore, the court reaffirmed the right of legislatures to impose penalties on individuals if their behavior violated public safety regardless of the causal influences of those actions.[28] Justice Black concurring in *Powell* noted that moral blameworthiness is but one consideration in law. Societal protections may override all other considerations:

> The accused undoubtedly commits the proscribed act and the only question is whether the act can be attributed to a part of "his" personality that should not be regarded as criminally responsible. Almost all of the traditional purposes of the criminal law can be significantly served by punishing the person who in fact committed the proscribed act, without regard to whether his action was "compelled" by some elusive "irresponsible" aspect of his personality. . . . [M]edical decisions concerning the use of a term such as "disease" . . . bear no necessary correspondence to the legal decision whether the overall objectives of the criminal law can be furthered by imposing punishment. . . . [M]uch as I think that criminal sanctions should in many situations be applied only to those whose conduct is morally blameworthy, I cannot think the States should be held constitutionally required to make the inquiry as to what part of a defendant's personality is responsible for his actions and to excuse anyone whose action was, in some complex, psychological sense, the result of a "compulsion."[29]

The law recognizes the need to take into account a host of factors in deciding the issues surrounding a particular case. Although evidence that addresses the causal influences of a person's behavior may be based on highly advanced and valid scientific evidence, courts must weigh that against other factors such as societal values and protection. Furthermore, because of the great number of crimes in which it is implicated, courts will naturally be hesitant to allow alcohol to be seen as an excuse for a person's behavior.

Apart from the issue of genetics and excuse, there is precedent for genetic factors to be weighed in mitigation at sentencing. For example, the California Supreme Court considered two separate disbarment cases in which the petitioners had not contested the fact that they misappropriated client funds. Both lawyers attributed their actions to substance abuse. One lawyer was disbarred and the other was placed on probation. The court found the latter's case more compelling because he was unaware of his "genetic predisposition to addiction." Because of that lack of awareness, he was presumably not able to make a conscious choice about abstaining from addicting substances.[30]

More recently, a Dutch study described a large kindred that contained several males who displayed borderline mental retardation, aggression, and violence. The investigators showed that these males had a selective deficit of the enzyme monoamine oxidase A, which is known to mediate aggression. They also identified a point mutation in a gene that regulated that enzyme.[31] Despite the fact that this study has subsequently come under considerable scrutiny and criticism and has even given rise to disclaimers by its author, the Dutch study was cited in a capital murder trial in Georgia. A Georgia superior court denied the defense's request to have genetic and blood testing done on the defendant to look for evidence of neurochemical imbalance or genetic abnormalities. The court's rationale was that the current level of scientific acceptance was not sufficient to justify admission of such evidence.[32] The individual was sentenced to death and on direct appeal to the Georgia supreme court, the death sentence was affirmed.[33]

Thus law, being a normative discipline, sets standards of behavior for people's actions. It presumes that people act freely and are held morally responsible for their behavior. On the other hand, it acknowledges that an individual's behavior may be influenced by environmental and social conditions as well as biological and genetic factors. The legal system makes judgments by proceeding "as if" one's actions are not determined but allowing for exculpating defenses or mitigating circumstances when doling out punishment.[34] As Herbert Packer said, "The law treats man's conduct as autonomous and willed, not because it is, but because it is desirable to proceed as if it were."[35]

## Juvenile Culpability

Juvenile courts are tending to move more toward retribution as a form of punishment despite some U.S. Supreme Court opinions that youths may be less culpable in certain situations because of developmental considerations. Psychological theory and research tend to support some of the Court's rulings.

### *Legal Considerations*

Grisso noted that because of the increasing rate of adolescent violence in the 1980s, legislatures have moved away from the traditional rehabilitative policies of juvenile court and have been creating a "fully retributive" system of punishment.[36] Fueling this is the belief that rehabilitation does not work for youth who commit serious crimes. This retributive focus is also occurring as a result of society's fear and anger and the belief that those who are endangering the public safety should reap their "just desserts." Thus delinquent children and adolescents who used to be seen as victims of harmful or inadequate social and environmental influences are increasingly being held morally blameworthy for their delinquent behavior. However, despite the general mood to hold juveniles more accountable, the U.S. Supreme Court has a history of recognizing childhood and adolescence as a time in one's life in which a person, for developmental reasons, may potentially be seen as less blameworthy. For example, Justice Stevens, addressing capital punishment of a 15-year-old, in a majority opinion wrote,

> Less culpability should attach to a crime committed by a juvenile than to a comparable crime committed by an adult, since inexperience, less education, and less intelligence make the teenager less able to evaluate the consequences of his or her conduct while at the same time he or she is much more apt to be motivated by mere emotion or peer pressure than is an adult.[37]

In a similar vein, Justice White, concurring in a case denying juveniles the right to a jury trial, wrote,

> Reprehensible acts by juveniles are not deemed the consequence of mature and malevolent choice but of environmental pressures (or lack of them) or other forces beyond their control. Hence . . . his conduct is not deemed so blameworthy that punishment is required to deter him or others. . .[38]

The juvenile justice system is in the throes of trying to balance society's desire for retribution and incapacitation with the juvenile courts' traditional emphasis on paternalism. During this process, important questions will need to be answered. For example, what are the relative benefits and costs of the increased emphasis on punitive actions? What is the fairest way to decide which youths are handled with leniency and perhaps diverted to treatment and which youths are sent to trial and perhaps incarcerated?

### *Psychological Perspective*

Evidence from the psychological literature supports the contention that, in general, juveniles differ from adults on issues related to social and cognitive developmental capacities. Grisso noted that adolescents choose to engage in more high-risk or illegal behavior when compared to adults, and those poor choices tend to decrease as one reaches adulthood. Evidence exists that supports the contention that adolescents between the ages of 15 and 17 (and, by implication, younger juveniles) differ from adults on decision-making capacities that are particularly relevant to problem solving. Furthermore, those choices appear to be related to social and cognitive developmental variables, which in turn influence problem-solving processes. Grisso concluded that ''substantial support is found for a developmental logic that renders many—perhaps most—adolescent violent offenders less blameworthy than adult offenders, and more greatly at risk of injustice if tried in criminal courts.''[39]

## Juveniles, Culpability, and Genetic Considerations

Considering that adult courts are hesitant to recognize a biological or genetic defense for antisocial or violent behavior, why would juvenile court be any different? And if it would act differently, under what circumstances? In general, given the current political climate and state of scientific knowledge, I do not think that juvenile courts will act any differently than adult courts when addressing the issue of culpability and genetics. There might be a basis for treating juveniles differently if the biological or genetic issues were different for the two groups. Indeed, some genetic evidence could be interpreted to justify treating juveniles more harshly than adults. A study by Lyons et al. indicates that juvenile antisocial traits may be less familial than adult antisocial traits.[40] Nonetheless, considering the relative nascence of the field of behavioral genetics and the fact that at this stage results of studies may raise more questions than they answer, the law should naturally hesitate to make decisions based on speculation that a child's genetic characteristics were in some way contributory to his behavior.

This is not to say that a juvenile's criminal or violent behavior is not influenced to some extent by their genetic makeup. However, several problematic issues are raised when the courts deal with issues of genetics and criminal behavior. First is the problem of multifactorial explanations. Researchers representing diverse viewpoints generally agree that both environment and genes influence behavior.[41] Particularly important to recognize is the fact that biology is "probabilistic." Current theories of neural plasticity posit that our brain's development occurs within a self-correcting and self-adapting system, which is not strictly genetically programmed.[42] Thus the bodily structure of people with identical genotypes develops in nonuniform manners. This effect is accentuated when we speak of behaviors. Wasserman discussed this issue in terms of "contingency":

> Because any genetic effect on behavior will be mediated by an array of amniotic, somatic, and environmental variables, as well as interactions with other genes, the actual commission of criminal or violent acts by a person with a given genetic constitution is highly contingent. A person predisposed to crime or violence may never commit a criminal or violent act. . .[43]

Thus our behaviors depend on a host of influences, all of which interact in often quite unpredictable ways. Even if we were able to say that a particular youth had a genotype that appeared to be related to factors that could influence criminal or violent behavior, we could not say that the genotype was in any way directly related to his or her particular criminal or violent act.

The second problem for courts is linking a specific crime to a youth's genotype. As noted earlier, no one purports to be able to hope to identify one gene, or even a group of genes that can be linked to a specific crime. But causality is a crucial issue for the courts. They are not concerned with a person being an "impulsive" individual or with a person's relative "biochemical imbalance" or genetic condition, but rather with a particular guilty act or crime. Therefore, unless a gene or gene product actually caused a particular action on the youth's part, the courts would likely not be interested in exculpation.

A third problematic issue is that of defining a threshold above which one would impute decreased responsibility to the genetic component. That threshold would be quite arbitrary and would likely depend on multiple factors, including the mechanism by which the behavior operates.[44] Certain crimes might not be amenable to a "genetic defense" if the mechanism of the genes expression was unrelated to the crime. For example, if the gene was related to a youth's impulsive behavior, an instrumental crime would not be defensible on such a basis.

Nonetheless, the juvenile court still might address the matter of genetics and culpability. This could conceivably occur for two reasons. First, the courts would want to explore all possibly relevant issues in the process of deciding the outcome of a particular juvenile's case. Juvenile court was developed with the flexibility to consider each person's case in an individual and creative manner. The whole basis of the juvenile court system is to look at the youth in the context of factors that contributed to his or her deviant path. Thus to the extent that genetic factors might be potentially exculpatory or mitigating, the court would want to know. A second reason that the juvenile court might address the matter of genetics and culpability when the adult courts would not is that the former has a different framework within which to view culpability in general. As we have seen, developmental psychological research and Supreme Court decisions have acknowledged that juveniles may have relatively undeveloped decisional capacities that would make them less blameworthy for their poor choices. Because of that recognition, any factors that might influence those decision-making capacities, such as genetic vulnerability of some sort, might be relevant in the particular case.

Assuming that the court did look at these issues (and that is a large assumption indeed), what might the court do with the information? In the juvenile court, genetic characteristics may become relevant to issues regarding blameworthiness, determining appropriate dispositions, and prevention. Blameworthiness and mental state tend to center around issues of insanity and *mens rea.* Because of the juvenile court's history of paternalism and the contention that youth cannot form the intent to commit a crime, one might assume that issues of *mens rea* and insanity would not be relevant in juvenile court. In fact, juvenile courts have no clear consensus with respect to recognizing an insanity defense for juveniles. Several states have statutory language providing for such a defense. Case law varies widely among jurisdictions, with some allowing for the defense and others specifically disallowing it. For example, in a case in Maryland, the Court of Appeals recognized that there is an element of culpability present for youths adjudicated in juvenile court.[45] Because of this implicit blameworthiness, the juvenile was entitled to mount an insanity defense. Issues of insanity and *mens rea* generally arise in the context of severe mental illness and mental retardation. Despite psychiatrists recognizing conduct disorder as a valid and potentially treatable disorder, courts universally (yet informally) exclude this as a disorder on which one might base an insanity defense. This is analogous to how adult courts treat persons with antisocial personality disorder.

A second area in which a juveniles' genetic characteristics may be relevant to the court is in the area of mitigation or dispositional alternatives. The seriousness of the crime or degree of violence associated with it would be inversely proportional to the willingness of the court to consider genetic factors in mitigation or at the time of a transfer decision. Indeed, for violent offenders evidence of a genetic link to their criminality might be an excuse for the court to view these individuals as less amenable to treatment, more dangerous, and particularly deserving of the stiffest sentences. On the other hand, if it could be shown that there were effective interventions available for such youth, a strong case could be made for more benevolent consideration of those whose criminal or violent acts were "proven" to be genetically linked.

Research in this area is poor. Most inquiry has shown that programs that have tried to treat violent or criminally recidivistic youth have had a poor outcome. To a certain extent, such data reflects a problem with the research methodology and outcome measurements of the individual programs. More recent research has shown that specific targeted interventions can reduce recidivism of youths who break the law.[46] Of course, any previous research concerning effectiveness of treatment alternatives might not be relevant to persons whose criminal behavior was to some extent a result of their genetic characteristics. New research would have to be done on this subgroup of offenders to see which, if any, interventions would be discriminably effective. Possibilities may encompass any of a wide range of treatments, including family therapy, group treatment, restorative justice-based interventions, or medications. Offering participation in medication therapy or some more invasive interventions as an alternative to secure incarceration would be an ethically questionable practice. Thus genetic factors might be an appropriate consideration at the time of disposition or transfer. However, such consideration would be contingent on several factors. For example, appropriate interventions should be available, ways need to be found to identify which youth would benefit from these programs, and money would have to be found to provide the services. Furthermore, issues of coercion need to be dealt with sensitively.

One final place where evidence of genetic characteristics that contribute to criminal or violent behavior would be relevant would be in the area of prevention. Society would have a strong incentive to identify those persons who would be most at risk for committing future crime. Unfortunately, such identification is unlikely, and even if it were, the implications would be potentially Orwellian. As Mark Small descriptively portrayed in the preceding chapter, clinicians assessing risk must overcome the formidable barrier of base rates. A base rate is the frequency that a specified behavior occurs in a population. Because the number of juveniles who are of most concern to the courts is quite low, even with a very good screening test, the ability to identify those individuals of greatest concern would be quite limited or prohibitively expensive. A further problem would be the incorrect identification of many individuals as "at risk" who were, in actuality, not at increased risk (i.e., false positives). The fact that the issues under consideration are inherently multidetermined contributes to the inaccuracy of the assessment of potential risk. However, one could make the argument that the cost would be minimal compared to the benefit to society of treating these potential offenders before they could exact harm on others. The best way to most efficiently test potential offenders would be to identify segments of the population who were most "at risk" for crime and screen them for the relevant genetic marker. Because of statistical considerations and the multiple factors that contribute to behaviors, there would still be many false positives. An additional problem with targeting segments of the population for a screening test is that that approach would tend to single out the most socially disadvantaged among us and lead to further societal stigmatization of these groups. Furthermore, would such screening be mandatory? That would most likely not pass constitutional muster. And if it were voluntary, what incentive would there be for a parent to have his or her child tested when the result of a "positive" test would be stigmatization despite the fact that, because of the unavoidable high rate of false positives the child would, in actuality, be at no higher risk for future criminality or violence than the average child.

## Conclusion

As our juvenile courts increasingly move toward punishment and retribution, we naturally are finding youth more responsible and blameworthy. Because social attitudes are a large influence on which specific laws are written and how they are eventually interpreted, this paradigmatic shift is merely a reflection of society's normative influence. Social attitudes, in turn, are influenced by a wide variety of factors, a major one being scientific knowledge. To the extent that our future understanding of human behavior is informed by scientific advances, this may be reflected in new laws that hold juveniles less or more culpable. Genetic understanding is particularly relevant to such issues. Advances in genetics that identify youth who may exhibit criminal or violent behavior could lead to eventual interventions of the juvenile court in areas of treatment, rehabilitation, and even prevention. However, these rapid advances in genetic and other scientific knowledge are not likely to challenge one of the fundamental assumptions of our legal system—namely that individuals act by their own free will and are morally responsible for their behavior. Courts realize that many factors are responsible for a person acting in the way he or she does and recognize that heritability plays some part. In the final analysis though, even if we had a perfect understanding of the reasons that people break the law, courts would likely not hold individuals less culpable. Although juvenile courts are based on paternalistic premises, society is most interested in keeping safe. Thus the legal system's notion of culpability and individual responsibility is not likely to change unless interventions are available that would ensure that public safety.

## Notes

1. Theoretical information regarding theories of punishment is based on P. W. Low, J. C. Jeffries, Jr., and R. J. Bonnie, *Criminal Law: Cases and Materials,* 2d ed. (Minneola, NY: Foundation Press, 1986).
2. *Id.*
3. *Id.*
4. *Id.*

5. *Id.*
6. *Id.*
7. In cases of strict liability, a *mens rea* element need not be proven. These types of cases are relatively rare.
8. Model Penal Code § 2.02 (1962).
9. Low, et al., *Criminal Law.*
10. R. C. Boldt, "The Construction of Responsibility in the Criminal Law," *University of Pennsylvania Law Review* 140 (1992):2245–2332.
11. S. H. Dinwiddie, "Genetics, Antisocial Personality, and Criminal Responsibility," *Bulletin of the American Academy of Psychiatry and the Law* 24 (1996):95–108.
12. Norval Morris, *Madness and the Criminal Law* (Chicago: University of Chicago Press, 1972).
13. D. H. Fishbein, "Biological Perspectives in Criminality," *Criminology* 28 (1990):1–72.
14. I. I. Gottesman and H. H. Goldsmith, "Developmental Psychopathology of Antisocial Behavior: Inserting Genes Into Its Ontogenesis and Epigenesis." In C. A. Nelson, ed., *Threats to Optimal Development: Integrating Biological, Psychological, and Social Risk Factors* (Hillsdale, NJ: Earlbaum, 1994).
15. G. B. Melton, J. Petrila, N. G. Poythress, and C. Slobogin, *Psychological Evaluations for the Courts* (New York: Guilford Press, 1987).
16. *Diagnostic and Statistical Manual of Mental Disorders,* 4th ed. (Washington, DC: American Psychiatric Association, 1994).
17. Dinwiddie, "Genetics, Antisocial Personality, and Criminal Responsibility."
18. *DSM-IV.*
19. D. W. Denno, "Human Biology and Criminal Responsibility: Free Will or Free Ride," *University of Pennsylvania Law Review* 137 (1988):615–671.
20. P. A. Brennan, S. A. Mednick, and B. Jacobsen, "Assessing the Role of Genetics in Crime Using Adoption Cohorts." In *Genetics of Criminal and Antisocial Behaviour* (Ciba Foundation Symposium 194) (Chichester, England: John Wiley and Sons, 1995); K. S. Kendler, "Genetic Epidemiology in Psychiatry: Taking Both Genes and Environment Seriously." *Archives of General Psychiatry* 52 (1995):895–899.
21. R. J. Cadoret, W. R. Yates, E. Troughton, G. Woodworth, and M. A. Stewart, "Genetic-Environmental Interaction in the Genesis of Aggressivity and Conduct Disorders," *Archives of General Psychiatry* 52 (1995):916–924.
22. D. Skeen, "The Genetically Defective Offender," *William Mitchell Law Review* 9 (1983):217–265.
23. D. Suzuki and P. Knudtson, *Genetics: The Clash Between the New Genetics and Human Values* (Cambridge, MA: Harvard University Press, 1989).
24. Denno, "Human Biology and Criminal Responsibility."
25. M. P. Coffey, "The Genetic Defense: Excuse or Explanation?" *William and Mary Law Review* 35 (1993):352–399; W. R. LaFave and A. W. Scott, Jr., *Criminal Law,* 2d ed. (St. Paul, MN: West Publishing, 1986).
26. Melton et al., *Psychological Evaluations for the Courts.*
27. Powell v. Texas, 392 U.S. 514 (1968).
28. Coffey, "The Genetic Defense."
29. Powell v. Texas, 392 U.S. 514 (1968).
30. Coffey, "The Genetic Defense"; R. C. Dreyfuss and D. Nelkin, "The Jurisprudence of Genetics," *Vanderbilt Law Review* 45 (1992):313–348.
31. H. G. Brunner, M. Nelen, X. O. Breakfield, H. H. Ropers, and B. A. van Oost, "Abnormal Behavior Associated With a Point Mutation in the Structural Gene for Monoamine Oxidase A," *Science* 262 (1993):578–580.
32. D. W. Denno, "Legal Implications of Genetics and Crime Research." In *Genetics of Criminal and Antisocial Behaviour* (Ciba Foundation Symposium 194) (Chichester, England: John Wiley, 1996).
33. *See* Summer's elaboration of chapter 6, this volume.

34. Coffey, ''The Genetic Defense.''
35. Herbert Packer, *The Limits of the Criminal Sanction* (Stanford, CA: Stanford University Press, 1968).
36. T. Grisso, ''Society's Retributive Response to Juvenile Violence: A Developmental Perspective,'' *Law and Human Behavior* 3 (1996):229–247.
37. Thompson v. Oklahoma, 487 U.S. 815 (1988).
38. McKeiver v. Pennsylvania, 403 U.S. 528 (1971).
39. Grisso, ''Society's Retributive Response to Juvenile Violence.''
40. M. J. Lyons, W. R. True, S. A. Eisen, J. Goldberg, J. M. Meyer, and S. V. Faraone, ''Differential Heritability of Adult and Juvenile Antisocial Traits,'' *Archives of General Psychiatry* 52 (1995):906–915.
41. D. E. Comings, ''Both Genes and Environment Play a Role in Antisocial Behavior,'' *Politics and the Life Sciences* 15 (1996):84–86.
42. M. Rutter, ''Discussant to: Legal Implications of Genetics and Crime Research.'' In *Genetics of Criminal and Antisocial Behavior* (Ciba Foundation Symposium 194) (London: John Wiley, 1996).
43. D. Wasserman, ''Genetic Predispositions to Violent and Anti-Social Behavior: Responsibility, Character, and Identity.'' Paper presented at the University of Maryland conference on Genetics and Criminal Behavior: Scientific Issues, Social and Political Implications, Queenstown, MD, August 1996.
44. Rutter, ''Discussant to: Legal Implications.''
45. In re Devon T., 584 A.2d 1287 (Md. App. 1991).
46. S. Henggeler, G. Melton, and L. Smith, ''Multisystemic Treatment of Serious Juvenile Offenders: An Effective Alternative to Incarceration,'' *Journal of Consulting and Clinical Psychology* 60 (1992):953–961; E. Mulvey, M. Arthur, and N. D. Reppucci, ''The Prevention and Treatment of Juvenile Delinquency: A Review of the Research,'' *Clinical Psychology Review* 16 (1991):133–167.

# PART IV

# Conclusions and Recommendations

# Chapter 9
# A BRAVE NEW CRIME-FREE WORLD?

Mary Coombs

The title of this book, and the workshop from which it derives, are reminiscent of an earlier conference that the National Institutes of Health supported. In this chapter, I want to use that conference and the controversy surrounding it as a jumping-off point to examine the concerns at issue regarding the politics of genetics and crime and the extent to which they reappear in the context of this volume and its related subject matter. I will conclude that, although the shift in focus in this volume reduces the direct relevance of those controversies, they nonetheless must be surfaced and considered in assessing the worth of this or any academic study of "genetics and crime."

The real significance of genetics and crime is likely to occur in criminal justice policy making, not in the trial of particular cases. The proponents of applying genetic knowledge are likely to be those seeking a solution to crime by effective crime control. The objects of that control will be the usual suspects: disadvantaged groups, primarily African American men. The public's perception of a genetic explanation for crime will serve not to exculpate individuals but to justify the control of populations.

## Genetics, Crime, and Race: A Long and Inglorious History

In 1992 a conference was scheduled to take place, to be titled "Genetic Factors in Crime: Findings, Uses and Implications." The conference organizer, David Wasserman of the University of Maryland's Institute for Philosophy and Public Policy, had received a grant from the Human Genome Project of the National Institutes of Health to support the conference. The conference brochure came to the attention of Peter Breggin, an independent researcher and gadfly on issues of mental health and race. Partly at his urging, opposition developed across a wide spectrum of the African American community, and the NIH withdrew funding. After appeals by the university, funding was provided for a new conference. The new conference, held in September 1995, had a somewhat different focus, evident from its new title, "The Meaning and Significance of Research on Genetics and Criminal Behavior." This conference included discussions of then-current scientific research but also discussions of the potential uses and misuses of that research and of similar research in the past.[1] Some opponents found the reconstituted conference still too dangerous or insensitive to the racial implications of such research, and the conference—though moved to a site in rural Maryland—was invaded briefly by protesters chanting "Maryland conference, you can't hide. We know you're pushing genocide."[2]

The original protests were triggered in part by the contemporaneous comment of Frederick Goodwin, then director of a division of NIMH, comparing inner-cities to jungles and their young male inhabitants to monkeys.[3] The comments were associated with the apparent plans of his and other government agencies to launch a federal violence initiative, which would examine violence as a biologically rooted phenomenon with a concomitant commitment to biologically based solutions, such as drug therapies.[4] The genetics and crime conference was seen as linked to a similar commitment to genetic explanations and medical–therapeutic solutions to the crime problem.

Protesters saw two fundamental problems with such an approach. Whether intentionally or not, it fed into images of crime as a problem of Black communities and located the problem as one within Black offenders and a result of their "genetic defects." This belief system in turn seemed to provide a veneer of objectivity and scientific legitimacy to racist attitudes and responses.[5] It also justified a shift from social programs to medical–psychological approaches, or even "benign neglect," by making the problem of violence a matter of genetic predisposition and thus seemingly making environmental change irrelevant.[6] These concerns were themselves part of the text of the reconstituted Maryland conference, though participants disagreed on whether the changes had gone far enough.

That quasi-abortive conference and the one that led to this volume are among the more recent manifestations of interest in theories of biology and human behavior. Such theories seem to arise, assert scientific legitimacy for claims that explicitly or implicitly valorize the genetic makeup of the theorist, and justify the domination of other groups. Then they are discredited, lie dormant, and arise in new forms. In one of the earliest such theories, three quarters of a century ago, Cesare Lombroso claimed that certain people were biologically predestined to be criminals. These criminals could be readily recognized by their distinctive long arms, sloping foreheads, and jutting jaws, and Lombroso's publications included photographs intended to illustrate the criminal physiognomy.[7] Not surprisingly, these natural-born criminals came disproportionately from the poorer classes.

In the first half of the twentieth century, the eugenics movement had a biological explanation for "feeble-mindedness" and "viciousness."[8] In a form of biologized Social Darwinism, it was assumed, first, that these undesirable traits occurred disproportionately in certain populations and, second, that the explanation was their heritability. Proponents of eugenics advocated both a positive and a negative program. Those with the best genes—white Anglo-Saxons of the upper classes—were encouraged to reproduce. Those with less desirable genes were to be excluded by immigration controls and, if already in the United States, encouraged not to dilute the future gene pool with offspring.[9] For those whose genetic inadequacies were most manifest, reproduction was to be prevented by compulsory sterilization. Laws authorizing compulsory sterilization of mental defectives or habitual criminals were passed in twenty-four states, and thousands were sterilized under such laws.[10]

Compulsory sterilization programs were examined by the United States Supreme Court twice. In *Buck v. Bell,*[11] Justice Holmes found no due process violation in the sterilization of Carrie Buck. The state had followed its own procedures for determining that she met the statutory criteria; and the legislature could reasonably accept the conclusions of eugenic science as the basis of its substantive choice. There was no due process right for someone like Carrie Buck not to be sterilized. If the best citizens can be called on to sacrifice their lives in war, Holmes suggested, surely the state could demand these "lesser sacrifices . . . in order to prevent our being swamped with incompetence." In a peroration that forever casts a shadow on Holmes's legacy, in the view of many, he asserted, "Three generations of imbeciles are enough."[12]

In *Skinner v. Oklahoma,*[13] the Supreme Court found that Oklahoma's compulsory sterilization program was a violation of the equal protection clause. The holding is narrow; the law at issue failed because it distinguished between those guilty of larceny and those guilty of embezzlement, subjecting only the former to sterilization. Whatever the heritable trait might be, the Court said, it was unlikely that nature followed the technicalities of the common law. (The Court chose not to point out explicitly the class prejudice apparent in the statute.[14]) However, the Court also held that the right to reproduce was "one of the basic civil rights of man," indicating the need for a high degree of scrutiny. By 1942 the misuse of eugenics by the Nazis was well-known, and the Court noted that "in evil or reckless hands [the power to sterilize] can cause races or types which are inimical to the dominant group to wither and disappear."[15]

More recently, sociobiology and its various bastard children have been used to assert and to explain biological differences among demographic groups. One of the most widely read and discussed examples is *The Bell Curve.*[16] Intelligence, the authors of the book claim, is measurable and largely a product of one's genes. Those who are cognitively disadvantaged are most likely to engage in socially undesirable behavior—illegitimacy, unemployment, and crime.[17] Although acknowledging that there are unintelligent White people, the book's graphs seem to indicate that the Black population of the United States is disproportionately concentrated on the dull-witted side of the bell curve. In an echo of the eugenicists of the 1920s and 1930s, the authors lament that the genetically disadvantaged are reproducing at a faster rate than more intelligent people (such as their readers). Some ameliorative measures are possible, such as making criminal law simple enough for the small-brained criminal class to understand,[18] but the authors assume that more coercive measures will also be needed to keep the dumb and dangerous away from the smart and productive.[19] The acceptability of this vision of an irredeemably dangerous class can be seen in the popularity of the book, as well as in the success political advertising relying on Willie Horton and welfare queen imagery and the increasing development of guardhouses, gates, blocked streets, and other techniques for walling off "us" from "them."[20] As Charles Murray, one of the authors, said, his book sells because it "make[s] well-meaning whites [who] fear that they are closet racists . . . feel better about things they already think but do not know how to say."[21]

The risk to racial justice from *The Bell Curve* or the proposals of the eugenics movement seem obvious. But there are substantial risks even from more nuanced analyses such as those in this book or at many of the panels at the Maryland conference. The potential harm reflects the confluence of a number of factors. Scientists, optimistic about their work and its potential, tend to overstate the scientific and social significance of their findings. The popular press and the general public, in turn, ignore nuances and read scientific findings as clearer and more policy determinative than is sensible. This is particularly so where the science appears to reaffirm existing prejudices. And it is even more so when that science provides an apparent justification for otherwise desired cutbacks in spending on social justice programs.

In its current, as in its past, manifestations, biological science has an optimism about its own value. There is a desire to find certain answers and an assurance that with sufficient effort they will be found. As Grob noted, psychiatrists, like members of other disciplines within Western structures of knowledge, are "committed to a search for final and irreducible truths."[22] The director of the Human Genome Project, James Watson, has said, "Our fate is in our genes."[23] Science, its proponents seem assured, will provide not only knowledge but solutions to social problems. "I would be surprised if there were no biological remedy" for violence, said one of the participants at the Maryland genetics and crime conference.[24] Even when scholars recognize the highly limited relevance of genetics to criminal law, traces of optimism remain. Botkin has said that genetic analysis and interpretation "may become an integral feature of the criminal and juvenile justice systems."[25] Science is knowledge and knowledge is by definition a human good.[26]

Scientists are not alone in their optimistic belief that science will provide a key to human behavior. The popular culture also delights in science, especially when it promises a solution for what have seemed irresolvable problems. Scientists and academics sometimes fail to recognize that nuanced reports of genetic links to behavior and other research become far more simplistic as reported to and understood by the general public.[27] The nontechnical book *Born to Crime,* for example, pervasively adopts language of genetic determinism.[28] Drug therapies seem to provide a certain and relatively inexpensive cure for mental illnesses that psychotherapy could not alleviate.

The authors of this volume indicate that genetics provides only a modest explanation for criminal behavior. Such a nuanced view, however, is likely to be radically oversimplified when it reaches the ''science'' section of the mass media and the local news reports. The public ''sees scientific information, regardless of the soundness of the methods, as powerfully legitimizing'' and assumes that ''genetic findings . . . are immutable.''[29] ''Genes Cause Crime''; ''Defective Gene Found Linked to Violent Behavior''; ''Fated to Kill.'' In headlines such as these, the Jukes and Kallikaks are likely to be resurrected in the popular imagination. Because such public misuse of genetic science is historically pervasive and likely to continue, it cannot be ignored. ''The history of fascism in science and scholarship demands that the researchers react vocally to claims about the implications and uses of their pure research.''[30]

The causes of violence and antisocial behavior are surely a complex mix of biology, environment, and human choice. Yet there is a high risk that policy makers and the public at large will opt for a simple, unitary explanation, and that explanation will be biological. Solutions, if they exist at all, will be individualized and medicalized.[31] The explanation, ''It's all caused by environmental pressures,'' would be equally simple (and equally wrong). It would also be less attractive. Environmental determinism posits the significance of such factors as diet, quality of schooling, family support systems, and availability of employment for the defendant now and for his parents while he was growing up. Reducing crime, then, would presumably involve efforts to change these environmental factors. As a society, however, we have largely given up on rehabilitation as a response to crime (even by young teenagers), and on crime reduction as a rationale for improving the lot of poor, inner-city children. We would rather not spend the resources, and we would rather not feel guilty for not spending the resources. ''Internalizing the source of criminality to the individual's genetic makeup,'' Crossley noted, ''relieves society of any responsibility for influencing individuals' actions.''[32]

A biological explanation, abetted by the prestige of science, seems to justify reducing resources to deter crime and to rehabilitate those who have broken the law. Note that there is no necessary relationship between whether a given condition is biological or environmental and whether that causal factor can be either prevented or treated.[33] Some biological characteristics, which may be untreatable, are in turn the result, wholly or in part, of environmental and theoretically treatable conditions. For example, a lack of prenatal care can lead to low birth weight and its associated negative consequences.[34] Conversely, the manifestation of genetic conditions could be controlled or the conditions themselves eliminated. Society as well as the individual could bear a responsibility for avoiding environmental triggers to the expression of genetically linked behaviors. If treatment via drugs or genetic surgeries were developed, there could be a social obligation to provide these to persons in need.[35] At the level of collective psychology, however, the thrust of a genetic explanation is to reduce the felt need for noncoercive means of crime prevention. ''[A] mood for slashing social programs can be powerfully abetted by an argument that beneficiaries cannot be helped owing to inborn . . . limits. . . .''

Let me be clear. I am neither suggesting that all those who examine the question of genetics and crime are consciously racist nor that such research should be forbidden. But work carried out in good faith by people of good will can nonetheless be put to evil uses. Historical humility should remind us that those who studied eugenics in the past genuinely believed that science led them to the conclusions of inborn racial hierarchy, though we might with hindsight see the causation running in a different direction. That studies have been done only on White populations also does not mean that the results of those studies cannot easily be used to justify racist conclusions and policy prescriptions.[37] Particularly when the subject is crime, all results will be considered in the context of a universe in which African Americans constitute 12% percent of the U.S. population but nearly 50% of its prison inmates.

In addition, genetic knowledge can provide a basis for genuine progress in public health and welfare. Studies such as the ones referred to in this book should not be forbidden, nor should public funding be a priori denied. What is needed is a caution born of the knowledge of how results have been and can be misused. "Scientists . . . need to show greater vigilance and responsibility, especially concerning public perception of their work."[38]

## What's New in this Volume

The focus of this book is distinct from, though related to, prior studies of genetics and social behavior or genetics and crime. The focus is on the possibility of using genetic evidence as a defense to criminal charges and on whether and to what extent genetic conditions can serve as a defense to criminal punishment. To develop this discussion, the authors must examine two questions. First, does a person's genetic makeup influence his or her likelihood of committing crime, and if so, how? Second, what difference would a positive answer to the first question make under existing theories of criminal responsibility or plausible extensions thereof?

The first question is a factual one: To what extent are we the playthings of our genes? The authors in this volume seem generally to stake out positions somewhere between cautious and skeptical. Twin studies, family studies, and adoption studies, they indicate, suggest there is a genetic component to certain mental illnesses, including schizophrenia, bipolar disorder, and antisocial personality disorder.[39] Similarly, alcoholism and drug addiction seem to have a biological basis. Whether genes predispose certain people to criminality or violence, apart from their influence operating through these specific conditions, is less clear. Even where a correlation is shown, it is never so high that one could confidently predict that an individual would develop one of the conditions by knowing she had a particular genetic makeup.[40] Furthermore, the mechanism by which genes affect these conditions is unknown and not readily discoverable; the one thing that seems quite certain is that there will not be some single magic genetic key that will allow us to recognize who has these defects or will develop the associated conditions.[41]

If the answer to the first question, then, is at best "maybe" and "only indirectly," the answer to the second is something very close to "no." The authors in this volume recognize the extremely limited impact that genetic knowledge is likely to have on the outcome of criminal cases.[42] If the defendant suffers from a mental illness and meets the jurisdiction's criteria for insanity, he should be found not guilty by reason of insanity, whatever the etiology of the mental illness.[43] At most, the effect of bringing in genetic evidence may be to make his case more psychologically persuasive to a jury. Laypersons may assume that geneticists are scientists, presenting objective truths, whereas psychiatrists may be seen merely as hired guns.[44]

What if the defendant, because of a genetic condition but without having a mental condition that would lead to a finding of insanity, commits a crime she arguably would not have committed if her genetic makeup were different? As the authors of this volume in general recognize, the genetic basis of a condition is unlikely to be relevant under current legal doctrine. Substance abuse, antisocial personality disorder, or the MOAA defect found in a Dutch family do not seem to affect a defendant's cognitive capacity.[45] The claim, rather, is that directly or through the mechanism of antisocial personality or substance abuse, the defendant's genes create a condition that makes it much more difficult for her to resist the temptation to engage in criminal acts. As a scientist, one might describe the genetic condition as *causing the behavior,* or, more precisely, *as among the behavior's multiple causes.* (Even when the genetic variation is expressed as phenotype it seems never to create an absolutely irresistible impulse.[46]) The criminal law, however, is largely uninterested in the causes of human behavior: "Biological causes of behavior are not grounds per se to excuse."[47]

The law assumes that defendants have free will. Exceptions to this maxim are extremely narrow. Some entities—young children, animals, and the insane—are seen as not, or not adequately, capable of the characteristics of thought and impulse control we associate with humanity, and are thus excused from blame. Similarly, there are a narrow range of situations, such as having a gun held to one's head, that are viewed as compulsion; in those situations persons otherwise responsible are seen as mere pawns and similarly excused.[48] Everywhere else, the doctrine of free will applies. This metaphysics of culpability does not deny that people differ in the ease with which they can obey the law; rather, it makes those differences essentially irrelevant to the question of criminal responsibility. The forces that pressure some people to commit crimes can be biological or environmental. In either case, they provide no defense.[49] As Schopp says, "Nothing about the different sources of the desires entails that the individuals are any more or less capable of refraining from acting on those desires."[50] Although it might be better if the law were more willing to recognize and provide protection in its doctrines for those who find it extraordinarily difficult to avoid criminal behavior,[51] the trend is toward a harsher doctrine, in which diminished capacity defenses are narrowed or eliminated[52] and insanity defenses drastically restricted.[53]

The more complex and subtle question is when and whether biological pressures might be relevant as *mitigation* of culpability. Where a jurisdiction allows for a diminished capacity claim,[54] it would be plausible to extend the doctrine to include diminished capacity stemming from genetic factors. At least one jurisdiction has considered the defendant's genetic predisposition to alcoholism, of which he was unaware, in imposing a lesser penalty in a bar disciplinary proceeding based on conduct related to the alcoholism.[55]

One might also imagine a new doctrine that would permit a genetic claim of reduced ability to control impulses to serve as a form of mitigation of murder to manslaughter. Typically, such mitigation requires a showing of circumstances that would lead the reasonable person to become extremely distraught, and no such defense is available for the defendant who is unusually jealous or thin-skinned. One rationale for limiting the doctrine in this way is that defendants are responsible for their character, and thus a man who flies into a murderous rage when the average person would not is culpable. One could argue that the defendant who is genetically programmed to be enraged easily and to find self-control difficult is not as culpable, for he was not wholly free to create a better character. This argument, however, first assumes both a certain kind of genetic defect for which there is not yet clear evidence and proof that this particular defendant suffers from it. Second, it would logically also apply to the defendant whose weak character was formed in an abusive family and a deprived neighborhood. Existing law, however, makes it clear that "rotten social background" is neither defense nor partial excuse. Third, the doctrines of excuse are, I suggest, designed to reduce punishment for those acting "out of character"; not for those whose characters, for whatever reason, make them unusually prone to commit antisocial acts. (The one exception is the excuse we allow for the insane, whom we lock up in any event.[56])

There is, however, one situation in which the defendant may introduce into evidence a wide array of facts about her history, her background, and her family situation: the penalty phase of a capital case.[57] Defense counsel may and should seek to explain how the defendant became the person who committed this crime, in the hopes of persuading the jury that she does not deserve the ultimate sanction. Although there is no good reason to grant genetic evidence more weight than other facts that influenced the defendant's character and her ability to control antisocial impulses, there is similarly no reason to grant it less relevance.[58] The *Mobley* case is thus an outlier in terms of the relevance of genetic evidence to criminal law. The decision to deny resources for genetic testing in that case cannot be justified as an abstract principle, though it may reflect a judgment that the likelihood of discovering relevant evidence through testing was sufficiently minute that the motion could justly be denied.

## What Might Policy Makers and the Public Make of the Studies in this Volume?

The previous discussion suggests that the impact of genetic knowledge on criminal trials will remain insignificant even if science gives us far better proof than we now have of a strong relationship between "genetics" and "crime."[59] The law's reluctance to acknowledge as relevant causes that do not amount to compulsion is unlikely to change because certain such causes are genetic.[60]

Then is this all a tempest in a teapot? I suggest that it is not. To understand the potential sociocultural significance of genetic research to criminal law, the existence of a substantial book on the topic is as important as the cautious conclusions drawn herein. Scientific research is likely to continue and, as it does, public perception of a genetic link to crime is likely to become increasingly widespread.

Many of the policies that may flow from those perceptions are the same as these that have been the subject of past controversies over research in genetics and violence, discussed previously. The focus on criminality as such, however, raises in an especially vivid form the specter of incapacitation and preventive detention.

In particular criminal trials, the claim of genetic causation will rarely be raised and generally rejected, for it is extraordinarily difficult, if not impossible, to show that individual X, in committing act A, had no choice but to act in a way dictated by her genetic makeup. Furthermore, the claim as excuse/mitigation, I suggest, may seem most plausible, both to the defense attorney and the fact-finder when the criminal behavior seems unlikely given the defendant's position in the social structure. To put it bluntly, successful defensive use is particularly unlikely for minorities, whose criminal proclivities are constructed as normal for them.[61]

What if a defendant does persuade the court that she suffers from a genetic defect? As Perlin noted, insanity acquittees, seen as dangerous though not culpable, are frequently subject to lengthier institutionalization than the noninsane who have engaged in the same behaviors.[62] Similarly, those who have manifested their genetic predisposition to engage in violent, antisocial behavior will be seen as in need of long-term institutionalization. If they can be predicted, with what is deemed an adequate degree of certainty, to be sufficiently likely to commit future criminal acts, quarantine would seem the most practical solution.[63]

Quarantine sounds like a nineteenth-century public health measure, and the commitment of insanity acquittees has always been justified, at least in part, by their need for treatment. My concerns, then, might seem somewhat paranoid. But recent events make clear that the notion of long-term detention of the criminally dangerous is not merely a right-wing fantasy.[64] It is, rather, a very slight variant of a reality found constitutionally permissible in 1997 by the U.S. Supreme Court in *Kansas v. Hendricks*.[65] *Hendricks* was a constitutional challenge to the Kansas Sexually Violent Predator Act. That statute provided for the involuntary commitment of those who (a) had been charged with predatory acts of sexual violence and convicted or found not guilty by reason of insanity or found incompetent to stand trial and (b) had a "mental abnormality" or "personality disorder." Once committed, such persons would remain committed until their abnormality or disorder had so changed that it was safe to permit the person to be at large. The statute applied to persons, such as Hendricks, whose criminal acts had all occurred prior to the statute's passage, which would make the statute unconstitutional if it were deemed punishment.[66] The Court found that it was not and, furthermore, that it was permissible under the limits imposed on involuntary civil commitment by the due process clause.

Kansas could decide that persons within the Act's contours, though not mentally ill as that term had previously been defined for involuntary civil commitment, were likely to engage in further acts of predatory violence and could delimit the detainable class as the statute had done. Their classification as antisocial personalities or mentally abnormal created "a limited subclass of dangerous persons" and thus fit them within the rationale of prior cases that involuntary civil commitment could be applied to "those who suffer from a volitional impairment rendering them dangerous beyond their control."[67] While the Supreme Court majority construed the legislation as based on an assertion by the legislature that treatment was available and that such treatment would be provided, it also suggested that the possibility of effective treatment was not necessary.[68] Even if the Kansas Supreme Court were correct in concluding that there was no known treatment for the pedophilia that Hendricks suffered, the statute remained legitimately civil, because "we have never held that the Constitution prevents a State from civilly detaining those for whom no treatment is available, but who nevertheless pose a danger to others."[69]

In effect, *Hendricks* authorizes life-long civil commitment for those who are dangerous, so long as the state has some means of defining a subclass, rather than applying its statute to all who have "a mere predisposition to violence."[70] Surely a class defined by their genetic makeup would provide the "plus" factor as well as one defined by their "mental abnormality or personality disorder."[71] *Hendricks* readily supports a program of involuntary, indeterminate detention of all those "genetic defectives" who have already engaged in an act of serious violence for which they have been or could have been punished.

The even more disturbing question is whether the state may engage in preventive detention of those whose genetic traits make them likely to commit acts of violence but who have not yet done so. *Hendricks* does not reach this question.[72] The prosecutor, Horton, recognized that the logical implications of proof of a strong genetic predisposition to crime might well include preventive detention of the class of persons at risk.[73]

In addition to the due process-like concerns raised by such potential indefinite, involuntary commitment on the basis of predispositions, such a program inevitably raises profound questions of racist application of the law. The groups subject to such social controls will be those that are seen as breeding grounds of substance abuse, antisocial personality disorder, and violence. In contemporary America, these are likely to be subsets of the same populations we now incarcerate so readily: African American (and, to a lesser extent, Latino) men. In theory, genetic defects could be scattered throughout the population;[74] in practice, we will look for them where we expect to find them.

The coercive responses suggested in these scenarios may not come to pass. The cost of locking up large numbers of people because some of them may commit (further) violent crimes in the future may be one society is unwilling to bear,[75] though in the context of not guilty by reason of insanity commitments, we seem much more comfortable with false positives (keeping people who are no longer dangerous institutionalized) than false negatives (releasing people who are still dangerous).[76] Furthermore, the group from which the institutionalized genetically defective will be drawn is demographically distinct from the decision-making group. Thus the costs of false positives are likely to be discounted. Coercive responses to a perceived link between genetics and crime are, I fear, far more plausible than the use of genetics as a defense in criminal cases. I do not believe it is a future that my coauthors welcome any more than I do. But it is a future that we may make more likely if we talk about genetics and crime without recognizing its shadow. One is reminded of the old Tom Lehrer song about the atom bomb. "'Once the rockets go up, I don't care where they come down. That's not my department,' says Wernher von Braun."[77] It is all of our departments to keep the bomb of genetic racism from coming down.

## Notes

1. Natalie Angier, "Disputed Meeting to Ask If Crime Has Genetic Roots," *New York Times,* September 19, 1995, C1; Peter Maass, "Conference on Genetics and Crime Gets Second Chance," *The Washington Post,* September 22, 1995, B1; "Amid Protest, Crime Gene Meeting Held in Maryland," *Biotechnology Newswatch* 1 (1995):1–4.
2. Peter Maass, "Crime, Genetics Forum Erupts in Controversy," *The Washington Post,* September 24, 1995, B3; *See also* "Conflict Marks Crime Conference," *Science* 269 (1995):1808–1809.
3. Peter R. Breggin and Ginger Ross Breggin, *The War Against Children* (New York: St. Martin's Press, 1994).
4. *Id.*
5. As New York University professor Dorothy Nelkin put it, "It's a dangerous kind of research. This is a way to mask racism as scientific reality." Quoted in Maass, "Crime, Genetics Forum Erupts in Controversy." at B3. *See also* "Amid Protest, Crime Gene Meeting Held in Maryland."
6. *See, e.g.*, Leslie Crook, "Protesters Crash Aspen Conference," *Queen Anne's Record-Observer,* September 27, 1995, 1.
7. Cesare Lombroso, *Crime: Its Causes and Remedies* (Boston: Little, Brown, 1918). *See also* Gina Lombroso-Ferrero, "The Born Criminal." In *Biology, Crime & Ethics,* Frank H. Marsh and Janet Katz, eds. (Cincinnati, OH: Anderson, 1985), 37, 38.
8. Edward J. Larson, *Sex, Race and Science: Eugenics in the Deep South* (Baltimore: Johns Hopkins University Press, 1995):21, 26–27.
9. Although the primary target of this wave of genetic enthusiasm was White ethnics, they were, in effect, racialized as non-White. Daniel J. Kevles quoted then vice president Calvin Coolidge: "America must be kept American. Biological law shows . . . that Nordics deteriorate when mixed with other races." Daniel J. Kevles, *In the Name of Eugenics* (New York: Knopf, 1985), 97.
10. Kevles, *In the Name of Eugenics,* 111. *See also* Larson's elaboration of chapter 1, this volume.
11. 274 U.S. 200 (1927).
12. *Id.*, 207. As it turns out, *Buck* was not only unjust but based on factual error. There were not "three generations of imbeciles." *See* Kevles, *In the Name of Eugenics,* 112; Paul A. Lombardo, "Three Generations, No Imbeciles: New Light on *Buck v. Bell.*" *New York University Law Review* 60 (1985):30–62.
13. 316 U.S. 535 (1942).
14. The felonies statutorily excluded from eligibility for compulsory sterilization were violations of prohibition, tax evasion, embezzlement, and political offenses. *Id.*, 537.
15. *Id.*, 541.
16. Richard J. Herrnstein and Charles Murray, *The Bell Curve: Intelligence and Class Structure in American Life* (New York: Free Press, 1994), 240.
17. They suggest that the cognitively impaired are more likely to commit crime because they are

unable to reason abstractly about the costs of crime, to engage in moral reasoning, or to understand the point of deferred gratification. *Id.* Compare the suggestion of a Michigan neurologist that "children of limited intelligence tend to become violent when they are treated as equals." Peter R. Breggin, "Campaigns Against Racist Federal Programs by the Center for the Study of Psychiatry and Psychology," *Journal of African American Men* 1 (Winter 1995/1996):3–21, citing comments of Ernst Rodin.

18. Herrnstein and Murray, *The Bell Curve,* 541–544.
19. *Id.*, 523–526, discussing the custodial state, which will keep the stupid and dangerous away from the smart and productive.
20. *See* Edward J. Blakely and Mary Gail Snyder, *Fortress America: Gated Communities in the United States* (Washington, DC: Brookings Institution, 1997).
21. Jason deParle, "Daring Research or 'Social Science Pornography'?" *New York Times Magazine,* October 9, 1994, 48, 50.
22. *See* chapter 1, this volume.
23. *See* chapter 3, this volume.
24. Lori Montgomery, "Scholars to Explore Link Between Crime, Genes," *Philadelphia Inquirer,* September 23, 1995, A2, quoting Gregory Carey, a behavioral geneticist at the University of Colorado. One participant went further, suggesting that *only* "genetic or other biological factors" could provide a means "to control violence." Douglas Birch, "Crime, Genetics Link Not Found," *Baltimore Sun,* September 25, 1995, 2A, quoting Adrian Raine, a psychologist at the University of Southern California.
25. Introduction, this volume.
26. Consider, for example, the editorial of Daniel E. Koshland, Jr., "The Rational Approach to the Irrational," *Science* 250 (Oct. 12 1990):189. All the data Koshland cited relates to genetic bases for mental illnesses, and even that is still highly preliminary and limited. Yet he concluded that these tools will "help in reducing crime" and that we may soon be able to "provide predictive diagnoses to distinguish those who are severely ill from those who merely represent harmless aberrations from the norms of society." The implicit denial herein of the possibility that knowledge and action predicated on new knowledge may be a net human evil is the subject of a new book, which uses genetic science in particular as one of its examples of dangerous knowledge. *See* Andrew Delbanco, "The Risk of Freedom," *New York Review of Books* (Sept. 25, 1997):4–7, reviewing Roger Shattuck, *Forbidden Knowledge: From Prometheus to Pornography* (New York: St. Martin's Press).
27. Rochelle Cooper Dreyfuss and Dorothy Nelkin, "The Jurisprudence of Genetics," *Vanderbilt Law Review* 45 (1992):313, 320; Maureen P. Coffey, "The Genetic Defense: Excuse or Explanation," *William and Mary Law Review* 35 (1993):353, 361, both describing phenomenon; John Horgan, "Eugenics Revisited," *Scientific American* 268 (June 1993):123–131.
28. Lawrence Taylor, *Born to Crime: The Genetic Causes of Criminal Behavior* (Westport, CT: Greenwood Press, 1984), an "individual is, in a sense, genetically 'programmed' before birth to commit criminal acts"; *id.*, 93, PMS sufferer who commits crime "may simply be acting out a genetically determined role."
29. "Crimes Against Genetics," *Nature Genetics* 11 (Nov. 11, 1995):223, 224. The ready acceptance of biology as a primary determinant of criminal behavior has always puzzled me, given the fatal weakness of any such assertion once crime is viewed dynamically rather than statically. Incidences of genetic markers in a population can change only very slowly absent human intervention of a kind not yet available. But crime rates have changed enormously over far shorter time periods. *See* Department of Justice, *Uniform Crime Reports: Crime in the United States* (Washington, DC: U.S. Government Printing Office, 1982), 36; and *id.* (1991), 7, both showing substantial changes in crime index over periods of less than a decade. *See also* "Crime and Punishment," *Scientific American* 273 (Dec. 1995):19–20.
30. Anita LaFrance Allen, "Genetic Testing, Nature, and Trust," *Seton Hall Law Review* 27 (1997):887, 891.
31. "If 'genetic markers' for aggressive behavior are found, then medical treatments could be devised

to counteract violent tendencies.'' Maass, ''Conference on Genetics and Crime Gets Second Chance.''

32. Crossley's elaboration, chapter 6, this volume.
33. For purposes of criminal law doctrine, it is important to distinguish between the possibility of prevention and of treatment. If a preventable condition, such as lead poisoning, highly predisposes the person to violence, society is culpable and should take responsibility for minimizing such a condition in the future. The already poisoned individual, however, remains a danger and the condition is unlikely to serve as a defense, if that would leave her free to commit further acts of violence. *Cf.* Deborah W. Denno, ''Human Biology and Criminal Responsibility: Free Will or Free Ride?'' *University of Pennsylvania Law Review* 137 (1988):615, 670. ''With the expected advances, we're going to be able to diagnose many people who are biologically brain-prone to violence. . . . I am encouraged by the opportunity to . . . screen people who might have high risk and to prevent them from harming someone else,'' said Stuart C. Yudofsky, chair of the Baylor College of Medicine psychiatry department, quoted in W. Wayt Gibbs, ''Seeking the Criminal Element,'' *Scientific American* 272 (March 1995):100–107.
34. *See* Kevles, *In the Name of Eugenics,* at 142–143. Some studies suggest that one of the clearest correlates of violence is high levels of lead in the bloodstream. *See* Denno, ''Human Biology and Criminal Responsibility,'' at 651–658 (reporting result of study indicating that ''lead intoxication'' was one of the factors significantly associated with adult criminality). Although lead poisoning of children can be avoided in the future, we currently have no treatment for those already so poisoned. *Cf. also* Malcolm Gladwell, ''Damaged,'' *The New Yorker* (Feb. 24, 1997):132, 140, reporting on research indicating that child abuse and neglect create changes in brain structures that in turn detrimentally influence behavior.
35. The recognition that treatment is available for biologically based social dysfunction is a two-edged sword, however. It could lead to provision of additional educational inputs to respond to biological limitations. It could also—and this is perhaps more likely—lead to Clockwork Orange-like coercive interventions to treat or medicate children for whom the genetic–psychiatric–medical disciplines deem to be at risk. One physician proposed ''the electrical stimulation of surgically implanted electrodes as a method of calming violent people.'' Breggin, ''Campaigns Against Racist Federal Programs by the Center for the Study of Psychiatry and Psychology,'' at 5, describing testimony of Dr. William Sweet.
36. Stephen Jay Gould, ''Mismeasure by Any Measure.'' *In The Bell Curve Debate: History, Documents, Opinions,* Russell Jacoby and Naomi Glauberman, eds. (New York: Time Books, 1995) 3, 4. *See also* Dreyfuss and Nelkin, ''The Jurisprudence of Genetics,'' at 346.
37. A number of the participants in the Maryland genetics and crime conference sought to rebut charges of racism regarding their work or the field in general by explaining that the research participants were all Caucasian. This is inapt, however. Results can be and predictably will be generalized to larger populations. Merely stating that one does not intend any racial implications is an insufficient prophylactic. *See, e.g.*, Angier, ''Disputed Meeting to Ask if Crime Has Genetic Roots''; Sam Vincent Meddis and Gary Fields, ''Genes, Crime Link Explored,'' *USA Today,* September 25, 1995, 5A. Indeed, one of the proponents, presumably reporting on similar ''White-only'' studies, asserted that attention deficit hyperactivity disorder was largely genetic in origin, and responsive to drug therapy for both White and Black patient populations. *See* Douglas Birch, ''Protesters Attempt to Halt Crime, Genetics Conference,'' *The Baltimore Sun,* September 24, 1995, 1C, 2C, discussing work of Dr. David Comings with hyperactive children. *The Bell Curve,* which was popularly viewed and responded to as a book about genetic differences between Blacks and Whites, discussed only intra-White differences for the first half of the book.

    ''When you consider the perception that black people have always been the violent people in this society, it is a short step from this stereotype to using this kind of research for social control.'' Anastasia Toufexis, ''Seeking the Roots of Violence,'' *Time,* April 19, 1993, 52–53, quoting Ronald Walters, a political science professor at Howard University.
38. 'Crimes Against Genetics,'' 224. In this regard, the Maryland genetics and crime conference had some encouraging results. One genetics researcher, Adrian Raine, said that he would ''do more to

'ensure that my findings are not misinterpreted in a way which could feed into the fears of the public.''' ''Conflict Marks Crime Conference.'' The sponsoring University of Maryland issued a statement that ''in our society, any research that links criminal behavior to genetic features may be mistakenly seen as implicating the black community and contribute to its stigmatization.'' Joyce Price, ''Crime-and-Genes Conference Resurfaces With New Focus,'' *Washington Times,* September 22, 1995, 5.

39. Such claims must be considered cautiously, given the frequent pattern whereby claims of genetic links made with great fanfare have failed the test of replicability. *See* Lori B. Andrews, ''Past as Prologue: Sobering Thoughts on Genetic Enthusiasm,'' *Seton Hall Law Review* 27 (1997):893, 898, n.48 ; Horgan, ''Eugenics Revisited.''
40. As Grob pointed out, ''The findings of modern genetics can only be stated in terms of probabilities, which hardly differs from the older concept of predisposition.'' *See* chapter 1, this volume.
41. *See* chapter 4, this volume. *See also* ''Address of Dept. of Health and Human Services Secretary Louis Sullivan to the American Academy of Child and Adolescent Psychiatry'': ''The very idea of a 'crime gene' is simplistic and misleading. . . . [T]here has been no human behavior for which any single gene has been found to be the cause.'' Quoted in Breggin and Breggin, *The War Against Children,* at 55.
42. *See* chapter 3, elaboration by Schopp and chapter 6, this volume.
43. Thus, for example, antisocial personality disorder, even if it is shown to be genetic in origin, will remain outside the definition of ''mental illness'' for purposes of the insanity defense. Stephen H. Dinwiddie, ''Genetics, Antisocial Personality, and Criminal Responsibility,'' *Bulletin of the American Academy of Psychiatry and Law* 24 (1996):95, 102–104.
44. As Perlin noted, ''some of the opposition to the insanity defense might be remediated . . . if we could pinpoint a visible gene or chromosome . . . that we could authoritatively say 'causes' otherwise-inexplicable aberrant behavior.'' *See* chapter 2, this volume. Crossley suggested that this greater public faith in genetics as ''real science'' may lead to an acceptance of genetics as an excuse where it is logically no different than non-genetic factors that do not excuse. *See* Crossley's elaboration, chapter 6, this volume. *Cf.* Wray Herbert, ''Politics of Biology,'' *U.S. News & World Report* 72 (April 21, 1997):72–80, in describing how lawmakers are more willing to acknowledge the reality of mental illness when shown PET scans, an advocate for the mentally ill says ''When they see that it's not some imaginary, fuzzy problem, but a real physical condition, then they get it. 'Oh, it's in the brain.'''
45. Even if these conditions do affect cognition, states need not allow juries to consider them in deciding if the defendant had the mental state otherwise required for conviction. *See* Montana v. Egelhoff, 518 U.S. 37 (1996) (finding constitutional Montana statute under which the jury was instructed ''that it could not consider respondent's 'intoxicated condition . . . in determining the existence of a mental state which is an element of the offense.''' 518 U.S. at 40, quoting MONT. CODE ANN. § 45-2-203).

    The prior brushfire of genetics-linked claims to explain and excuse criminal acts, surrounding the apparent relationship between XYY chromosomal abnormality and disposition to violence, was put out as a defense by the opinion in *People v. Yukl,* 372 N.Y. 2d 313, 319 (NY Sup. Ct. 1975), which stated that the abnormality could be treated as insanity and thus provide a defense only on a showing that ''the genetic imbalance must have so affected the thought processes as to interfere substantially with the defendant's cognitive capacity or with his ability to understand the basic moral code of his society.''
46. As Moldin noted, it is unlikely that we will ever conclude that a gene has such strong predictive power that one could say that its expression in behavior is determined or irresistible. *See* chapter 5, this volume.
47. Stephen Morse, ''Brain and Blame,'' *Georgetown Law Journal* (1996):527, 527; *See also* Michael S. Moore, ''Causation and the Excuses,'' *California Law Review* 73 (1985):1091–1149, discussing difference between causation and compulsion; *see* chapter 2, this volume.

    The Supreme Court has also noted this disjuncture when it referred to ''[t]he conceptual difficulties inevitably attendant upon the importation of scientific and medical models [of causation and free will] into a legal system generally predicated upon a different set of assumptions.'' *Powell v. Texas,* 392 U.S. 514, 526 (1968).

48. The extremely narrow scope of this doctrinal exception to free will is apparent in the Supreme Court's decision in *Powell v. Texas,* 392 U.S. 514 (1968). In rejecting an involuntariness defense to the charge of being drunk in public, the Court held that there was insufficient scientific evidence to support a finding that chronic alcoholics ''suffer from such an irresistible compulsion to drink and to get drunk in public that they are *utterly unable* to control their performance of either or both of these acts.'' *Id.*, 535 (emphasis added).
49. Brock and Buchanan asserted that genetic deviations, so long as they do not make criminal acts absolutely irresistible, but only more difficult to resist, are akin to environmental factors. *See* chapter 3, this volume. Though the authors suggested that we may have more control over our environment, that is only partially true. Those who have been abused as children are disproportionately more likely to be abusers when they grow up, but a child realistically has no more control over his subjection to childhood abuse than his genes.
50. Schopp's elaboration, chapter 3, this volume. Only Daniel Summer, Mobley's attorney, asserted that persons with a genetic trait that ''precludes or *to some extent* interferes with the ability to form the legal concept of criminal intent'' should be excused or have their punishment mitigated (emphasis added). *See* Summer's elaboration, chapter 6, this volume. If the genetic condition indeed made it impossible for the defendant to form the *mens rea* required for the crime charged, this would indeed ordinarily be a defense. There is no evidence, however, that genetic defects, except insofar as they lead to cognizable insanity claims, are associated with such true incapacity. Rather, they seem to be linked to conditions of impulsivity and lack of control.
51. *See* Parker's elaboration, chapter 3, this volume. Dresser suggested that the underlying structure of our criminal law would permit a defendant to avoid liability if he could show that because of a genetic condition he ''could not fairly be expected to avoid the criminal act,'' or, more narrowly, if the action were wholly unavoidable. However, as she also recognizes, the defendant can nonetheless be held liable for exposing himself to a situation in which such strong urges are likely to arise. *See* chapter 6, this volume. *Cf.* People v Decina, 2 N.Y.2d 133, 138 N.E. 2d 799 (1956) (defendant who had an epileptic seizure while driving could be convicted of criminal homicide because he took the chance of driving, knowing of his condition). *See generally* Mark Kelman, ''Interpretive Construction in the Substantive Criminal Law,'' *Stanford Law Review* 33 (1981):591, noting that the time-framing of the defendant's behavior is both crucial to the outcome and not determinable by any set of doctrinal rules.
52. *See, e.g.*, California Penal Code § 28(a) (1981). Under that statute, evidence of mental disease defect or disorder can be used to show the defendant did not have the required mental state but not ''to show or negate the capacity to form any mental state.'' Further, ''as a matter of public policy there shall be no defense of diminished capacity, diminished responsibility, or irresistible impulse in a criminal action or juvenile adjudication hearing.'' § 28(b).
53. *See generally* John Kaplan, Robert Weisberg, and Guyora Binder, *Criminal Law: Cases and Materials,* 3d ed. (Mineola, NY: Foundation Press, 1996), 745–749.
54. That is, a claim that the defendant as a result of an abnormal mental condition was at the time of the crime incapable of forming the mental state required for the crime charged. The issue almost always arises in the context of homicide, in which the practical effect of such a doctrine would be to reduce the crime of conviction from first-degree murder to second-degree murder, for lack of premeditation or from murder to manslaughter, for lack of intent to kill. An alternative view of diminished capacity sees the mental condition as partly excusing, without focusing on *mens rea* as such and is explicitly limited to the homicide situation. *See generally* Joshua Dressler, *Understanding Criminal Law,* 2d ed. (New York: Matthew Bender, 1995), § 26.01–.02.
55. Baker v. State Bar of California, 781 P.2d 1344 (Cal. 1990). *See also* Coffey, ''The Genetic Defense.''
56. As Perlin noted, the insanity defense is rarely raised, rarely successful (unless prosecution and defense concur), and leads to institutionalization for a longer period on average than would occur if the defendant had been found guilty. *See* chapter 2, this volume.

57. Similar considerations traditionally were used in arguing sentencing questions to the judge. The recent change to a sentencing guidelines system in the federal and many state systems, however, has made these particularities of the defendant's situation far less significant.
58. *See* Denno, ''Human Biology and Criminal Responsibility,'' discussing admissibility of evidence of brain deformation and CAT scans in penalty hearings.
59. Many of the chapters in this volume carefully deconstruct the false essentialism of the first term. Equal attention must be paid to avoid naturalizing or essentializing the concepts of *crime* or even *violence.* The meaning of these terms is not natural, but socially constructed. For example, Stephen Mobley's family apparently included both a number of people who had engaged in antisocial violence and a father whose aggressive business tactics—likely subject to social disapprobation in other cultures—had made him a millionaire. *See* Deborah W. Denno, ''Legal Implications of Genetics and Crime Research.'' In *Ciba Foundation Symposium 194: Genetics of Criminal and Antisocial Behavior* (London: Ciba Foundation, 1996), 248–264; Skinner v. Oklahoma, 316 U.S. 535 (1942) (no biological correlate of distinction made in sterilization statute between larceny and embezzlement). Even studies of the link between genetics and mental illnesses suffer similar difficulties as definitions of, for example, schizophrenia, change over time. ''Crimes Against Genetics,'' at 223. Furthermore, we must distinguish between *crime* and *violence,* not assuming that studies of causes of the latter are adequate explanations of the former. The conflation of the two in the popular imagination reflects an obsession with street crime—the acts of a violent ''other'' against us and our children—rather than white-collar forms of criminality.

    The essentialization of *crime* has a depoliticizing effect, demonstrated dramatically by the suggestion of three doctors that the 1967 Detroit riots might be attributable to brain disorders among the rioters. *See* V. H. Mark, W. H. Sweet, and F. R. Ervin, ''Role of Brain Disease in Riots and Urban Violence,'' *JAMA* 201 (1967):895.
60. Indeed, as Botkin noted, genetic evidence might conceivably be used against defendants in individual cases as evidence that the defendant had a particular character and, consistent therewith, did the charged act. *See* introduction, this volume. Although traditionally the state has not been permitted to introduce such character evidence, recent amendments to the Federal Rules of Evidence have modified that rule in certain situations. *See* FED. R. EVID. §§; 413–415. *See generally* Michael Graham, *Handbook of Federal Evidence.* 4th ed. (St Paul, MN: West Publishing Co., 1996), §§ 413–415.
61. In Summer's case histories, at least three, including his client, Mobley, are described as White, suggesting that genetic defect as a defense strategy is most plausible when the defendant is among a class of people we do not already stereotype as dangerous, violent, and subhuman. *Cf.* Dreyfuss and Nelkin, ''The Jurisprudence of Genetics,'' at 331, which acknowledges that ''hereditary traits . . . often reduce to ethnic group membership.''
62. *See* chapter 2, this volume.
63. At least one court has limited the traditional power of quarantine in accordance with the protections of the Americans with Disabilities Act. City of Newark v. J.S., 279 N.J. Super. 178, 652 A.2d 265 (1993). The dangerous condition in that case, however, was tuberculosis, clearly covered by the act. The limitation might not be available for allegedly dangerous genetic defects, particularly if they are manifested in conditions such as pedophilia or antisocial personality disorder. *See* 42 U.S.C. § 12211(b)(1) (specifically excluding pedophilia and other sexual disorders). *Cf.* Taylor, *Born to Crime,* at 147, 158–159, which asks rhetorically what ''society [can] do to protect itself'' from a 3-year-old ''who is scientifically shown to have a 78% chance that he will become a compulsive rapist or a 61% chance that he will eventually take at least one human life violently and without justification,'' and then quotes approvingly from Justice Holmes's opinion in *Buck v. Bell* [(276 U.S. 200 (1927)].

    Diane Fishbein, a criminologist at the University of Baltimore, has said that treatment should be mandatory for offenders. ''They should remain in a secure facility until they can show without a doubt that they are self-controlled'' and ''held indefinitely'' if no effective treatments are available. Quoted in Gibbs, ''Seeking the Criminal Element,'' at 107.

64. *Cf.* Herrnstein and Murray, *The Bell Curve,* at 523–526, discussing the custodial state, which will keep the stupid and dangerous away from the smart and productive.
65. 521 U.S. 346 (1997).
66. It would constitute both double jeopardy, because it applied in addition to the sentence Hendricks was already serving, and an *ex post facto* law.
67. 521 U.S., 356.
68. As the dissent notes, this point is dicta, because the Kansas statute specifically indicated the state's belief that conditions such as Hendricks's pedophilia are treatable. Even a requirement of "treatment" is cold comfort, however. In effect, it constitutionally authorizes the imposition of whatever treatment modalities a subgroup of the appropriate professional discipline think may be useful in changing the predispositions of the subject population. *See* C. R. Jeffery, "Criminology as an Interdisciplinary Behavioral Science." In *Biology, Crime and Ethics,* at 44, and Hugo Adam Bedau, "Physical Interventions to Alter Behavior in a Punitive Environment: Some Moral Reflections on New Technology." In *id.*, 202.
69. Kansas v. Hendricks, 521 U.S., 359. This conclusion was arguably prefigured by Justice O'Connor's crucial concurrence in *Foucha v. Louisiana,* suggesting that a state might continue to detain an insanity acquittee who was no longer mentally ill if "the nature and duration of detention were tailored to reflect pressing public safety concerns related to the acquittee's continuing dangerousness"; 504 U.S. 71, 87–88 (1992). *Hendricks* seems to have undermined the position, plausible after *Foucha,* that involuntary civil commitment can be justified only if it is therapeutically appropriate. Bruce J. Winick, "Ambiguities in the Legal Meaning and Significance of Mental Illness," *Psychology, Public Policy and Law* 1 (1996):535.
70. *Hendricks,* 521 U.S., 356. The subclass in *Hendricks,* as in previous civil commitment statutes, was defined based on mental conditions. Nothing in the language of the case, however, indicates that a subclass could not be defined based on a physical condition that impairs one's ability to avoid antisocial behavior. 1996 WL 469200 (U.S. Amicus Brief) *Kansas v. Hendricks* (U.S. Supreme Court Amicus Brief) August 16, 1996, *Brief for the American Psych Assoc. as Amicus Curiae in Support of Leroy Hendricks,* 1–64, The amicus briefs of both the Washington State Psychiatric Association and the American Psychiatric Association sounded this warning. The former noted that the criteria of "mental disorder" and "antisocial personality disorder" were so amorphous that they could "be stretched to fit all sex offenders," and thus "allow use of our mental institutions for warehousing all 'undesirables,' i.e., people who repeatedly show disregard for the rights of others." The American Psychiatric Association similarly suggested that a decision upholding the law would confirm" a general state power to confine people indefinitely based on predictions that they would likely cause serious harm.' 1996 WL 468611 (U.S. Amicus Brief) August 14, 1996, *Brief of Amicus Curiae Washington State Psych. Assoc. in Suport of Respondent,* 1–26.
71. The American Psychiatric Association amicus brief in *Hendricks* suggested that any rationale that would uphold this law would also justify detention "for any number of classes of individuals, such as alcoholics and substance abusers, whose characteristics are significantly associated with violence." Indeed, if one needs merely an innate factor significantly associated with criminal violence, could one civilly commit males who had engaged in such violence and were predicted to do so again?
72. The statute as it reached the Supreme Court was construed to require a conviction, with the narrow exceptions set out earlier. The statutory language, however, only requires proof that the defendant "has been convicted of or charged with a sexually violent offense." KAN. STAT. ANNOT. § 59-29a02(a) (1997). This possibility has not escaped the attention of some of the other contributors to this volume. *See* chapter 3, this volume ("It still may be necessary for society to protect itself from persons whose genes disclose them to be dangerously violent"); Wettstein's elaboration, chapter 4, this volume, also suggests that the state could preventatively detain individuals at high genetic risk of violence; and chapter 6, this volume, notes that if a genetic defense were recognized it would increase pressure "to restrain even people of that genotype who

have not yet committed any antisocial acts,'' quoting Dennis S. Karjala, ''A Legal Research Agenda for the Human Genome Initiative,'' *Jurimetrics Journal* 32 (1992):121, 162.

73. Horton's elaboration for chapter 6, this volume. Although one may expect a lawyer, rather than a psychiatrist, to be most likely to notice risks to civil liberties, it is striking that the prosecutor, rather than a defense attorney or a legal academic, included this point in his contribution. Only one of the authors seems to anticipate coercive measures against the genetically impaired as a desirable social policy. *See* chapter 4, this volume, noting that ''a substantial majority of all serious personal crimes are committed by individuals suffering from antisocial personality, alcoholism, or other substance abuse. If it were not for these three conditions, the crime problem in the United States would be vastly reduced.'' Although Guze does not discuss the means of eliminating these conditions, or those suffering from them, there currently is no effective, noncoercive means to do so.
74. In fact, it is unlikely the genes that control significant behaviors are distributed racially. A major study by population geneticists indicates that, apart from superficial characteristics such as skin color, the distribution of genes is almost identical across those population groups we call races. Luca Cavalli-Sforza, Paolo Menozzi, and Alberto Piazza, *The History and Geography of Human Genes* (Princeton, NJ: Princeton University Press, 1994).
75. The numbers of people who might be subject to such intervention is high. Irving Gottesman, a professor of psychology at the University of Virginia and author of some of the twin studies that have supported claims of genetic links to crime, stated that 7.3% of men and 1% of women suffer from antisocial personality disorder, ''which can lead to criminal behavior.'' David L. Wheeler, ''The Biology of Crime,'' *The Chronicle of Higher Education* (Oct. 6, 1995):3–4.
76. *Cf.* chapter 7, this volume, discussing the impact of public fear of crime on the preference for false positives over false negatives.
77. Tom Lehrer, ''Wernher von Braun.'' Broadcast on *That Was the Year That Was* (1955). WEA/ Warner Brothers (available on compact disc).

# Chapter 10
# CRIMINAL LAW

Leslie Pickering Francis

Both now and in the near future it is highly unlikely that genetic information will be available that will change our basic understandings of the criminal justice system. Even a vast expansion in the empirical information available about the causes of human behavior does not warrant abandoning metaphysical commitments to free will and responsibility. Nor does it recommend fundamental shifts in the retributivist paradigm of the criminal law. A number of contributors to this volume do suggest, however, that as genetic information mounts we may change our beliefs about the circumstances under which responsibility should be attributed to individuals—and to society—as well as about the appropriate social responses to criminal behavior. This chapter describes the principal recommendations for the criminal law reached by contributors to this volume. The need for caution is the most important recommendation, because genetic information is very likely to be misunderstood in the context of contemporary attitudes toward the mentally ill and those accused of crime. In this highly charged political context, courts and policy makers should proceed extremely gingerly with genetic information and work to become appropriately educated about the complexity and limits of the information that is available.

In broad terms, the criminal law defines offenses, establishes responsibility on the part of offenders, and mandates punishment or other remedies. Genetic information is potentially relevant to any of these decisions, and an examination of its significance must consider the issue context. For example, defenses to a criminal prosecution function either to exculpate entirely or to diminish the level of crime for which the defender is to be held responsible. Defenses tend to function in an all or nothing way, establishing if successful that the accused was "not guilty"; yet the genetic information that is likely to become available will be tentative and probabilistic at best. It is possible, therefore, that some of the information rejected as relevant in the defense context may be more helpful in the sentencing arena, where judgments can be more nuanced, at least if they are not subject to mandated guidelines.

Beyond the retributivist paradigm, the criminal law also serves a protective function, through deterrence and incapacitation. Even if genetic information does undermine notions of responsibility in limited cases, it will not abrogate the protective functions of the criminal law in these cases. At most, it would suggest new information about protective institutions and treatments, and the disaggregation of the retributive and protective functions of the criminal law.

Finally, important evidentiary issues can be expected to arise if genetic information is introduced into the criminal justice process. The recent changes in the standards under the federal rules of evidence for the introduction of expert scientific testimony can be expected to increase the likelihood that efforts to introduce genetic information will be successful. Yet the result may be the danger of significant misunderstanding and misuse. These new evidentiary standards increase the burden on legal professionals to acquire sufficient understanding of these complex scientific issues.

## Criminal Responsibility and Genetic Information

The standard retributivist paradigm of criminal punishment, based on the metaphysical understanding of ourselves as free and responsible agents, is that defendants should not be held responsible unless they are to blame for what they did. A common understanding of the meaning of this paradigm, developed in chapter 6 by Rebecca Dresser, is that no one should be held responsible when they could not have acted otherwise. As Dan Brock and Allen Buchanan argue in chapter 3, this volume, empirical information about particular explanations for human behavior does not imply a change in the metaphysical understanding of human freedom that underlies the retributivist paradigm. As empirical information mounts, however, it may change our understanding of particular cases; and in the long run the justifiable worry is that it might undermine the justification for conceiving ourselves as free and responsible agents.

Contributors to the volume conclude, however, that the limited and probabilistic nature of the genetic information becoming available is such that this retributivist paradigm is unlikely to be undermined, except in unusual cases. As outlined in Steve Moldin's contribution in chapter 5, layers of probability judgments are involved in any efforts to link genetic information and behavior. At a minimum, these layers include the identification of genes themselves, the identification of linkages between genotype and phenotype, and the understanding of the biochemical bases of any linkages between phenotype and behavior. It is thus very difficult at the present time to believe that there will be reliable linkages between genetic information and the causation of behavior. Yet such reliable linkages to behavior are what would be needed for exculpation on the retributivist paradigm. To be sure, our developing knowledge in genetics may be relevant to determinations of responsibility in particular kinds of cases, but these are likely to be the exception rather than the rule.

However, it is fair to say that the working group identified likely persistent disagreement about the underlying models of cause and responsibility that are relied on. For some, judgments that an individual was not responsible because of the role of disease in conduct would need to rely on clear cognitive or volitional impairment rendering the defendant genuinely unable to have done otherwise. For others, difficulty in resistance—the likelihood that it is simply harder, perhaps much harder, for an individual to conform his or her conduct to the requirements of law—would suffice for altered judgments of responsibility. As knowledge of physiology and behavior increases, it is likely that we will learn that people with certain biochemical makeups (perhaps even coded in an identified genetic way) demonstrate characteristic patterns of neural activity, and describe themselves as finding it more difficult to resist certain kinds of temptations than people without that biochemical makeup. The law will need to resolve whether such background stories exculpate in any way. With such information, we still would not know the actual causal links between neural activity and behavior; we would at best have established correlations between genotype, brain chemistry, and experiential reports about behavior. Some contributors to this volume argue that the criminal law should continue to set a very low threshold for responsibility, exculpating only when we know that a demonstrable impairment caused the behavior at issue. Others argue that judgments that the defendant was not responsible for what he or she did should extend more broadly to circumstances in which there is evidence of correlation between biochemistry and the description of behavior control as very difficult. One possibility is that if such evidence does not show cause in the way regarded as exculpatory, it might be regarded as relevant to decisions about sentencing rather than to decisions to exculpate.

At the present time, virtually no case law or statutes recognize the relevance of genetic diagnoses to the criminal law. A single case in Georgia, described by Daniel Summer in his elaboration of chapter 6, this volume, did allow a woman on trial for murder to plead not guilty by reason of insanity in virtue of evidence that she suffered from Huntington's disease. There have been several other attempts to use genetic information, particularly in sentencing but also to establish exculpation: All have been unsuccessful. These efforts are surely likely to increase, however, and it is important to understand what linkages might be suggested and why the available genetic information is unlikely to support them except in unusual cases.

## Potential Links Between Genetic Information and Criminal Culpability

There are at least three different ways links between genetic information and behavior might come into the criminal defense process. First, crimes are typically defined in terms of an act and a mental element. The genetics–behavior link could be relevant to whether there was an act at all, and so whether a crime occurred. A standard example would be a patient with a seizure disorder who, in the course of a seizure, strikes and injures another. If the blow was the result of an uncontrolled physical movement occurring in the course of the seizure, it would not be an act in the sense required for assault crimes. Genetic evidence might be relevant to the etiology of the seizure and thus to the defendant's argument that he or she did not act at all.

Second, genetic information might be regarded as relevant to a determination of what the offense at issue was. For example, homicide offenses typically are graded in seriousness, depending on the mental element of the offense. The difference between murder and manslaughter is established in this way. Genetic information might be relevant to understanding the capacities of the defendant and thus whether the defendant was capable of or did have the mental element required for the offense. This use of medical information about the defendant potentially diminishes responsibility by reducing the grade of offense but not exculpating altogether. Examples suggested by Summer, elaboration for chapter 6, this volume, include patients who have altered neurotransmittor receptors, thus altered behavior control mechanisms, and thus arguably diminished capacities to engage in planned rather than impulsive behavior. Genetic evidence might be relevant to establishing whether such a defendant might be guilty of manslaughter rather than murder.

Third, genetic information might be used to establish the insanity defense in one of its classic forms. The M'Naghten test for insanity requires proof that the defendant was unable to understand the nature and quality of his or her act or that it was wrong. An example of a defendant meeting this test might be a schizophrenic who is delusional and cognitively impaired as a result. The Model Penal Code test for insanity requires a showing that, because of a mental disease or defect, the defendant was substantially unable to conform his or her conduct to the requirements of law. An example of this might be a defendant who could show that because of altered brain chemistry, she or he was substantially less able to control her or his behavior in accord with law. This kind of causal showing, however, is unlikely to be available except in unusual cases.

## Current Legal Analogies to the Use of Genetic Information

In law today, diagnostic medical information is regarded as relevant to establishing the elements just outlined: whether the defendant acted, whether the defendant should be regarded as culpable of a diminished offense, and whether the defendant can establish a classic form of the insanity defense. If information about genetic diagnoses is analogous to information about other medical diagnoses, current policy may be judged inconsistent if it fails to recognize genetic information while continuing to rely on other medical information.

The analogy between genetic information and other medical information is not straightforward, however. Care needs to be taken in understanding what information is actually available. When medical information is incorporated in the criminal defense process, it generally is as evidence of a function-impairing condition. For example, evidence of a brain tumor that is symptomatic in affecting behavior might establish that the defendant was not guilty by reason of a mental disease. Evidence that the defendant is suffering from other neurological diseases, diabetes, psychiatric conditions, or mental retardation also functions in this way. Genetic information, however, might concern genotype, phenotype, or fully expressed condition with symptoms. For the analogy with other medical diagnoses to work, then, the genetic information would need to establish the presence of a function-impairing condition, rather than simply serving as an indication as to the cause of a condition. Evidence that the defendant carries the gene for Huntington's disease, however, is not sufficient to establish linkages with behavior that are analogous to the brain tumor example. What would need to be shown in addition is that the defendant evidenced altered behavioral function as a result of the progress of Huntington's disease at the level of a fully expressed condition.

To be sure, there are situations in which it is unclear whether behavior is a manifestation of a functional impairment. As information grows about links between genotype and behavior, genetic information might be helpful in establishing diagnoses of functional impairment. Ordinary descriptions of behavior may be equivocal about whether somebody was undergoing a psychotic episode, for example; genetic evidence could be relevant in establishing the likelihood that the accused really did undergo a psychosis. It may also be useful as evidence of a deteriorating, degenerative condition, if the act was the first manifestation of the condition. Although the information about the case in which a genetic diagnosis of Huntington's served to exculpate is incomplete, it seems that the genetic information enabled the trial court to understand the defendant was functionally impaired in a way that had affected her behavior. Thus it was the presence of the functional impairment, rather than the genotype by itself, that was regarded as exculpatory in the courts.

In this summary so far, medical information has been viewed as potentially exculpating. But medical information may also serve to inculpate, and analogies between currently used information and genetic information might be developed in this way as well. Predispositions to expressed conditions such as alcohol or drug abuse can function to inculpate if the defendant reasonably should have been on notice of the risk. For example, courts currently use probabilistic information about alcohol or drug abuse as the basis for concluding that the defendant should have known the risks of exposure to alcohol and taken steps to avoid the possibility of alcohol use. Predispositions rooted in genotype might be found analogous to predispositions to alcohol or drug abuse. In cases of such known genetic predispositions, the defendant might be regarded as having been on notice of the importance of avoiding triggering circumstances and as culpable for the failure.

## Genetic Information and Legal Ethics

Genetic information should be treated by lawyers just as other available information is handled. Lawyers' obligations of confidentiality apply regardless of the type of information involved. Prosecutors have the obligation to disclose potentially exculpatory evidence to defendants. If the prosecution possesses possibly exculpatory evidence as a result of genetic testing, disclosure to the defense will be required. Defense lawyers do not have the general obligation to disclose potentially inculpatory information (although they do of course have the obligation to respond to lawful demands for information and not to suppress evidence illegally). If the defense engages in genetic testing of the accused but the results are noninformative or possibly inculpatory, it will not be required to disclose the information, to the extent that it is not currently required to disclose other inculpatory evidence. It would be misleading, however, for the defense to suggest the possibility of exculpation when it has reason to believe it would not be borne out.

## Genetic Information and the Rules of Evidence

Like other scientific information, genetic information is difficult for courts to understand and to assess. In criminal law, courts have had longstanding difficulties in separating legal standards from psychiatric expert testimony and in understanding the expert testimony. Genetic information, which is highly complex scientifically, will add to these difficulties.

The rules of evidence are designed to ensure that misinformation and irrelevant information is not considered by the finder of fact (often the jury, but also the judge). The *Daubert* standard, currently prevailing for admissibility of scientific evidence in court, requires expert scientific testimony to be reliable and relevant to the issue at hand. Questions are certain to arise about whether genetic information can pass this test and whether this test is adequate to screen out pseudoscientific or otherwise problematic use of genetics.

Genetic information may be well-enough established as science to meet the reliability portion of the *Daubert* test. Although there will be questions about the reliability of the laboratories performing genetic testing, these problems are not more daunting in the criminal defense context than in other forensic contexts. One current use of genetic evidence, DNA fingerprinting, illustrates how these issues of reliability have largely been resolved. But there is a second aspect to the *Daubert* test: relevance. When genetic evidence is used for identification, the relevance of the DNA match is not at issue (except perhaps in the unusual case of identical twins). In the use of genetic information to establish defenses to crime, however, there is much more dispute about what, if anything, the evidence might show. The question is whether the genetic information is relevant or of any probative value to the criminal liability of the defendant. It can be expected, for example, that there will be questions about whether genetic information is admissible when linkages are tenuous at best between genotype and expression in functional impairment. Courts should exercise particular care in the scrutiny of the relevance of genetic evidence, with attention to whether it is being offered to show genotype, a link between genotype and phenotype, or a link to functional impairment and behavior, and whether it is then relevant to the point at issue. Members of the working group suggested courts should consider establishing guidelines for evidentiary hearings about genetic information.

Even when scientific evidence is of some probative value, moreover, this does not end the inquiry about its admissibility. The rules of evidence require courts also to consider whether the probative value of scientific evidence is outweighed by the likelihood that it will prove prejudicial. Courts will thus need to consider whether there are particular circumstances in which genetic evidence, though relevant, would be more prejudicial than probative, and should be ruled inadmissible on that basis.

## Future Directions

We can anticipate, from the debates to date, that many arguments will be made for the use of genetic information in the criminal justice process. Given the probabilistic nature of the information that will be available, and its tenuous connection at best to behavior in many cases, courts should scrutinize these arguments with great caution. It is certainly possible that these arguments will find open acceptance—in courts, by legislatures interested in reducing crime rates or incarcerating offenders who are believed to be especially dangerous, or in discussions in the media. The risks of a rush to judgment in such instances are significant. The working group emphasized the importance of attention to less obviously public ways in which the acceptance of genetic information might prove influential on legal policy and the criminal justice process.

One less public arena in which genetic information might be taken into account is in the discretionary exercise of the prosecution function. There is considerable discretion in prosecutorial decision making: in decisions about diversion from the criminal justice process, about whether to charge, and about the charges to be brought, among others. Yet the prosecution decision is largely unreviewable and plays a major role in outcomes of the criminal justice system. Questions may well arise about whether prosecutors should use genetic information in deciding how to charge. Genetic information may well be relied on by prosecutors intent on protecting (or being perceived as aggressive in protecting) the public against potentially dangerous offenders. Genetics aside, it is an important question of public policy whether prosecutors should make decisions about whether to prosecute based on estimates of an alleged offender's future dangerousness. Genetic information may compound this debate, in ways that the working group regarded as seriously problematic. Any links between genotype and dangerousness will be very tenuous; even if there are some linkages, it is likely that there will be many false positive predictions. Prosecutors may well overestimate the predictive power of genetic information. Finally, the concern should be raised that inappropriate reliance on genetic information might compound currently racist or otherwise unjust prosecutorial practices.

Another area of concern is that genetic information might be thought to provide added support for determinate sentencing or habitual criminal statutes. Such statutes are sometimes justified on the basis of the supposed dangerousness of the offenders in question. Overestimates about the predictive value of genetic information might lend support to these statutes on an unjustified basis.

On the other side, if there are cases in which genetic information is probative, then courts will need to consider whether there would be an obligation to pay for the testing for indigent defendants. With respect to other issues of mental status, once the defendant shows that there is a colorable question and that he or she is indigent, the court will pay for the exam. With genetic information, expensive testing may be needed to show that there is a colorable issue that the evidence is significant. If these questions are resolved against defendants, potentially helpful genetic information may not become available.

Finally, as genetic information is increasingly brought into the criminal law, as it inevitably will be, there are many risks for particular defendants, for others who might come into the criminal justice system, and for society as a whole. For defendants, the risks are particularly impressive and warrant special care in the use of genetic information. There are risks of stigmatization and demonization: the judgment that "he's a sexual predator," for example, or the potential for racism described by Coombs (chapter 9, this volume). Stigmatization, in turn, may result in decreased respect for autonomy and decreased self-control because of self-fulfilling prophesies. More severe prosecution, linked to perceptions of dangerousness (whether misleading or not) may also result when the defendant has been identified as possessing a genetic trait. There is also the possibility of indeterminate detention (perhaps even without treatment) of those regarded as a threat, either without conviction and sentencing or after a sentence has been served, as happens under the sexual predator statutes today. For others in the criminal justice system, the principal risk from genetic information is the possibility that they will be treated unfairly because of the misuse of genetic information in a prior case.

The risks to society of the misuse of genetic information are also apparent. There is the possibility of unjustified exculpation: of a defendant who should have been punished, or worse, a defendant who is released because of a failure to disaggregate judgments of responsibility from the protective functions of the criminal law. There are the risks of deeply problematic social policies, such as the eugenics movement that, historically, was used to try to reduce "undesirable" criminal behavior. Perhaps worst of all, genetics has a deservedly clouded reputation in the United States, and there are risks that genetic data will be used selectively, inaccurately, and inappropriately to reinforce racism or other prejudices.

Despite these risks, some possible benefits from genetic information can be identified, if the information is used cautiously and with understanding of what it does and does not indicate. For defendants, there are the possibilities of appropriate exculpation, not otherwise obtainable without genetic information. There is also the possibility of treatment that is tailored to the genetic condition, and of increased self-control as a result. For society, there are possibilities of better understanding of the causes of crimes of violence and of more efficient treatment and intervention. There is also the possibility of obtaining better information about release decisions and methods of managing probation. There may, in the long run, be the ability to refocus as a society on understanding the social as well as the biological bases of crime.

## Directions for Further Research

Because genetic information poses such extraordinary potential for misuse, and such apparent risks and benefits, research about its reception and understanding seems particularly important. A number of research directions are especially promising:

1. How do jurors respond to evidence about genetics? Do they regard it as exculpatory? How well do they understand it?
2. How much do judges know about genetic information? To what extent is it incorporated into judicial opinions, and is it done so with appropriate understanding?
3. What are the phenomenological experiences of people with genetic conditions expressed in behavior? To what extent, for example, do people with Tourette's syndrome describe themselves as incapable of controlling their behavior?
4. What is the relevance of such phenomenological understandings to defenses in criminal law? What kind of understanding of the causation of behavior is relevant to defenses in criminal law?

## Recommendations

Genetic information, as it becomes available, should be used with extreme caution, if at all, by courts and policy makers in the area of criminal defense. This might counsel admission in the courts if it is factually relevant on a case by case basis. Courts should be cautioned that they are likely to be regarded as standard-setters in this area by other institutions—for example, educational institutions. They should be careful in this regard, although it should also be noted that it is likely to be misleading for other institutions to model themselves on courts. These institutions will need to think through the relevance of genetic information in their own areas. There should be ongoing continuing education of judges and the prosecution–defense bar about the extent and limits of available genetic information. Understanding of probabilistic information is especially important and complex. Professional genetics organizations should consider developing standards in this area. If calls for the use of genetic information in the criminal law grow, there may be need for a new field of forensic genetic counseling.

# Chapter 11
# THE NEW GENETICS AND JUVENILE LAW

William M. McMahon

The preceding chapters provide no precise predictions nor easy solutions for problems that face our society as we grapple with new genetic information and implications for juvenile law. As this book goes to press, no genetic findings have yet forced us to radically reconsider our notions of the development of personal responsibility by juvenile offenders. If such future findings show that some detectable gene or genes contribute substantially to the development of criminal behavior, we will need an informed debate on the social policy implications. Anticipating that claims of such a gene are likely to occur, and that great potential exists for a misinformed, divisive, and counterproductive argument, this chapter synthesizes the prior chapters into a set of consensus statements, overriding conclusions and specific recommendations. Besides material in the previous chapters, this synthesis draws on discussion by the Working Group on Juvenile Law and on the clinical and research experience of the author in psychiatry and genetics.

## Development, Disorders, Delinquents, and Society

Human development is complex, multifactorial, multidimensional, and unfolds over time. Genetics and environment interact continuously, sometimes in highly specific ways at highly specific times (critical periods). Single gene effects are modified by multiple other genes and environmental conditions over time in the process of epigenesis. Single genes are unlikely to directly *cause* specific complex behaviors but rather act to set reaction ranges that define probabilities or predispositions. In other words, a mutation may make some general type of behavior more likely—for example, aggressive or impulsive behavior. But the effects of such a mutation are also likely to be dependent on other genes and a host of environmental influences. Furthermore, complex actions are likely to result from a complex chain of events, with each link in the chain being necessary but not sufficient to cause the final outcome.

Psychopathology in childhood and adolescence, like human development, is also complex, multidimensional, and unfolds over time. Genetic influences are likely to contribute to the etiology of many disorders with onset in childhood or adolescence. Current understanding of adolescent psychopathology is imperfect, but several diagnostic categories are associated with increased rates of conduct problems that may result in referral to the legal system. These include learning disabilities (especially reading disability or dyslexia), attention-deficit/hyperactivity disorder, substance abuse, and conduct disorder. Disorders traditionally described in adults, such as schizophrenia and bipolar disorder, may have onset in childhood or adolescence with age-related symptoms, such as behavior problems, that may be difficult to recognize as secondary to a psychiatric disorder prior to full unfolding of the characteristic picture in adulthood. For many diagostic categories, the boundaries between normal development and developmental psychopathology are difficult to define. Furthermore, the diagnosis of one disorder in an adolescent increases the probability that other diagnoses also apply to that adolescent. This is the problem of comorbidity, or co-occurrence of multiple disorders. For example, an adolescent diagnosed with conduct disorder is also more likely to have attention deficit/hyperactivity disorder, learning disability, and substance abuse.

Each diagnostic category may actually represent a collection of disorders, all with different underlying causes. Attention-deficit/hyperactivity disorder, as a case in point, may not be the unitary condition that is portrayed by some psychiatric literature and by most popular media. Different genes, environmental toxins, or environmental stressors may all cause conditions that look identical but may not have the same long-term outcome nor respond to the same treatments. For example, two putative causal genes have been identified in different samples of children with attention-deficit/hyperactivity disorder: the thyroid receptor and the dopamine transporter protein.[1] Neither gene may be a causal factor in the majority of cases, but may contribute to the causation in some subset of cases. Given the complex nature of attention and impulse control, it seems reasonable to expect the list of genes that may cause attention-deficit/hyperactivity disorder to expand greatly. Consider Leppert's example (elaboration for chapter 4, this volume) of the eye disease retinitis pigmentosa, which has ten different genetic causes: Should we not expect that substantially more than two genes will be found for attention-deficit/hyperactivity disorder?

In some rare cases, a yet undiscovered specific mutation that affects only one family may cause attention-deficit/hyperactivity disorder (probably in combination with other genetic and environmental factors). This could be analagous to the MAOA mutation found in the Dutch family, in that the mutation could be unique to a single family.[2]

Delinquency may be a manifestation of a psychiatric disorder, but it is complicated by a number of other phenomena. For example, many "normal" teenagers commit an illegal act but never get caught and do not develop a chronic pattern of antisocial behavior. On the other hand, a small proportion of all delinquents commit a large proportion of juvenile offenses. In addition, the trends toward increasing use of guns, increasing influence of organized gangs, and increasing substance abuse have caused juvenile crime to become a major concern for society. Drive-by shootings raise fears and create political pressure for ensuring safety.

Juvenile law has deemphasized rehabilitation and increased emphasis on due process, retribution, incapacitation, and transfer to adult court. Despite this change in emphasis, juvenile law is still more flexible than adult law in how offenders are managed. This flexibility allows for both positive and negative perspectives with respect to the new genetics. For example, the discovery of a gene that confers risk toward criminal behavior may be used to screen juvenile offenders. On the positive side, an offender with a positive screening test may benefit if the test is truly predictive and if there is an effective intervention that decreases that risk. On the negative side, a screening test with a high rate of false positives may result in unjustified labeling or institutionalization. Likewise, a positive test combined with no known effective intervention to decrease risk may result in a sense of hopelessness and prolonged incarceration.

Society as a whole will determine the influence of genetics on juvenile law. The concept of childhood is a social construction. Consensus about the needs, capacities, and treatment of children have changed greatly with time. Events such as genetic and other developmental discoveries, drive-by shootings, and media portrayals such as the movie *Sling Blade* shape public opinion. Thoughtful, extensive public dialogue is needed to raise public awareness and build consensus.

## Conclusion

Great skepticism and caution are warranted in evaluating any future claims that a specific gene causes or substantially contributes to any illegal or antisocial behavior, particularly in juvenile offenders. There is currently little evidence that a single gene plays any role in even a subset of antisocial acts carried out by juveniles. Although it is likely that genetic factors contribute to the development of several psychiatric disorders and personality traits, any relationship of genes to juvenile antisocial acts will include environmental and other genetic modifiers that will limit predictive validity of any screening or testing in the forseeable future. The current excitement about findings of single genes that predict such diseases as breast cancer and Huntington's may set inappropriate expectations for genes involved in complex behaviors. Current evidence supports the conclusion that social variables play the predominant role in causation of juvenile antisocial behavior. Furthermore, both genetics and environmental factors interact in a dynamic process, making prediction of complex behavior based on one factor unwise.

A second major conclusion is that there is a great need for systematic knowledge, comprehensive management, and multidisciplinary understanding of the biopsychosocial phenomena relevant to managing the juvenile justice system. Much is unknown about genetic and nongenetic influences on normal development and psychopathology, as well as the processes of the juvenile court and related agencies and the social construction of childhood by all relevant elements of society.

A third major conclusion is that care and management of juvenile offenders, with or without identified psychopathology, will always be needed, whether any single or multiple genetic influences contribute to antisocial behavior. At present, environmental factors appear greatly more influential in etiology. Even if genetic influences are found to be stronger for some individual or subset of cases, psychosocial interventions and care are likely to be the most practical and effective possible responses.

Our recommendations can be summarized in one statement: Society at all levels must engage in the process of understanding and coping with the genetic and nongenetic causes of juvenile crime. New findings must be subjected to both scientific and public scrutiny that transcends transitory headlines and sound bites. Thoughtful solutions to chronic problems require systematic effort over time. Following are some specific recommendations.

- Organizations representing professional geneticists, such as the American Society of Human Genetics, the American College of Medical Genetics, the American Association for the Advancement of Science, and other groups concerned with developing and interpreting genetic information should take an active role in translating scientific claims for the public. A proactive approach is needed on an ongoing basis, with educational outreach to segments of society including those listed below.
- Medical policy-setting groups such as the Institute of Medicine and the National Research Council should convene task forces to increase focus on ethical, legal, and social aspects of the so-called genetics revolution in medicine, particularly relating to child and adolescent development and psychopathology.
- Increased programmatic research is needed to delineate aspects of normal development, genetic and nongenetic risk, or protective factors, the validity and long term outcome of child and adolescent psychiatric disorders, and the understanding of youth violence and antisocial behavior.
- Juvenile justice education programs transmitting the complexity of genetics–behavior relationships and the need for caution in evaluating popular claims about the deterministic nature of genes should be directed at all levels of the juvenile justice system. Facilitation of such efforts should occur through such organizations as The National Judicial College, The National Center for State Courts, the Federal

Judicial Center, the State Justice Institute, and the Council of State Court Administrators.

- State government officials should join similar educational efforts that allow for more informed debate and policy decisions. Workshops should include executive and legislative branch officials through groups such as the National Governors Organization and the National Conference of State Legislators.
- Efforts of public advocacy groups (such as the National Alliance for the Mentally Ill) in educating the public about psychiatric disorders should be supported and expanded.
- Media portrayal of these complex topics should be improved so that oversimplification does not lead to a misinformed public. Writers, producers, and others who creatively portray the vital but complicated aspects of human nature should be honored for bringing the power of the mass media to inform the public dialogue.
- Private foundations that fund genetics research should be encouraged to increase priority for research on the ethical, legal, and social implications of genetic research.

## Notes

1. P. Hauser, A. J. Zametkin, P. Martinez, B. Vitiello, J. A. Matochik, A. J. Mixson, and B. D. Weintraub, ''Attention Deficit Hyperactivity Disorder in People With Generalized Resistance to Thyroid Hormone,'' *New England Journal of Medicine* 328 (1993):997–1001; E. H. Cook, Jr., M. A. Stein, M. D. Krasowski, N. J. Cox, D. M. Olkon, J. E. Kieffer, and B. L. Leventhal, ''Association of Attention-Deficit Disorder and the Dopamine Transporter Gene,'' *American Journal of Human Genetics* 56 (1995):993–998.
2. H. G. Brunner, M. Nelen, X. O. Breakfield, H. H. Ropers, and B. A. van Oost, ''Abnormal Behavior Associated With a Point Mutation in the Structural Gene for Monoamine Oxidase A,'' *Science* 262 (1993):578–580.

# Author Index

Numbers in italics refer to listings in notes and reference sections.

# Subject Index

# ABOUT THE EDITORS

**Jeffrey R. Botkin, MD, MPH,** is a professor of pediatrics and an adjunct professor of internal medicine in the Division of Medical Ethics at the University of Utah. He is a graduate of Princeton University and the University of Pittsburgh School of Medicine. He is a pediatrician with fellowship training in law, ethics, and health at the Kennedy Institute of Ethics at Georgetown University, and he is the director of the Genetic Science in Society (GENESIS) program at the University of Utah Center for Human Genome Research, a program devoted to education and research on the ethical, legal, and social implications of genetic research. His research and writing focuses on ethical and legal issues in prenatal diagnosis and genetic testing for cancer susceptibility. He is a member of the Committee on Bioethics for the American Academy of Pediatrics and is on the editorial board for the *American Journal of Medical Genetics.*

**William M. McMahon, MD,** is associate professor of psychiatry and adjunct associate professor of pediatrics and psychology at the University of Utah. He received his undergraduate and medical degrees as well as his first year of psychiatry residency at the University of Kansas, and subsequently served as a medical intern at Gorgas Hospital in the Canal Zone. He completed residency training in general psychiatry, followed by a fellowship in child and adolescent psychiatry at the University of Utah. He served as president of the Utah Psychiatric Association in 1981. His clinical duties have included directing an inpatient psychiatry unit for children and adolescents; directing a clinic for children with Tourette syndrome, learning problems, and autism; and serving as a court-appointed examiner for committment hearings. He was awarded the Karl Manwaring Memorial Award by the Utah Tourette Syndrome Association, and is listed in *The Best Doctors in America.* In 1992, he was awarded a National Institute of Mental Health grant for training in molecular and clinical genetics. His research is focused on genetic components of Tourette, autism, and other neurobehavioral disorders.

**Leslie Pickering Francis, JD, PhD,** is a professor of law, professor of philosophy, and an adjunct professor of internal medicine at the Division of Medical Ethics at the University of Utah. She received her PhD in philosophy from the University of Michigan in 1974 and her JD from the University of Utah in 1981. She was a law clerk to Judge Abner Mikva of the U.S. Court of Appeals for the District of Columbia Circuit in 1981–82. She specializes in health law, bioethics, and legal ethics, and is the author of a number of articles on issues in philosophy of law, health care, and professional ethics. She is currently a member of the American Law Institute, the American Bar Association's Commission on the Legal Problems of the Elderly, the Executive Committee of the Pacific Division of the American Philosophical Association, and the Utah state bar's Ethics Advisory Opinion Committee. She also chairs the American Philosophical Association's Committee for the Defense of the Professional Rights of Philosophers.